Systems Thinking and E-Participation: ICT in the Governance of Society

Jose Rodrigo Cordoba-Pachón
University of London, UK

Alejandro Elias Ochoa-Arias
Universidad de los Andes, Mérida, Venezuela

Information Science REFERENCE

INFORMATION SCIENCE REFERENCE
Hershey · New York

Director of Editorial Content: Kristin Klinger
Senior Managing Editor: Jamie Snavely
Assistant Managing Editor: Michael Brehm
Publishing Assistant: Sean Woznicki
Typesetter: Kurt Smith
Cover Design: Lisa Tosheff
Printed at: Yurchak Printing Inc.

Published in the United States of America by
Information Science Reference (an imprint of IGI Global)
701 E. Chocolate Avenue
Hershey PA 17033
Tel: 717-533-8845
Fax: 717-533-8661
E-mail: cust@igi-global.com
Web site: http://www.igi-global.com/reference

Library of Congress Cataloging-in-Publication Data

Systems thinking and e-participation : ICT in the governance of society / Jose Rodrigo Cordoba-Pachon and Alejandro Elias Ochoa-Arias, editors.
p. cm.
Includes bibliographical references and index.
Summary: "This book provides a systemic-based inquiry platform to explore boundaries, limits, and advantages of information and communication technology use in the public decision making processes"--Provided by publisher.
ISBN 978-1-60566-860-4 (hardcover) -- ISBN 978-1-60566-861-1 (ebook) 1. Internet in public administration--Cross-cultural studies. 2. Political participation--Computer networks--Cross-cultural studies. 3. Communication in politics. I. Cordoba-Pachon, Jose-Rodrigo. II. Ochoa-Arias, Alejandro.
JF1525.A8S98 2010
352.3'802854678--dc22

2009035156

British Cataloguing in Publication Data
A Cataloguing in Publication record for this book is available from the British Library.

List of Reviewers

Ahmed S Al-Maimani, *University of Hull, UK*
Alexandre F. Barbosa, *Brazilian Internet Steering Committee, Brazil*
Diego Barría, *Leiden University, The Netherlands*
Gustavo Campos, *University of Chile, Chile*
Juan-Gabriel Cegarra-Navarro, *Universidad Politécnica de Cartagena, Spain*
Rajneesh Chowdhury, *PricewaterhouseCoopers, India*
J. Ignacio Criado, *Universidad Autónoma de Madrid, Spain*
Jacinto Dávila, *Universidad de los Andes, Venezuela*
Eduardo Henrique Diniz, *Fundação Getúlio Vargas, Brazil*
Angela Espinosa, *University of Hull, UK*
Marcelo Fernandez, *Secretariat of Finance of State of São Paulo, Brazil*
Amanda J. Gregory, *University of Hull, UK*
Alvaro Junqueira, *Secretariat of Finance of State of São Paulo, Brazil*
Cecilia Loureiro-Koechlin, *Oxford University, UK*
Endrit Kromidha, *Military University "Skenderbej" Tirana, Albania*
Deepankar Medhi, *University of Missouri–Kansas City, USA*
Federico Monteverde, *Electronic Government Researcher, Uruguay*
Eduardo Araya Moreno, *University of Chile, Chile*
Irena Ograjenšek, *University of Ljubljana, Slovenia*
Kevin Orr, *University of Hull, UK*
Francisco-Javier Palm-Rojas, *University of Los Andes, Mérida, Venezuela*
María-Ángela Petrizzo-Páez, *ENDOLOGICA, Open Informatics and Professional Services Cooperative, Venezuela; SPIRALIA C.A. Networking Learning Solutions, Venezuela*
Elias Pimenidis, *University of East London, UK*
Otávio Prado, *FGV-EAESP, Brazil*
Ana Magaly Reyes, *Freelance Consultant, Venezuela*
Ioannis Savvas, *Agricultural University of Athens, Greece*
Alexander B. Sideridis, *Agricultural University of Athens, Greece*

Table of Contents

Section 2
Mediation in the Context of E-Participation

Section 3
Emerging Forms of Community Interaction: Groups and Tools

Section 4
E-Configurations, Citizens and the State

Detailed Table of Contents

subsystems can be intervened for change. The role of information and communication technologies (ICTs) to support this type of action is also discussed.

This chapter develops an argument of both technology and government as historically determined phenomena within what has been called liberalism. Technology helps government to make clear and attend individuals' choices. The author asks the question about the existence of an alternative ideological framework to guide the use and implementation of both government and technologies and defines possible ways forward for the pursuit of citizenship in societies.

This chapter proposes alternative ways to conceive of and manage e-government initiatives in the context of the information society. Three different patterns are defined and described; their application to study the case of the Colombian e-government "Gobierno en Línea" (online government) reveals the importance of sensitivity to and awareness of different and intertwined discourses about e-government projects, and how systems methodologies can help enquiry.

This chapter provides a combination of analytical tools to help those involved and affected by local e-government initiatives to make sense of complexities encountered and reveal the substratum of traditions that helps or inhibits e-government development.

Section 2
Mediation in the Context of E-Participation

This section of the book includes practical interventions of e-participation initiatives some of which embed the use of systemic principles. Reflections of practice show the complexity of initiatives and the importance of citizens' participation for their further improvement.

The chapter presents an in-depth study of an e-government multi-agency project in Brazil (Nota Fiscal Electrónica or NF-e), using the ideas of the project management book of knowledge framework (PMBOK) to assess the degree of formality of the project. Reflections and findings suggest that continuous involvement, informal communication, commitment and participation of different stakeholders compensate for the apparent lack of project formality. Moreover, these aspects helped people prioritise adequately certain areas of intervention (integration, cost, scope management) to help the project to succeed.

The chapter suggests that small and medium enterprises (SMEs) can benefit greatly from the provision of e-government services (e-services) if these are conceived of as part of processes of knowledge acquisition and renewal. Using concepts of exploitation of knowledge (EK) and organisational memory (OM) the authors develop a framework to guide the development of e-services. The role of systems methodologies in enabling knowledge-related activities leading to e-services provision is also discussed.

This section shows how information and communication technologies have contributed to generate new forms of interaction between people, and how learning can become a key element within citizens' communities. Learning can be enhanced by the appropriation of technologies and their social use, as well as by re-thinking fundamental concepts of community, citizenship and government.

The chapter describes the basic tenets of online communities, their typology and their fundamental elements with a view of informing ways in which people can get to participate online, something that those individuals or organizations dealing with e-governance initiatives also aim at. A key consideration is the importance of socializing as an ingredient that motivates individuals and which can offer them a degree of freedom to discuss government related matters.

put forward and explored in the case of Valencian Comunidad Autonómas (autonomous community). The chapter shows how the interplay of institutional practices, the development of new forms of organization to provide e-government services and the unfolding of regional policies within Spain and Europe all contribute to make e-government what is now in the region of study.

Using the ideas of Kingdon's model of multiple streams, the author proposes a framework to assess the development of e-government initiatives which can then be used by e-government actors to reflect on and act upon e-government implementations. The experience of the Uruguayan case shows how e-government unfolding happens within certain windows of opportunity. The implications for policy and practice are also discussed in this chapter.

This chapter provides a framework used to historically analyse the development of e-government (e-gov) in Brazil. The framework allows for the exploration of different conditions (not only technological) that have contributed to e-gov development. The authors identify a number of factors and relations between different stakeholders which influenced stages of e-government up to now. In addition, it suggests further avenues to enhance the understanding of e-gov development in future studies.

Albania is an example of how policies for e-government in the context of the information society have been taken forward. This chapter discusses the implementation of policies in the Albanian context and identifies a number of shortcomings of traditional evaluation methods use for e-government initiatives. It suggests focusing on evaluation of e-government as a way to facilitate inclusion of different issues and people and thus to improve initiatives.

Preface

In September 2004 and thanks to a small grant offered by the University of Hull in the United Kingdom, an interdisciplinary research group called "E-government and Local Integration with Sustainability" (ELIS) was set up with colleagues. Its original aim was to look at how we could make e-government more inclusive and participative. At that time, e-government had been defined as an ideal to be achieved in many countries. As such, it was involving the use of technology by citizens and government officers. In the initial meeting we exchanged ideas and possibilities. There was a group of academics from Spain, Venezuela, Poland and the UK, and a council officer (Shaun Nicholson). In one way or another, we have continued working together. Despite geographical and ideological distances have inspired the development of this book which brings the best of both systems-thinking and e-government.

The ELIS group met again in Hull in 2008, and an idea of a collection of chapters to critically assess e-government with systems thinking was put on the table. Alejandro Ochoa and myself took up this challenge and began spreading the word from our bases in UK and Venezuela. As a result, we got the interest of people in several countries and the theme of e-participation. Systems-thinking has become vehicle through which we have travelled together in this journey. It has enabled us to pull contributions together, and to make our voices heard now that we see many democracies trying to emerge whilst others have not fulfilled entirely the aspirations of citizens. As the group of contributors, we want to highlight that whatever conditions surround a particular society, citizens find ways of using technologies to make their voices heard. The book is our way to show how we can move forward any effort to facilitate societal participation with information and communication technologies in mind *and acknowledging that technologies are becoming critical in the conduction of society as a whole.*

We aim to provide a series of perspectives on how we see the uptake of electronic resources (systems, technologies, websites, open software tools, etc.) in civil societies *and the state*, and how this can be conceptualised with the help of systems thinking, a collection of ideas and concepts to help us select appropriate methods to pursue societal improvements. In the book we have strived to provide a comprehensive set of perspectives which show how electronically mediated systems of participation could provide answers or even more questions to current issues. We have coined the term e-participation to account for all these perspectives.

Since the original idea was on the table we have seen that what we see as the official face of e-participation (e.g., electronic government or e-government) has spread across the world and continuously every day. New systems are continuously created to facilitate meeting of government requirements by citizens. Despite a degree of sophistication in the technologies being used, as well as in the integration of data sources from government offices or citizen's mobile devices, we still have not seen other forms empowerment than those pre-determined by design and justified by security and confidentiality concerns. This leads us to think carefully before claiming a positive degree of success in the implementation of electronic government. In the academic quarters related to e-participation (or to e-government as its officially adopted name), it makes us think on how we need to make sense of a diversity of experiences around the world, and how we can be more explicit about underlying theories used as well on rigorously obtained and useful recommendations for action (Heeks & Bailur, 2007).

A key issue derived from what we believe should be a healthy degree of scepticism in relation to e-participation is that of societal *governance*, and how as an attribute it is to be exerted by people. To many, modern societies are just surveillance camps where, in the name of democracy the use of technology has spread across areas of intimate life, touching upon our identities, consumption habits and even political affiliations. This is only one possible side of the story of governance. The other is one in which we involve ourselves and the rest of our fellow human beings in the pursuit of better societies for all, which we call *improvement*. The chapters of the book reflect accounts that might lead us to lean more towards one side or the other.

We leave it to the reader to do so, although *the editors* share the view that it is important to step aside from official versions of electronic participation and engage in constructing our own one(s). As systems thinkers, we also advocate a critical stance towards the design and implementation of electronic systems that forgets about the social context in which such systems are to be used. The form of this stance can take many avenues, each being helped by different systems-oriented approaches to planning and evaluation.

A GUIDE FOR THE READER

To help the reader we have divided our book in the following main sections:

Section 1. Systemic conceptualizations of e-participation, e-governance, e-government, technology or society. We include here the use of systems ideas to interpret phenomena related to e-participation and to open up new avenues of enquiry into this phenomenon. Córdoba and Ochoa-Arias begin by presenting two perspectives on e-participation (official, unofficial) and to open up the discussion on possibilities. Gregory follows on and discusses the problem of governance using the ideas of autopoiesis and complexity theories. This is a provoking account that sets the tone of systemic discussion in the book and helps us raise awareness about the role of citizens in societies, as well as the use of information systems and technologies in it. Ochoa-Arias then looks at technology and government as being historically conceived phenomena which together reinforce a way of looking at the world as a tool for control and use of resources. With his account, Ochoa-Arias is suggesting alternative interpretations of government. Along similar lines, Córdoba presents a number of patterns to help us understand and deal with e-government with a view that these patterns could be used in different social contexts to guide enquiry and action into electronic government implementations. Finally, Córdoba and Orr present a combination of analytical tools to develop alternative conceptions in relation to local government; it is a combination of the ideas of systems boundary critique and traditions, which generates an interesting synergy to facilitate a deeper degree of critique into local e-government projects.

Section 2. Mediation in the context of e-participation. This section aims to raise awareness about the type(s) of (democratic) citizen-based systems in which uses of technologies emerge, as well as on how other elements of such context can be addressed in the practice of setting up electronically mediated systems. Using systems thinking, Chowdury and Mehdi present a framework to facilitate the implementation of a public health information system in India and its access by illiterate people. Their framework proposes a number of principles which find support in the commitments of critical systems thinking as previously defined by Midgley (1996). The authors also propose an architecture to handle the diversity of requests from users and thus reflect their commitment to promote diversity and social inclusion. Their architecture is in some respects similar to the one proposed by Savvas, Pimenidis and Sideridis in the following chapter. These authors envisage electronic government systems as composed by different levels. Flexibility of and communication between these levels can help attend a variety of audiences with appropriate information services.

Leaning more towards the organisation of electronic participation systems, Espinosa and Al-Maimani give us an account of a research project in Oman in which they look at the organization of national e-government projects, and how it could be improved using ideas of Beer's Viable System Model (VSM) and other systems-based methods. They present us with an example of how one of these projects is being improved with the help of other systems techniques including Checkland's soft systems methodology root definition. The unfolding of complexity in e-government projects can be well managed by VSM according to the authors; an adequate degree of autonomy can be developed. Autonomy is also something that Junqueira, Diniz and Fernandez highlight in their account of the electronic invoice project "Nota Fiscal Electronica" (NF-e) in Brazil. They reflect on this experience using some of the key project areas defined by the project management book of knowledge (PMBOK) to this project. The project, as the authors say, was managed in a somehow informal but impressive way regarding its results. With the experience there are also many challenges ahead, one of them being the development of successful (e.g., autonomy based) working environments between organisations concerned with implementing an electronic service for the public.

In the realm of small and medium enterprises (SMEs), Córdoba and Cegarra present a framework to assess and further develop the provision of electronic services to this type of business. They argue that current provision needs to consider how SMEs acquire and manage knowledge, in a way that new electronic services can be integrated to such processes. The authors draw on the use of systems methodologies to facilitate the development of knowledge processes and with it the improvement of existing services. They also interpret some preliminary results of using ideas from this framework in the Spanish telecommunications industry. Results indicate a number of opportunities to conceive of and implement e-government services to help businesses improve.

The above accounts offer us a stance that makes emphasis on the exploration of the context of electronically mediated participation systems and reflect different ways to go about such exploration However, and as is often the case, technologies move very dynamically and we have then to consider how best to use their potential and create appropriate environments for their use. With new technological arrays mediating between citizens and governments, new forms of organisation emerge. Electronically mediated systems have spread very rapidly and have contributed to new forms of relation between people, in many cases replacing old ways of relation.

Section 3. Emerging forms of community interaction: Groups and tools. A broader notion of governance that is not (in principle) related to that of formal government helps us to acknowledge the existence of new forms of interaction between people; in those technology is still a mediating element, but people are not pre-determined in how they can use it or constrained by existing forms of organisation. Loureiro-Koechlin begins this section by focusing on the dynamics of online communities and how such communities fulfil certain purposes. She then discusses their features and ways to facilitate their development, and suggests connections between the dynamics of online communities and what is being understood as *governance* in the public sector domain. The notion of community is a powerful one that we have found when looking systemically at e-participation. Such a notion can pervade geographical locations, jobs or sources of expertise and even the political sphere of societies.

Petrizzo and Palm give their account of what it means to politically engage through the Internet. They introduce the notion of e-citizenship to show how people can use the Internet to participate politically in their societies, and how their development of social networks is a key goal as well as a medium to facilitate deliberation. Their chapter also brings a notion of proactive citizenship which in our view is holistic and can inform the improvement of relations between citizens and their government institutions.

In conjunction with such notions of 'e' citizenship and governance, governments are also reviewing their own governance practices. This is shown by Dávila and Reyes' chapter. In it, they propose a form of *articulated planning* for government affairs in Venezuela to be best developed through information and communications technology (ICT) use. The authors portray an example of how their proposal works in practice, showing also how citizens were engaged from the grassroots, and how they interacted with different government organisations with friendly and simple to use ICT tools. The new configurations of people and technology do not have to be sophisticated but meaningful and easy to 'use'. At the heart of their chapter Davila and Reyes aim to rescue some ideas of *socialism* which they see appropriate to help citizens face contingencies and share their knowledge. Their proposal challenges what we see as a dominant view of e-government which aims to gradually transform government institutions which often results in missed opportunities.

Section 4. E-configurations, citizens and the state. Moreover, a form of 'gradual change' to e-participation does not sit easily when governments have to accommodate their offline and online operations, in other words the 'old' and the 'new'. Criado begins this section by providing a thorough and thoughtful account of the development of e-government in the Spanish "comunidades autonomas" under the lenses of administrative modernization. This is an account which shows the multifarious nature of e-government which requires equally multi dimensional perspectives to study it in its richness, as it generates dynamic assemblages and arrays of technology, people and institutions. Two further chapters reinforce this point. Firstly, Monteverde shows this more clearly by reviewing the development of e-government in Uruguay. In his chapter and using the ideas of Kindgon he shows how different 'streams' (political, technological, etc.) historically converge into *windows of opportunity* to give possibilities for the improvement of participation. Monteverde concludes by suggesting that the framework developed can help an exploration of conditions surrounding e-government prior to its implementation. The framework makes it clear that within what appears to be 'anarchical' situations, there is scope to act. Secondly, Barbosa et al show the development of e-government in Brazil, and account for a number of factors which usually do not belong to the technological domain but which contribute to create a path through which influences the adoption of e-government policies and plans. These factors range from previous projects (the millennium bug) to leadership of key individuals in taking ideas forward. These factors generate conditions and with them possibilities for action.

In a similar way to a possibility of 'scoping to act' on e-participation that is derived from the above accounts, Kromidha and Córdoba review the unfolding of e-government in Albania, making emphasis on the importance of evaluating e-government projects in a non-traditional way. Their discussion provides an overview of dominant e-government evaluation approaches (e.g., cost-benefit based), and the challenges ahead if alternative evaluation approaches are adopted. The case of Albania shows a number of possibilities to make use of existing e-government configurations through e-government evaluation. Speaking also of different configurations, in their contribution Araya, Barria and Campos show how the use of Internet based technologies has pervaded the political arena in Chile. There is a variety of technologies being used within and outside websites of political parties. Interestingly, these and other communities created around the use of technologies for political purposes can provide much more open spaces for citizens. The authors also make the point that technologies can help people to operate 'outside' traditional systems. This confirms what many other book contributors have raised and what intends to be a key message of this book: *the phenomena of e-participation goes beyond technology use, and invites us to re-consider how we engage with it.*

To make this point more clear and conclude our book, Ochoa-Arias and Petrizzo disclose a meaning of citizenship by addressing some of the issues confronted when e-democracy and e-management are conceived to be closely related in spite of analytical efforts to keep both fields separated. This opens up a number of possibilities for the future of e-participation, and the authors reformulate an 'official' (stage-based) model of e-government within a new conception of citizenship.

With these insights, we reconsider what has been said in relation to e-government research by Heeks and Bailur (2007). They aim to enrich the discussion by proposing that e-government research should be more explicit about the theories used, with new and innovative methods to collect data, and with clearer implications for action from studies on this area. They also advocate adopting a more inter-disciplinary approach to research, and to value existing research in e-government as an emerging area. We think that this book contributes to enrich theoretically and practically informed discussion and also to provide a 'system' of accounts on e-participation. By doing the latter and by showing how systems thinking can inform practice and reflection, the book can help researchers and readers to review their own assumptions in research or intervention; to be critical about e-participation phenomena; and to appreciate the complexities derived from the use of information and communication technologies in this domain. The door remains open to continue engaging with e-participation and to pursue societal improvements with the help of information and communication technologies.

José-Rodrigo Córdoba-Pachón
Alejandro Ochoa-Arias

REFERENCES

Heeks, R., & Bailur, S. (2007). Analyzing e-government research: perspectives, philosophies, theories, methods and practice. *Government Information Quarterly, 24*, 243-265.

Midgley, G. (1996). What is this thing called CST? In R. L. Flod & N. Romm (Eds.), *Critical Systems Thinking: Current Research and Practice* (pp. 11-24). New York, Plenum Press.

Acknowledgment

Our thanks go to our contributors, to members of the Centre for Systems Studies at Hull University in the UK, to our current employers (Royal Holloway in the UK, Universidad de los Andes in Mérida - Venezuela respectively) for giving us time for this project, and to our families. We would like to thank Juan-Gabriel Cegarra and Irena Ograjensek from the ELIS group for their help with some of the chapter reviews.

And to Joel Gamon from IGI Global and his collaborators for their patience!

Section 1

Systemic Conceptualizations of E-Participation, E-Governance, E-Government, Technology or Society

Chapter 1
Introduction:
Searching upon the Limits of E-Government and E-Participation— A Systemic Appraisal

José-Rodrigo Córdoba-Pachón
University of London, UK

Alejandro Elias Ochoa-Arias
Universidad de los Andes, Mérida, Venezuela

ABSTRACT

This chapter gives an overview of the phenomenon of e-government within the context of the information society. The importance of values, critical thinking and boundaries is raised for the study of this phenomenon. The chapter offers a critical appreciation of the dominant discourse of e-government to open up new possibilities for enquiry, intervention and citizenship.

INTRODUCTION

Those of us who have had the fortune of witnessing the last years of the twentieth century and the beginning of the new millennium would say that the pace of change is still increasing. Computers are now essential part of our daily lives, and so is electronic information. Worldwide we see that we have become a global information society, and as such, it is time to think about how the ways in we relate to each other have unfolded.

This book is about the phenomenon of electronically mediated participation (e-participation) in society, which it has been commonly associated with electronic government (e-government). One could say that e-government seems to be an explicit or visible manifestation of e-participation in the same way in which one could say that the internet is an explicit manifestation of the information society. In both phenomena we have to say that neither e-government nor the internet as visible manifestations can account for all past, present and future developments. Our task is then to go beneath the surface of these phenomena, and try to understand the thinking behind their development so we can contribute in theory and practice to better it. We do this to see if as policy makers, practitioners or citizens we need to positively change the ways in which both e-participation and the information society are being developed.

DOI: 10.4018/978-1-60566-860-4.ch001

This chapter presents an introduction to the topic of e-participation and the book in general. We contextualize it within the information society, a discourse that can help us make sense of the changes in societies that we have been witnessing and will continue to be either witnessing or intervening in. Our initial contextualization aims to dispel a common idea about this phenomenon as something completely revolutionary or emerging out of the blue. With this conceptualization, we offer two main alternatives to talk about e-participation. We hope to give the reader some elements that they could explore when reviewing the collection of essays that the book contains.

The chapter presents then two main perspectives to understand e-participation by looking at e-government in official and non-official ways. The first one refers to what has dominated the development of e-government in the context of the information society. The second one emerges as a response to the need to go beneath the surface of the official version and complement or challenge what we see is in place in e-government initiatives. The emerging nature of this second perspective also gives us the opportunity to include developments in the use of technologies by people in society. At the end of the chapter we propose some ways forward to inform the thinking about e-participation.

To start our discussion we put ourselves in the picture of the network idea of society.

SOCIETY AS A NETWORK OF INFORMATION

Individuals who ventured to make sense of the changes happening in our societies in the 80s and 90s talked about the emergence of a new paradigm for society: That of a network. What they meant was that a society (national, regional, local or global) could be conceived of as a network of flows of information. Such network was de-centering traditional centers of power in organizations. It was supposed to open space for the emergence of new forms of organizations. De-centering of power was possible because information could now flow instantly through different and geographically spread physical locations, and ultimately through different groups of individuals.

Through history we are used to think of society in terms of flows of resources, influences, relations with governments and power. However and recently, information and communication technologies were greatly contributing to a shift in our understanding of society as a network of information. Technologies have contributed to enhance information exchange as well as generate economic opportunities. With them not only there would be the possibility for societies to facilitate the creation and exchange of information anywhere as a capability to leverage economic growth. Technological change then leads to change the nature of job structures, consumer behavior and ultimately institutional arrangements (Perez, 1983)

The information-based network idea of society presupposes that social change is to be a partial result of technological one and conceived as a societal improvement in itself. Little is explicitly said about how social changes are to take place, or the sort of institutional impacts that are to unfold at different levels in society. We only know that there would be exclusions when traditional nodes of power (e.g. institutions) would be–in principle displaced to give way to 'hubs' or particular locations. Such hubs would attract talented individuals and leave behind those not so-talented or with little possibilities to skill themselves to participate in this network of flows of information and knowledge. Institutional resistance was considered but with time overcome due to the inevitability of technological adoption. How institutions could make use of technology to perpetuate them-selves at the expense of society was considered part of such change (Toffler, 1992).

Thus, within the network idea there was consideration of changes at high cost for those not being prepared to adapt to them. Information was to become the new strategic asset of organizations and governments which needed to be protected; those with access to it would be at advantage from others. This sort of exclusion was perceived by some as a threat to society. To this possible outcome of the development of the network society, Castells (2001) says:

'Why don't you leave me alone?! I want no part of your Internet, or your technological civilization, of your network society!' Well, if this is your position, I have bad news for you. If you do not care about the networks, the networks will care about you, anyway. For as long as you want to live in society, at this time and in this place, you will have to deal with the network society (p.282).

With this, Castells signals the inevitability of change, and the need of individuals to take action on the face of it. He shows how in a number of cases how people can take collective action with the help of technologies so as to redress existing imbalances of power as well as protect (or rescue) their own values, interests and ultimately concerns for their societies. Although this can help us make sense of the change that takes place, we need as individuals to go beyond the possibility of 'reactive' or 'one sided' (e.g. government-led) action, and involve both as part of a whole phenomena which we call e-participation. Critical thinking is needed not only to engage people but to review the assumptions and limitations of the network idea and its implications for people's participation in societal affairs. In this regard, it is important to consider that although individual potentiality increases in the information-based network idea of society, there is still a need of creating and developing collective groups in order to have a sustainable impact in society if it is considered to be relevant.

VALUES, CONCERNS AND CRITICAL THINKING

The development of the information society has also brought its critics who argue that in this radical transformation the idea of community is being undermined if not lost (Afele 2003; Mansell, 2002; May, 2002; Menzies, 1996; Wresch, 1996). The voices of marginalized groups or countries can still be heard in this regard, but often these voices have been (hopefully partially) silenced by other discourses (environmental protection, economic crises, terrorism, etc). Such voices tell us that people are being uprooted from their communities, and they need to go back to it in some way or another. Critics of modern society discourses do not favor the radical nature of transformations and prefer a more gradual process (Webster, 2002) in which people could still maintain their social relations (Brown & Duguid, 1999) and decide on which values they want to keep for themselves and future generations. This claim also suggests that for new forms of interaction (e-participation included) needs consideration of the social context in which people are as a substratum that shapes change in society. It also needs active engagement with such social context and its particular forms of participation, community involvement, and information technology use. Unfortunately, this consideration is being replaced by forms of consumer-like engagement in the information society (Mattelart, 2003).

Maintaining the above consideration helps us to give momentum to the idea of collectively driven action that arises when some of the 'pioneer' information societies have not achieved a number of desired benefits (inclusiveness, participation, cohesiveness). In the words of Wickam (1997), for instance we have early evidence about the information society in Ireland. At that time, he acknowledged that despite the pervasiveness of information and communication technologies,

Individuals buy and sell across the world, but are unable to sustain relationships of mutual trust with their neighbours. (p.289)

This and other claims (Van Audenhove, Burgelman, Nulens, & Cammaert, 1999) speak about the social nature of changes and how we still need the social fabric even to promote radical transformations (Webster, 2002). There are to be intermediaries that ensure that people do not remain marginalized from accessing information and technologies (Madon et al, 2004). This can also mean that practitioners and policy makers or individuals with expertise in the management of information systems and technologies need to mediate between revolutionary possibilities for the development of societal change and the social fabric. Those advocating new forms of electronically mediated participation could benefit from looking at how a better degree of understanding of the context and conditions for such participation can be gained through theory and practice.

For the information society, rescue from this potential risk of individual isolation (or exclusion) comes through critically thinking about the network paradigm. There are two aspects we need to consider in order to review the revolutionary and inevitable nature of the network paradigm in society. Firstly, it is the idea that we are not dealing with a completely revolutionary idea. Some critical thinkers argue that the network idea was not new. Mattelart for instance (2003) argues that since the nineteen century and with a concern to improve human communication in societies, the idea of a big communication network has been in the minds of many individuals. A network would facilitate the satisfaction of individual needs if we are able to highlight these needs and be supported by other people in the pursuit of needs' satisfaction. Behind this idea is the notion that mankind is in the pursuit of realization and that societies are to help individuals fulfill such goals. Paramount to the idea is the possibility of human communication directed to identify and address such needs, but change takes place within a social context. As simply put by May (2002):

...The vision of an information society itself often takes the character of an all-encompassing story about this new age...when we look at what allows some of us to become rich and the rest of us merely to get by on our pensions, this still has something to do with who owns what. (pp.1-2)

Secondly and in support of the importance of considering the social context in which e-participation is to take place, there is a systemic view of society within the idea of a network. In the twentieth century we see again that some researchers aim to model the behavior of human, animals and organizations, and that communication processes can be improved so as to enable those parties involved to grow together (Turner, 2006). The idea is more evident in the case of Japan, where a plan was drawn to develop the Japanese Information Society. According to Masuda, one of the key writers of the information society in Japan (Masuda, 1980):

The hypothesis can be formulated that the future information society will be a 'highly organismic society' resembling an organism. What I am referring to will probably be a multi-centered complex society in which many systems are linked and integrated by information networks. Moreover, this society will have the dynamism to respond more quickly and more appropriately to than contemporary society to changes in the external environment... (pp.58-59)

The above two aspects (consideration of the social context and the systemicity of the new information-network based paradigm) give us the possibility to consider different aspects that are to be interacting with each other at different levels (individual, collective) in the information society. The phenomena of e-participation seems to adhere to the information-based network idea of society, but now we need to consider the social conditions that can help its development as well as how to mitigate potential exclusions that result from it. Thinking in terms of 'systems' (e.g. social, technological, intra and inter-organizational) can help us to mediate in the dynamics of change. Using the ideas and language of systems, we can then explore the implications of their unfolding, as well as make use of the systemic nature of change to improve e-participation and with it the society as a whole. Even more, adopting a systemic perspective helps us to critically address the way in which e-participation and information society are becoming social constructs upon which social change is grounded.

There is another challenge to face. The pervasiveness of technologies in business organizations influences the ways in which we adopt them when it comes to use technology in the realm of the 'public'. The literature of management talks about the 'value' that technologies can leverage for customers, and of the importance of aligning corporate plans to those that aim to implement systems and technologies, all in the name of competitive advantage (Ward & Griffiths, 2002). This partly explains why now in the realm of public administration and for many societies, we now have a number of accounts that portray new forms of organization and management of the technology function within government (Contini, 2009; Dunleavy, Margetts, Bastow, & Tinkler, 2006), all of them driven by the notion of 'value' and 'customer-oriented' service. What is even more concerning is that the adoption of technologies for public administration or service delivery is said to be on the name of 'citizen-centred' perspectives of such services or functions. An example is the European strategy for e-government, whose policy documents reveal such a vision as a way of improving participation (European Commission, 2007).

This is also why we believe that e-government has been strongly associated with e-participation, and why we now make a plea for more socially accountable approaches to develop and assess e-government service performance. It is time we broaden our boundaries and look underneath the surface of adoption of technologies and explore with a more critical eye what we mean by e-participation. Is it something that should be 'aligned' to public service delivery? What value do we (or should we) associate e-participation with? How is that we are to consider the social context in which e-participation is said to flourish? How will people (us) be able to produce alternative notions of e-participation, and use technologies to enable such notions to co-exist (or fight against) existing notions?

To help the reader in seeing how e-participation is being looked at with the help of systemic ideas, we now present what we see as an official (and so called revolutionary) perspective of this phenomenon. We take this as a point of departure for critical and systemically oriented thinking so as to consider the social context of e-participation.

THE VISIBLE FACE TO E-PARTICIPATION: ELECTRONIC GOVERNMENT (E-GOVERNMENT)

As an initial and provocative element of the book, we now propose a perspective of the visible and widely used face of e-participation (electronic government) which seems to be the official one (to be) adopted by societies worldwide. It is a perspective that embeds revolutionary change. Interestingly, this perspective does not provide a unique definition of the term e-government. Bannister (2007) acknowledges this when he suggests that there is a spectrum of types of applications that are termed under the banner of e-government. West (2004) sees e-government as referring to the delivery of government information and services online through the internet and other digital means (p.16). What we can discern from these definitions and the development of applications related to 'government' is that they involve the use of ICT to deal with public (administration) matters and that require the concourse of citizens at some stage, very often at the stage of 'use'.

Interestingly, this perspective is one that sees a historical and linear progression in the interplay of technology and public service delivery and administration domains. The story behind it promises to deliver a number of benefits, one of which is to contribute to "build better relationships between the government and the public by making interaction with citizens smoother, easier, and more efficient" (Layne and Lee, 2001, p.123). It is a story with many manifestations worldwide and although it might be difficult to encapsulate it in a short ways, some stages can be identified.

The story begins when public administrations began using information technologies for their own administrative tasks. With the emergence of the web, the information to be offered to citizens in order for them to fulfill their part of the tasks was put into governmental websites. This is a stage called 'information', and it is still arguably the most popular and recognized by citizens in relation to their governments (Carter & Bélanger, 2005; West, 2004). It is included as part of the story to illustrate the nature of following changes as 'revolutionary'.

In a similar vein to the spread use of information systems for competitive advantage in organizations, the next stage has become the proliferation of transactional facilities that government offices offer to citizens. Thus the stage is called 'transaction'. Facilities include payment of taxes, renewal of licenses or acquisition of permits and certificates. E-government initiatives (and now the term appears too) focus on connecting an internal government system to online interfaces and allowing citizens to transact with government electronically (Layne & Lee, 2001:125). With automation, governments pride themselves in having achieved important cost reductions and efficiencies which the citizens (should) value as they also increase the transparency and accountability of government methods and processes.

The transactional proliferation leads then to the unification (or integration) of interfaces, platforms and technologies. Governments embark in the construction of sophisticated portals which although connecting legacy systems from different government institutions (via technologically remarkable platforms or architectures) offer one single type of environment (e.g. a single electronic interface) to citizens. The expectation is that with friendliness and efficiency, citizens will then become more prone to use online services. We see this stage in the form of national portal projects that take the form of one stop online shop for citizens (customers). A recent policy document of the European Union in relation to e-government describes this transition as:

Using what is termed a 'life-event' approach, citizens will be able to access government services using a web portal to select their desired life event, such as 'I want to get married' or 'I want to drive a car', without having to worry about the services, organizations or processes involved underneath (European Commission, 2007). (p.6)

This is what is termed 'citizen-centred' electronic government. Its purpose is to include different citizen stages. This type of discourse assumes that citizens are to be served throughout their whole life. It also assumes that governments are to integrate horizontally their services, so that citizens can do many things in one-stop online shops (similar to what one could do in a banking branch: buying insurance, a mortgage, and retirement products to cover the whole of one's life).

More often than not, horizontal or vertical[1] integration of services in portals seems ahead of the regulatory and appropriate environments for the sharing of electronic information (Contini, 2009), and lacks any explicit theoretical justification (Heeks & Bailur, 2007) . Despite these difficulties, the story continues progressing, now with consideration of issues of information privacy and confidentiality. Policy makers and adminis-

Figure 1. Stages of e-government (adapted from Layne and Lee, 2001)

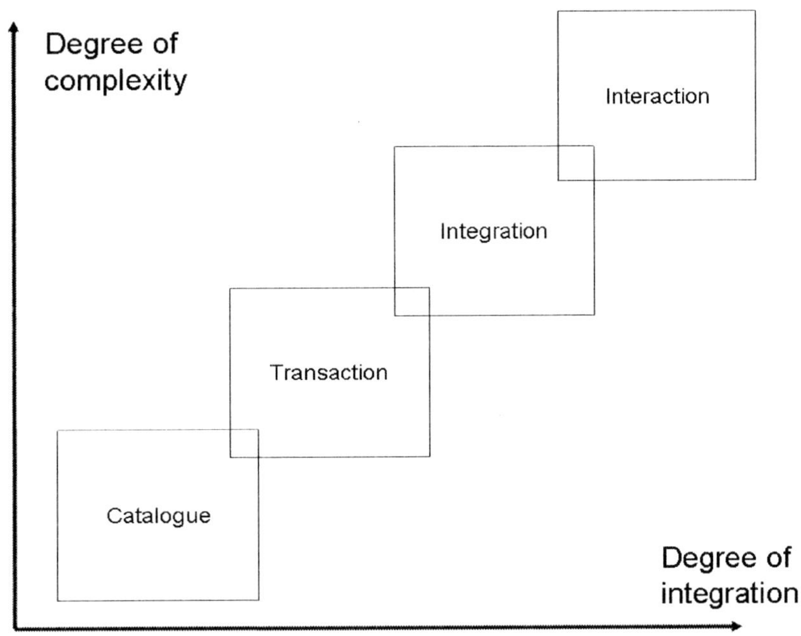

trations struggle to come to terms with possibilities and realities of information redundancy and ownership. Cases of lost of data have come to the fore, and the tensions between two different domains become evident: The technological and the governmental (Ciborra, 2001; Dunleavy et al., 2006). The initial success and possible integration of e-government into a technological infrastructure now generates constraints and potentially new problems for governments who want to innovate and get closer to their citizens. These problems are being seen as related to institutional factors, conflict between groups, culture and political behavior among others (West, 2004).

At this point the official story seems to acknowledge the need to consider the social context in which the visible face of e-participation is unfolding. Unfortunately, the focus is still on the progression towards transformation and managing the complexity due to arrival of new 'actors' (Dunleavy et al., 2006). Despite these minor 'social' hiccups, the official story suggests that there is a last stage to be fulfilled. It is the stage of *online interaction* with citizens. In such a stage, it is assumed that governance systems are in place, meaning that government systems are transformed for the benefit of the public; information is securely shared; trust is gained and maintained across government departments at different level (locally, regionally, nationally and internationally). Any barrier of communication is overcome, and it could look like all a citizen has to do is click and type (or even speak) to interact with their governments and be listened to or fulfill their duties. Safe and secured democracy now takes place in an electronic type of setting, market or one-stop shopping portals. Moreover, existing systems (stemming from government websites, possibly with mobile, and virtual access to them) move beyond service delivery to 'system-wide political transformation' (West, 2004). In the case of websites, they offer options for 'personalisation' according to one's own interests so that one can receive relevant information. People can provide feedback and exert some degree of accountability through two-way communication. They can engage in conversation and exert influence on government matters.

Technological design can make this type of engagement friendly so that confidence in government can increase. However the degree of influence of citizens in governing their own affairs via the government is an area which still is not discussed, partly because the challenges of institutional integration and transformation seem to be take most of the effort of those involved (policy makers, technology suppliers, government officers). Citizens and their needs take a secondary role in the process. This role is relevant for us though, because it could provide us with a space where systemic inquiry could reveal the becoming of a new form of society and of requesting political support and responsibility.

Within the official story, systems-thinking is starting to make an impact. Some of the existing 'institutions' (law, public administration) are seen as self-produce at the expense of others (Brans & Rossbach, 1997; Luhmann, 1996) at the expense of other systems (e.g. participatory systems). Outside the 'official' version of e-government, the use of systemic thinking could enable the identification of self-producing behavior for change. It can help us to see how we can as individuals provoke change and include several aspects of it. In this book we aim to show how the official and widely used story is only part of a whole phenomenon of participation in society, and how such a phenomenon has different manifestations in the social, economic, political and communitarian spheres of society. To be able to show this holistic nature of e-participation, we need to evaluate the impact it has had so far in societies and in different spheres, as well as how it is being impacted by them.

This story is allegedly biased towards what we consider a very linear (not systemic) way of thinking whose hiccups have to do with the context of people in which it is developed. The evaluation of the degree of success of e-government systems is limited to performance when not to the number of services being delivered online (Dunleavy et al., 2006). The story also makes it homogeneous the deployment of systems under the discourse of transparency and accountability, something to which citizens are supposed to have been consulted. This 'exclusion' is reinforced at the policy level. In Europe for example, the priority of eEurope initiatives are to first bring 'everyone' into the digital age and online; this can then pave the way to literacy and social inclusion (European Commission, 2007).

Moreover, the story generates a low degree of uptake of e-government services. Benchmarking is used as a practice to ensure homogeneisation and penetration of services 'online'. Benchmark shows that on the one hand, governments get better at what is to be measured. On the other, priorities efforts are determined by what is to be benchmarked, often leaving out consideration of aspects to be addressed if citizens are to be involved. As Bannister (2007)) says:

The result [of benchmarking] is that countries build clever and impressive looking systems that people rarely use (p.181) (brackets added).

EMERGING ALTERNATIVES: E-GOVERNMENT ASSEMBLAGES AND CONTEXT RE-INTERPRETATIONS

The official story on e-government thus begins to be reviewed. From a variety of perspectives, there is concern about how the challenges of integration between technologies, organizations and people are really dealt in practice. Contini (2009) shows how in reality some e-government systems are deployed with particular reference to the judicial realm. Using the concept of *assemblages* as collections of institutional and technological components which tend to maintain their specificity, he proposes a new 'model' for e-government systems. This model involves the following stages:

a. Identification of e-government services that can "reasonably be handled online"
b. Identification of simple *procedures* embedded in such services which can be functionally simplified, and of sub procedures that can be easily translated into strict cause and effect, streamlined and inscribed into an electronic (online) system. For instance, when making a claim against the judicial system, the payment is part of the procedure of completing a claim and it can be selected for automation.
c. Selection of appropriate gateways and interlocking mechanisms between online and offline procedures to support their smooth flow, even then these go beyond the limits of online proceedings or when it is appropriate to switch between online/offline. This includes for instance the use of existing payment facilities (credit and debit card) or the use of email systems to communicate with citizens and other stakeholders involved in a procedure.
d. Identification of organizations involved in the innovation process (some third parties included, and organizations that need to regulate the 'new' process)
e. Statutory changes to ease translation of procedures, so that in this case, online payment and electronic signatures are considered 'legal'.

Contini (2009) also suggests two principles can apply when developing effectively an e-government system. The first one refers to focusing on a specific user group to whom the existing infrastructure adds value via simple adaptation. This signals that e-government is to start from some feasible possibilities and use some technology. The second one refers to avoiding getting 'locked in' by the installed technological base so that dependencies from third parties, their norms and ultimately their interests are to be avoided. These two principles acknowledge the importance of what is in the context and what is feasible to accommodate given technologies and users.

Still, the focus on the interplay between concerned actors is limited to those being part of formal institutions. However, and between these two principles the role of a mediator gains importance. Such a role consists of (Contini, 2009):

[facilitating] communication between the parties to assist them in focusing on the limited compatibility between the elements to be assembled and in unraveling the techno- juridical knots in a workable way and, as a consequence, to design and set up robust assemblages. (p.268)

This speaks about the importance of being able to identify certain opportunities in the process of e-government transformation (Lyytinen & Newman, 2008; McLoughlin & Cornford, 2006), as to operate within constraints and possibilities that an e-government service can offer to users.

A perspective like Contini (2009)'s has similarities with those developed by Heeks and collaborators (Heeks, 2005; Heeks & Stanforth, 2007) to gradually manage implementation of e-government systems. Heeks (2005) identifies the existence of different (social) contexts: Design, implementation and use. In each, there is a variety of assumptions that influence the deployment of systems and which is necessary to identify, challenge and/or accommodate. Heeks et al suggest a piecemeal e-government development approach to indicate a more and context sensitive approach to its deployment. In such approach there is also involves consideration of people's values and concerns. Such concerns might be embedded in sub-networks of relations which also act as 'gates' that facilitate the adoption or rejection of information systems (Introna, 1997). There could be a variety of such sub-networks being developed through time; some of these are excluded or marginalized within the process of e-government systems definition and implementation (see the chapter by Cordoba and Orr in this book). It becomes essential to be able to identify and study the dynamics of such sub-network; in there we might find the seeds of new, innovative and more empowering forms of relations between individuals, some of which speak more truthfully about how they govern themselves and others than the official sub-networks of government in their social context.

The emerging nature of these and other approaches suggests that now people are to be at the front of e-government development, not necessarily at the front of its official manifestations. If consideration of the social context is taken into account, then the focus on e-participation systems (not only e-government ones)' deployment is shifted. Within such a context (and now with citizens fully included) there are a number of possibilities. We can improve the uptake of e-services (those to be deployed) but we can also question their nature and their implications for the relationships between individuals.

Moreover and using the ideas about the information society, we can then look beyond institutions and consider a more holistic type of development and evaluation of e-government. Complexity is not there to be tamed but to be operated within. Systemic thinking can reinforce the possibility of mediation in different types of contexts. It can also help us to critically review the notions that have been associated with e-participation and the context where it is supposed to take place (citizenship). We now conclude with some remarks about the role(s) that we attribute to systemic thinking in dealing with e-participation and the uses of information and communication technologies in societies worldwide.

CONCLUSION: THE ROLE OF SYSTEMS THINKING

Systems-thinking has been regarded at the same time as a discipline and as an operational arm of other disciplines that aim to pursue societal improvements. Confusions might still exist as to what is its true nature and purpose and as we see it, this is part of the continuous dialogue within and beyond the systems community. To us, systems thinking is an ally that will help us clarify existing conceptualizations about e-participation and e-government, and will also help us appreciate different ways of making these two phenomena more 'citizen' friendly. By t focusing on exploring these phenomena we also hope to inform discussions in systems thinking.

Moreover, adopting a systemic view will also enable us to include other perspectives and to enrich our current understanding of the phenomenon of e-government beyond the assumed limits of being a technological provision for running public affairs. We expect that such insights could reach to all of us interested from whatever perspective on he subject, and for the reader, to appreciate their history, similarities and differences with what s/he could be experiencing now as a citizen, technology practitioner, government officer or member of a community.

In this chapter we have started a disclosure and openness of the social scene in which e-government takes place as a technological answer but also as a social challenge. In following a path of a systemic inquiry it should become paramount that what has been a dominant (official, widely used) perspective of e-government requires to be understood and surpassed by a will for gaining comprehension on the issue beyond the dimension of the citizen as a user or customer. In such a strive lies the challenge of considering the effort contained in this book as a step towards a critical systemic account of e-government, e-participation and the configuration of a new social context in which we are already living without realizing the advantages and disadvantages that are self-contained in such a process. Therefore, it is not possible to arrive at a better conclusion now, except to indicate that the challenge to inquiry even further and more critically about the phenomenon although has been accepted here is far from being completely solved.

REFERENCES

Afele, J. S. (2003). *Digital Bridges: Developing Countries in the Knowledge Economy*. Hershey PA: Idea Group Inc.

Bannister, F. (2007). The curse of the benchmark: an assesment of the validity and value of e-government comparisons. *International Review of Administrative Sciences*, *73*(2), 171–188. doi:10.1177/0020852307077959

Brans, M., & Rossbach, S. (1997). The autopoiesis of administrative systems: Niklas Luhmann on public administration and public policy. *Public Administration*, (75): 417–439. doi:10.1111/1467-9299.00068

Brown, J., & Duguid, P. (1999). *The Social Life of Information*. Harvard (Mass): Harvard University Press.

Carter, L., & Bélanger, F. (2005). The utilization of e-government services: citizen trust, innovation and acceptance factors. *Information Systems Journal*, *15*(1), 5–25. doi:10.1111/j.1365-2575.2005.00183.x

Castells, M. (2001). *The Internet Galaxy: Reflections on Internet, Business and Society*. Oxford: Oxford University Press.

Ciborra, C. (2001). *From Control to Drift: The Dynamics of Corporate Information Infrastructures*. Oxford: Oxford University Press.

Contini, F. (2009). ICT, assemblages and institutional contexts: understanding multiple development paths. In F. Contini, & G. F. Lanzara (Eds.), *ICT and Innovation in the Public Sector* (pp. 244-271). London: Palgrave.

Dunleavy, P., Margetts, H., Bastow, S., & Tinkler, J. (2006). *Digital Era Governance: IT Corporations, the State, and E-Government*. Oxford: Oxford University Press.

European Commission. (2007). *Information Society and Government: Linking European Policies*. Brussels, Belgium: European Commission.

Heeks, R. (2005). e-government as a carrier of context. *Journal of Public Policy*, *25*(1), 51–74. doi:10.1017/S0143814X05000206

Heeks, R., & Bailur, S. (2007). Analyzing e-government research: perspectives, philosophies, theories, methods and practice. *Government Information Quarterly*, *24*, 243–265. doi:10.1016/j.giq.2006.06.005

Heeks, R., & Stanforth, C. (2007). Understanding e-Government project trajectories from an actor-network perspective. *European Journal of Information Systems*, *16*(2), 165–177. doi:10.1057/palgrave.ejis.3000676

Introna, L. D. (1997). *Management, Information and Power: A Narrative of the Involved Manager*. Basingstoke: Macmillan.

Layne, K., & Lee, J. (2001). Developing fully functional E-government: A four stage model. *Government Information Quarterly*, *18*, 122–136. doi:10.1016/S0740-624X(01)00066-1

Luhmann, N. (1996). *Social Systems*. (J. Bednarz, & D. Baecker, Trans.). Stanford, CA: Stanford University Press.

Lyytinen, K., & Newman, M. (2008). Explaining information systems change: a punctuated socio-technical change model. *European Journal of Information Systems*, *17*(6), 589–613. doi:10.1057/ejis.2008.50

Mansell, R. (Ed.). (2002). *Inside The Communication Revolution: Evolving Patterns of Social and Technical Interaction* (1st ed.). Oxford: Oxford University Press.

Masuda, Y. (1980). *The Information Society as Post-Industrial Society*. Tokyo: Institute for the Information Society.

Mattelart, A. (2003). *The Information Society: An Introduction.* (S. Taponier, & J. Cohen, Trans.). London: Sage.

May, C. (2002). *The Information Society: A Sceptical View*. Cambridge, UK: Polity.

McLoughlin, I., & Cornford, J. (2006). Transformational change in the local state? Enacting e-government in English local authorities. *Journal of Management & Organization, 12*(3), 195–208.

Menzies, H. (1996). *Whose Brave New World? The Information Highway and the New Economy*. Toronto: Between the Lines.

Perez, C. (1983). Structural change and assimilation of new technologies in the economic and social systems. *Futures, 15*(5), 357–375. doi:10.1016/0016-3287(83)90050-2

Toffler, A. (1992). *Power Shift: Knowledge, Wealth and Violence at the Edge of the 21St Century*. London: Bantam Books.

Turner, F. (2006). *From Counterculture to Cyberculture: Stewart Brand, the Whole Earth Network, and the Rise of Digital Utopianism*. Chicago: University of Chicago Press.

Van Audenhove, L., Burgelman, J., Nulens, G., & Cammaert, B. (1999). Information Society in the developing world: A critical assessment. *Third World Quarterly, 20*(2), 387–404. doi:10.1080/01436599913811

Ward, J., & Griffiths, P. (2002). *Strategic Planning for Information Systems*. Chichester, UK: John Wiley and Sons.

Webster, F. (2002). *Theories of the Information Society* (2nd ed.). London: Routledge.

West, D. (2004). E-government and the transformation of service delivery and attitudes. *Public Administration Review, 64*(1), 15–27. doi:10.1111/j.1540-6210.2004.00343.x

Wickham, J. (1997). Where is Ireland in the global information society? *The Economic and Social Review, 28*(3), 277–294.

Wresch, W. (1996). *Disconnected: Have and Have-nots in the Information Age*. New Brunswick, NJ: Rutgers University Press.

ENDNOTE

1 Layden and Lee (2001) distinguish two stages within the stage of transaction in e-government. They argue that vertical integration takes place when a service delivery requires the concourse of different government institutions. Once this is automated, horizontal integration of services takes place to cover the whole 'citizen life cycle' services under a single interface or transaction.

Chapter 2
The Problem of Governance from an Autopoietic Perspective (And a Critical Comment on the Role of ICT)

Amanda J. Gregory
University of Hull, UK

ABSTRACT

The aim of this chapter is to examine the role of Information and Communication Technology (ICT) in the governance of modern society from an autopoietic perspective. For the main part, discussions of ICT use by government have focused on either practical issues, dissemination issues, or the assessment of perceived benefits of e-government initiatives; in this chapter a more critical approach will be adopted. This chapter will adopt an autopoietic view in seeking to address the role of government as a steering mechanism in modern society and the use of ICT as an instrument of governance by administrators.

INTRODUCTION

The aim of this paper is to examine the role of Information and Communication Technology (ICT) in the governance of modern society from an autopoietic perspective. For the main part, discussions of ICT use by government have focused on either practical issues, such as the development of the necessary technological infrastructure to support e-government initiatives, dissemination issues, such as concern about trust in the security of the medium (see for example, Grest et al, 2005), or the assessment of perceived benefits of e-government initiatives, such as the ease of buying a vehicle tax disc on-line. In this paper a more critical approach will be adopted. This paper will adopt an autopoietic view in seeking to address the role of government as a steering mechanism in modern society and the use of ICT as an instrument of governance by administrators.

HOW DID WE GET TO THE STATE WE ARE IN?

In order to understand the role of government in modern society and the potential use of ICT as a tool of governance, it is first necessary to under-

DOI: 10.4018/978-1-60566-860-4.ch002

stand how society came to be organised as it is. Addressing the question of 'how did we get to the state we are in' requires reflection on earlier societal arrangements and how they have changed. There are many different competing theories on the causes of periods of change and stability in societal systems but that provided by systems theory, particularly complexity and autopoiesis, are highly persuasive.

In the next section, complexity theory will be used to explain the change from pre-industrial society to industrial society. The argument is advanced that the industrial revolution marked a bifurcation point from pre-industrial society (characterised by the integration of economic activity with social, religious and political functions) to industrial society (characterised by the differentiation of economic and other functions into separate sub-systems). Following this, the theory of autopoiesis will be used to help explain the steady state represented by industrial society.

The Industrial Revolution as a Bifurcation Point

According to Swedberg and Granovetter (2001), in the 1950s Karl Polanyi addressed the application of economic theory to preindustrial and industrial society and, in so doing, established that the industrial revolution marked a distinct point of change in social history. Such a distinct point of change may be seen to represent a bifurcation point. In a system that is at equilibrium or near equilibrium, there exists a steady state that is dependent on the value of certain control parameters. If the value of these parameters is exceeded through, for example an inflow of energy, behaviours are disrupted and a cycle of positive feedback may be initiated which serves to further amplify the effects of the surge in energy throughout the system. Consequently, the behaviour of the system grows increasingly erratic until the threshold of stability is reached which is marked by a bifurcation point. At a bifurcation point the nature of the whole system may change especially as, unlike systems that are at equilibrium or near equilibrium, far-from-equilibrium systems are highly sensitive to external conditions. According to Prigogine and Stengers (1984), "far from equilibrium, new types of structures may originate spontaneously. In far-from-equilibrium conditions we may have transformation from disorder, from thermal chaos, into order. New dynamic states of matter may originate states that reflect the interaction of a given system with its surroundings. We have called these new structures dissipative structures to emphasize the constructive role of dissipative processes in their formation" (p.12).

At the point of change, or the bifurcation point, the system may have several different paths open to it; the choice of path is essentially random and therefore unpredictable. Thus, "The indeterminacy at the bifurcation points and the 'chaos-type' unpredictably due to repeated iterations both imply that the behaviour of a dissipative structure can be predicted only over a short time span. After that, the system's trajectory eludes us." (Capra, 1996, p.178). Whilst prediction of long-term system behaviour may be problematic, Prigogine and Stengers emphasise the pattern that underlies these change processes, "We expect that near a bifurcation, fluctuations or random elements would play an important role, while between bifurcations the deterministic aspects would become dominant." (p.176). This pattern of behaviour is clearly reflected in recent social history and, having discussed the change from pre-industrial to industrial society, in the following section the steady state represented by industrial society, characterised by functional differentiation, will be addressed.

Industrial Society: A System Near Equilibrium

Industrial society is characterised by functional separation. Luhmann (1995) looks at society as closely approximating a self-producing or autopoietic system that is comprised of six subsystems: the economy, politics, law, science, religion and education (hence each sub-system, except education which imparts ideas and concepts from other subsystems, is itself autopoietic and each is environment to the others).

The theory of autopoiesis evolved from Maturana and Varela's (1980) exploration of what distinguishes living systems from non-living and how living systems persist despite changes in structure and components. Maturana and Varela proposed that the fundamental characteristic of autopoietic systems as living systems is autonomy which is realised through the self-production of the component parts. Indeed, Maturana (1975) states "...autonomy in living systems is a feature of self-production..." (p. 313). Hence, an autopoietic system is made up of networks of recurring interactions of the production of component parts.

The application of autopoiesis in the social domain has always been a contentious issue; Maturana expressed reservations about its direct applicability to such systems and a summary review of the literature reveals that theorists have adopted different stances (see for example, Gomez and Probst (1989), Robb (1989a and b), Mingers(1989, 1995)) and the question of what constitutes the component parts of such systems is a contentious issue. In his most influential working of autopoietic theory, Luhmann viewed communications as the elements of reproduction of society's sub-systems (1995) and later he viewed decisions as the elements of reproduction of organizations (Hernes and Bakken, 2003, referring to Luhmann, 2000). Decisions are communicated through a form of communication that is specific to that sub-system. For example, in the case of the political system decisions are communicated through assertions of power. It is important to recognise two things here. Firstly, when Luhmann uses the term 'communication' it is in a technical manner; he is not referring to the ongoing chatter that takes place between individuals in the organizational context since this is relegated to the realm of environment. Rather, drawing on Shannon and Weaver, Luhmann (1995) regards communication as comprising of: information (what the message is about), utterance (form in which the communication is expressed) and understanding (meaning that the communication creates). Secondly, 'for Luhmann, 'power' is not a 'repressive' or 'ideological' force, it is simply a functional resource facilitating the implementation of some decisions and courses of collective action over others and for moving society forward.' (Rempel, p. 64).

According to Dunsire (1996), the dynamic of social subsystems is driven by a binary relation: having a property or not having it. The binary relation in the political sub-system is that of holding versus not holding political office. According to Rempel (1996), "The first side of that binary, holding office is what yields the formal authority to communicate and act in politics. In other words, it comprises the basis of the authority to use the generalized medium of power to execute collectively-binding decisions. Furthermore, "the binary splitting of the code in this way...creates elements that only have meaning in relation to the other of each pair...they belong only together..." (Dunsire, p. 304). Hence, "A conservative party is only necessary because a progressive party has emerged...The mutual check-and-balance of these coupled actors in their niche regulates from the inside what cannot, in autopoietic theory, be regulated from without." (Dunsire, p. 305).

Controversially, it has been suggested (Vanderstraeten, 2005) that there is a tension within the theory of autopoiesis between the notions of self-production and the closure of the system with the coupling of the system to the sub-systems that form its environment; this suggestion does not sit easily though. Bailey (2005) argues that "Luhmann (1989, 1995) decried such over-reliance on internal models, saying that the real value of social systems theory was in the study of flows across system boundaries... He also expanded the traditional model by recognizing that not only do holistic systems interact with their environment but also that their internal system components can have their own independent interactions with the environment." Hence Bailey appears to applaud Luhmann for shifting away from the traditional emphasis on internal operations to emphasising the potential for coupling between sub-systems. It should be recognised that Luhmann rejects the traditional organic view that regards systems as being environmentally determined. From an autopoietic stance, a system changes in response to perturbations from other sub-system but this is not in a strict tit-for-tat fashion. The autopoietic system is structurally constrained to react in a manner amenable to the maintenance of its own autopoietic state. Hence the autopoietic system responds to a perturbation by producing a feasible set of responses (communications/decisions), consistent with its own prior responses. Therefore, the autopoietic system is neither determined by its environment nor its internal operations alone; it is a product of the interaction of the two and it is said that through structurally coupling the two co-evolve.

According to Hernes and Bakken (2003), Luhmann's autopoiesis represents "a correction of Parsonian structural functionalist theory" (p.1512). For Parsons, separation of specialist functions within society introduced the problem of integration whereas, from Luhmann's perspective, the problem is one of there being a lack of differentiation (Rempel, p.59). Any overlap between systems would mean that they were not sufficiently differentiated from other sub-systems in their environment and hence vulnerable to being overcome by environmental complexity. Further, with regard to the environment, according to Vanderstraeten, "In important regards, the current form of functional differentiation narrows society's attention span. Because every function system is solely and completely responsible for its own function, it tends to observe its environment only in as far as this environment is relevant within its programmatic perspective.". Further, the different sub-systems adopting their own sub-system specific medium of communication is a way of attenuating environmental variety in that they cannot talk to one another

Clearly, autopoietic systems are predisposed to looking after their own interests but when this is to the cost of the environment that sustains them then they are said to have become pathologically autopoietic (Beer, 1985). To avoid this, some kind of co-ordinating and control function to balance the demands of the different sub-systems and punish those that err towards pathological behaviours is required. But, as Vanderstraeten recognises, "In modern society, there is no supervising reason, no centre, no apex that would allow steering of the development of society and its connections with the environment."

Even if there were such a guiding mechanism, its attempts to regulate societal sub-systems and organizations would meet with resistance. Chapman (2004) views this resistance in terms of resilience, thus "Viewed from this perspective the resistance to change exhibited by many organisations is not because of bloody-mindedness on the part of the individuals involved, although that may be a contributing factor. The resistance to change is actually a measure of an organisation's ability to adapt; it is a measure of its resilience. This resilience is therefore expected to be greater the longer the institution has existed and been required to adapt – which is broadly the case." (p. 53). Given the resilience of such systems, the important question is whether this is desirable or not. It should not be assumed that all theorists who have argued for the existence of the autopoietic organization are claiming that the existence of such organizations is desirable. According to Beer (1975), quoted by Mingers (1989), "... any cohesive social institution is an autopoietic system – because it survives, because its methods of survival answer the autopoietic criteria, and because it may well change its entire appearance and its apparent purpose in the process." (p. 172). This ability to persist despite, as Mingers puts it,

"...deliberate and sustained attempts to destroy them..." (p. 172) surely introduces doubt about whether or not such organizations can be managed or directed given that Mingers cites Faucheux and Makridakis (1979) as arguing that they are characteristically autonomous. Indeed, Robb (1989b) declares that, "To those who would see the achievement of autopoietic organization as a desirable objective in organizing, I warn that such an aim may result ultimately in the subordination of all human aspirations and ambitions, values, and welfare to the service of preserving the unity of such systems, and not to any human end. Once formed such organizations appear to be beyond human control, indeed to be real-world systems." (p. 348). Robb sounded an early warning signal stating that "The received wisdom (e.g. Peters and Waterman, 1982) that we shall always be able to make interventions which will loosen up organisations and induce cultural changes so as to direct the organisations activities to serving human purposes is very much open to question." (1989a, p. 250).

GOVERNMENT AND GOVERNING

In the previous section, autopoiesis in general and Luhmann's work on the application of autopoiesis to explain society were discussed. In the context of this paper, this work is important because it throws considerable light on the issue of the government's ability to govern and in this section we will focus particularly on this.

The political subsystem is the subsystem with the function of making collective decisions binding on society. However, the binary dynamic of the political subsystem means that it is only the political party that holds office which yields the formal authority to 'form a government' in order to execute collectively-binding decisions. This notion of 'forming a government' is important because the additional self-description reinforces the unity of the political system. As Luhmann (1995) states "Orientation to the state enables the closure of self-reference...and couples it with matters to be decided, interests, and structural changes in the political system's environment. Thus here too self- and other- reference are processed simultaneously, so that order is continually reproduced on the basis of order and disorder." (p.463).

Whilst 'the government' has the authority to execute collectively-binding decision it should be recognised that, according to the logic of authopoiesis "the self-referential and organizationally closed nature of other subsystems means that it cannot regulate them." (Dunsire, 1996, p.300-301). This limitation, though, does not mean that it does not have ambitions to regulate other sub-systems, "The state or government is but one corporate actor among many; but, hubristically or not, it conventionally takes upon itself responsibilities for at least maintaining the society's integrity against external threat or internal disruption and claims legitimacy for the pursuit of other electorally ratified goals." (Dunsire, p.307). If systems are to maintain their autpoieisis then "the possibilities of one social subsystem (the political) governing another, or steering the society as a whole, look remote" (Dunsire, p. 308). Dunsire goes on to elaborate on this theme by quoting an example from Teubner (1992): "if the state decides to impose a price freeze on the economy, the state can act only by creating such noise in the environment of the legal system that the law learns to change its internal order by such cognitive and normative operations as will produce a price control law, which can do no more than produce further legal operations. That law by itself does nothing but make environmental waves for the economic system, such that (if it has to) it, in turn, learns to adapt internally..." (Teubner, 1002, p. 622, quoted in Dunsire, p. 308). It can be seen that ICT may play an important role in this creation of 'noise' and hence influence.

We have reached an important point in this paper. It has been established that autopoiesis places severe constraints on attempts by one subsystem to steer another, or society as a whole. Hence autopoiesis opens up the opportunity to look at constraints on the role of government in a different way to the dominant narrative which is influenced by notions of public choice. An alternative perspective is reflected here in looking at possibilities for governance, political intervention in the operations of other social subsystems, compatible with their self-referentiality and the raising of important questions such as 'does it make sense to talk of government governing or is this mere hubris'?

According to Dunsire (1996), 'autopoietic theory presents social actors who seek to discharge societal responsibilities by exerting governance (administrators) with problems but the possible solutions to such problems work by finding a way of tapping into their selfgovernance and harnessing it to public purposes by using the familiar tools of government'. Dunsire (1996) goes on to suggests three main possibilities (subsidization, neocorporatism, and reflexive law) to which he adds a fourth strategy that he entitles, co-libration:

- **Subsidization:** Social systems require energy inputs such as financial subsidies. Attaching conditions to such grants, loans, or subsidies even unspoken conditions is subject to all the problems of attempting to regulate them: The target system either does not receive the message or it tries to adapt and endangers its own self-reproduction (it is "colonized" by the political system) (Habermas, 1981, p. 542). Hence the effective use of this strategy relies on the selection of a social actor or institution which is already performing what the administrator regards as a public service to support.
- **Neocorporatism:** Political actors enter into a form of coproduction with economic or other social actors by common membership of or communication with intermediary bodies operating in the 'public interest').
- **Reflexive law:** Reflexive law is more of an application of pure legal instruments to impose constitutional or procedural terms upon the internal operations of the target institution. Basic identity is preserved as this form of intervention modifies only structural relations, such as modes of decision making, without affecting organizational closure.
- **Co-libration:** Co-libration is an intervention into the dynamics of the reproduction of structural relations that makes use of the built-in checks and balances of a particular kind of social subsystem or action arena and introduces a bias or compensator affect such that it arrives at a steady state when otherwise it might not.

Having outlined the main strategies which administrators, as government agents, may use to intervene in the operations of other social subsystems, Dunsire (1996) goes on to evaluate their steering capacity:

- The pure subsidization method scores least on policy influence, unless accompanied by conditions whose enforcement would negatively affect integrity; and the only sanction is withdrawal of the subvention, which impinges on what steering capacity there is
- Partnership with intermediaries is more promising because there is a trade-off situation. Intermediary and administrator come to share a common view as they become mutual beneficiaries
- The reflexive law strategy promises great steering capacity because it has the power of social pressure behind it; for the very purpose of encouraging subsystems to externalize more, to give meanings to their environment that match its demands. However, reflexive law does not confer steering over substantive decision making.

- Co-libration works on the dynamics of the situation: fine-tuning by trial-and-error, changing a little at a time, letting things settle, seeing whether a bit more change is required. This can be a discourseless process: more focusing on changing the the conditions of engagement than heavy negotiations or explicit pressure.

In this section, the work of Dunsire on outlining and evaluating the various strategies available to enable administrators, as agents of the political system, to steer other sub-systems and society at large, has been described. The potential role of ICT in this will be taken up in the next section.

ICT ENABLED GOVERNING

In the previous section Dunsire's discussion of the relative steering capacity of the different strategies was summarised. It may be argued, though, that Dunsire's elaboration of these strategies implicitly assumes the traditional instruments of governance that administrations have in their armory such as bureaucratic and legal measures. ICT potentially represents a 'new' instrument of governance which may work to particularly promote the effectiveness as co-libration as a steering strategy by enabling improved communication between adminstrators and other parties.

This notion that the putting in place of an ICT infrastructure may be insufficient to bring about effective e-governance is reflected in a recent study. A study by Gartner Inc. Analysis in 2002 found that more than 60 percent of e-government initiated fail or fall short of their objectives and one of the reasons cited for the high failure rate was a lack of adequately skilled personnel (Matthews, 2002). Interestingly, though, it may be that ICT changes not only imply certain skills requirements but also certain personality types: 'Traditionally, governments have designed their human resources practices to attract people who like to work in a secure, stable and often less competitive workplace. For e-government, they need people who not only adapt to change quickly, but also drive change.' (Judith Carr, vice president and senior program director for Gartner's Executive Programs, quoted in Matthews, 2002). Autopoiesis opens the opportunity for such change "...our knowledge and understanding change over time: we are capable of *learning* with others. According to Maturana (1988), language has emerged as the by-product of the autopoiesis of individual human beings not only act, we also strive to coordinate our actions. Language helps in this process: it allows us to coordinate our coordination of actions. Over time, a particular use of language to co-ordinate co-ordinations of actions may become more and more elaborated, allowing people to exist in very subtle, well-coordinated relationships – and these can be distinguished in conversations. Therefore, whole rational domains of interaction, relevant to human activity systems, are created in language." (p. 1068).

The use of ICT as a tool to facilitate communication between administrators, organizations and actors is an important one. ICT can amplify the ability of the government, through administrators, to gather massive up-to-date information about the other and to present that information in the most meaningful way thus facilitating mutual understanding and influence between administrators and the organizations and actors they are seeking to steer. As Córdoba and Midgley (2006) recognise "...people should be able to raise concerns that they see as relevant to the improvement of their lives as constituted by the different domains of action in which they participate (family, business, community, etc.) (Córdoba and Midgley, 2003). An ideal situation in planning involves reviewing one's own assumptions and listening to other people's; taking account of similarities and differences in the concerns expressed; and looking for synergies that can be expressed in actions." (p. 1069). Hence ICT use by government makes it easier for relevant parties to communicate their concerns and interests but is the mere development of an ICT equipped government administration sufficient to bring *understanding* about?

According to Córdoba and Midgley (2006) the theory of autopoiesis can bring a sense of ethical awareness which implies that "As individuals, we should recognize the inevitable limitations of our own knowledge that stem from our local histories and positionings, and take responsibility for the consequences of our actions. In planning, people should therefore be encouraged to reflect on and review their own assumptions and values in relation to situations. Second, there is the possibility of *creating new understandings* about a situation through conversations with others. When we encounter divergent views about a situation, we should assume an attitude of openness and respect for the concerns of others, and consider possibilities for creating new domains of action in which multiple concerns can be addressed." (p. 1068). To reflect on the obligation of government administrators to be ethically aware is an interesting thought experiment. In the UK, the television sitcom 'Yes Minister' plays a strong influence; the show having a recurring theme of government administrators sacrificing any altruistic desires to do 'the right thing' to their self-interested needs to secure some kind of political advantage enabling elevation to higher office.

In the light of the above, it may be concluded that developments in the use of ICT are necessary to making co-libration a more effective strategy for enabling government to govern. It is easy to focus on what else needs to happen to further enable this development, such as skills enhancement, without stopping to think about whether the use of ICT as a tool in the arsenal of government administrators to steer other societal sub-systems is a good or bad thing. The use of ICT under various e-government initiatives is assumed to deliver assorted benefits to agencies and the public, it should also be consider though that it may rather less serve empower citizens and more to empower government and the administrators through which it acts.

CONCLUSION

In this paper the argument was advanced that industrial society is characterised by the differentiation of economic and other functions into separate sub-systems. The self-produced nature and necessary closure of each sub-system places severe constraints on any attempts by one subsystem to steer another, or society as a whole. Hence whilst the political system is responsible for the making of collective decisions, it can only aspire to the regulation of their implementation by other sub-systems and the steering of society at large. Autopoietic theory is a problem for governmental intervention and in seeking to address this problem, a range of strategies and tools have been developed including various ICT applications collectively referred to as e-government. Whilst the use of ICT is commonly regarded to be a mechanism for empowering citizens a long term evaluation of its use may reveal it to more empower government and the administrators through which it acts.

REFERENCES

Bailey, K. D. (2005). Beyond system internals: Expanding the scope of living systems theory. *Systems Research and Behavioral Science*, *22*(6). doi:10.1002/sres.659

Beer, S. (1985). *Diagnosing the System for Organizations*. Chichester, UK: John Wiley and Sons.

Capra, F. (1996). *The Web of Life: A New Synthesis of Mind and Matter*. London: Harper Collins.

Chapman, J. (2004). *System Failure: Why Governments Must Learn to Think Differently (2nd Ed.)*. London: Demos.

Córdoba, J.-R., & Midgley, G. (2006). Broadening the boundaries: an application of critical systems thinking to IS planning in Colombia. *The Journal of the Operational Research Society*, *57*, 1064–1080. doi:10.1057/palgrave.jors.2602081

Dunsire, A. (1996). Tipping the balance: Autopoiesis and governance. *Administration & Society*, *28*(3), 299–334. doi:10.1177/009539979602800302

Gomez, P., & Probst, G. J. B. (1989). Organizational closure in management: A complementary view to contingency approaches. *Cybernetics and Systems*, *20*, 311–320. doi:10.1080/01969728908902210

Grest, D. W., Taylor, M. J., & Lunn, J. (2005). Legal consideration in cyber-activism. *Systems Research and Behavioral Science*, *22*(6), 565–570. doi:10.1002/sres.625

Hernes, T., & Bakken, T. (2003). Implications of self-reference: Niklas Luhmann's autopoiesis and organization theory. *Organization Studies*, *24*(9), 1511–1535. doi:10.1177/0170840603249007

Luhmann, N. (1995). *Social Systems.* (J. Bednarz, Jr. & D. Baecker, Trans.). Stanford, CA: Stanford University Press.

Matthews, W. (2002). E-gov falters easily. *Federal Computer Week*, *13*, 43.

Maturana, H. R. (1975). The organization of the living: a theory of the living organization. *International Journal of Man-Machine Studies*, *7*, 313–332. doi:10.1016/S0020-7373(75)80015-0

Maturana, H. R., & Varela, F. J. (1980). *Autopoiesis and Cognition: The Realization of the Living.* Dordrecht, Holland: Reidel Publishing Co.

Mingers, J. (1989). An introduction to autopoiesis – implications and applications. *Systems Practice*, *2*(2), 159–180. doi:10.1007/BF01059497

Mingers, J. (1995). *Self-Producing Systems: Implications and Applications of Autopoiesis.* London: Plenum.

Prigogine, I., & Stengers, I. (1984). *Order Out of Chaos: Man's New Dialogue with Nature.* New York: Bantam Books

Rempel, M. (1996). Systems theory and power/knowledge: A Foucauldian reconstruction of Niklas Luhmann's systems theory. *The International Journal of Sociology and Social Policy*, *16*(4), 58–90. doi:10.1108/eb013250

Robb, F. F. (1989a). The limits to human organisation: The emergence of autopoietic systems. In M.C. Jackson, P. Keys & S. Cropper (Eds.), *Operational Research and the Social Sciences.* New York: Plenum Press.

Robb, F. F. (1989b). The application of autopoiesis to social organization – A comment on John Mingers' 'An introduction to autopoiesis – implications and applications.' . *Systems Practice*, *2*(3), 343–348. doi:10.1007/BF01059979

Swedberg, R., & Granovetter, M. (2001). Introduction. In M. Granovetter & R. Swedberg (Eds.), *The Sociology of Economic Life (2nd Ed.)*. Oxford, UK: Westview Press.

Vanderstraeten, R. (2005). System and environment: Notes on the autopoiesis of modern society. *Systems Research and Behavioral Science*, *22*(6). doi:10.1002/sres.662

Chapter 3
Technology and Government:
The Pursuit of Governmental Technologies in the Present

Alejandro Elias Ochoa-Arias
Universidad de Los Andes, Mérida, Venezuela

ABSTRACT

This chapter develops an argument of both technology and government as historically determined phenomena within what has been called liberalism. Technology helps the government to make clear and assist with individuals' choices. The author asks the question about the existence of an alternative ideological framework to guide the use and implementation of both government and technologies and defines possible ways forward for the pursuit of citizenship in societies.

INTRODUCTION

One of those many challenges currently posed to almost every sector of society is an increasing complexity associated with information flows and communication channels in which society is currently embedded.

The ever increasing capabilities of information processing and communication has been associated to a widespread belief that there is universal access to information and communication technology (ICT) for everybody in society. Therefore, it was concluded that it is a matter of developing appropriate technical skills in order to constitute a platform for public decision making and running of public affairs based on ICT.

This chapter unveils some of those hidden grounds upon which relationships between government and technology could be understood as being driven by a will to power that inevitably knots together technology and government. The assumed neutrality and transparency based on the ICT incorporation in running public affairs requires to be assessed from a more critical standpoint which could address technology as a socio-historical outcome of a process of political and cognitive struggle concentrated in the dominance of the world.

In order to do that, a brief account of current society from a technological-driven perspective and the way in which society can be ruled provides

DOI: 10.4018/978-1-60566-860-4.ch003

an ideal type of society which will be contrasted with current affairs and tendencies regarding the incorporation of ICT in government. The contrasts will allowed us to reveal how far the process of a technological driven society is opening up a new way of ordering society as a whole. Concluding remarks refers to some of the side-effects of such processes in the present.

TECHNOLOGY IN THE PRESENT

There is not doubt that technology plays a key role in the conduction of our current affairs. It could be argued that technology allowed us to grasp a new way of being human. Technology makes us free from external constraints and internal shortcomings to deal with the world. In this sense, technology can be understood as the constitution of a "world of second order" based on certainty and predictability. Indeed, the claim of technology as a vehicle for alienation of human beings from their own contexts of meaning or life, has been a common complaint among social theorists who considers the whole process of technology as being beyond the control of human being, at least in the current state of affairs. Of course, there is another group of advocates who claim that technology is an inevitably process in which human being is engaged in the realization of a natural condition for human being to become "homos faber" (Arendt, 1958).

However, these claims are misleading. On the one hand, technology has been considered as leading towards a determinism of society in which power is channeled and exercised by a very exclusive elite based on technical reasons. In this sense, scientific knowledge and technology have been a birth mark of authoritarian processes based on power-knowledge as it was explained by Foucault. The process of linking knowledge as a constitutive part of the rationale upon which society is designed and governed implied a growing role for the experts and an ever diminishing role for the multitude or masses. In this regard, technological determinism would drive human being to a comfortable life but without freedom. It could be understood as a totalizing life in which human beings become stranded from themselves in order to adapt to a complex technical world. A second possibility opens up a way towards an understanding of technology as providing an ever changing arena for the debate and collective construction of the future by considering technological devices and rationale as the most advanced stance of human beings in building up their own constitution as "homo sapiens". (see Feenberg for a complete account on this perspective) (Feenberg, 2002)

The wide range of possibilities upon which we can consider technology as an "iron cage" based on instrumental rationality or the most advanced ground for a debate about the civilization processes in which human being engages is just a revealing sign of the holding sway of technology and discourses associated to it in current times. It ranges from a despair based on an inevitable destiny driven by a rationality beyond human control to a hope based on a capability of correcting and improving the human products by humankind itself. Therefore, it seems that technology is inevitably tied to a conception of power. In a nutshell, technology potentiates human beings in a web of relations in which actions are possible due to a will to power.

Briefly, it implies that a way in which technology holds sway in human action is based on a will of control and predictability. It is important to remark that technology is considered here as a complex socio historical product in which discourses, devices and modes of enquiry are intertwined and bring together a unity called "technology". There are two particular dimensions of technology upon which this chapter is grounded. They are the role of discourses and modes on inquiry in the historical development of a link between technology and government.

The notion of a discourse of technology is understood here as a set of arguments presented in order to argue about the process through which human being is capable of ordering and manipulating things and people in order to accomplish the conduction of them as desirable and to be considered as good in itself. In this line of reasoning, it is important to remark that it is not a technical discourse. Instead it is a set of beliefs, approaches and claims about the behavior of the world that ensures a space for the conduction of the human and natural affairs alike by the development of a capability of prediction and control.

This discourse is inevitably tied to a way of inquiring into the world. In this regard, it should become apparent that there is a notion of regularity and independence from the subject who inquiries, and the subject which is studied. In this sense, the separation of both as independent parts plays a key role in ensuring that conduction of things and people can be performed from rational grounds and without any attachment to a specific set of attributes regarding the subject. As soon as the world is independent of a specific will then, in this same proportion, it will be subject of control and prediction by getting the appropriate laws and patterns of behaviour and to gain systematic knowledge about them. It suggests that the driving force is to enhance power by disembodying it from a specific agent or actor. The will to conduct will be powerful if it does not depend on a specific will but instead, of a given set of norms that regulates such a behaviour. The strength of power lies on its ability to hide (see Foucault, 1991).

Therefore, the pursuit of a knowledge systematic and rationally acceptable and regulated implies a pursuit of a framework in which the validity is not contingently tied to processes involving human will. A reductionistic approach will render that framework possible. That is the process in which knowledge is separated from its historical conditions. This made possible a way of developing technology and the ordering of the world were considered as being driven by an instrumental interest (Habermas, 1972).

From above, it should become evident that technology is grounded upon a point of view in which the world is considered to be a dispositive to be used. Such viewpoint is self-legitimising. Such an approach was considered as the most legitimate form of gaining knowledge and control over the world. However, there is a twist that we require to announce now and to deal with it later on. It is the power of information and its processing as a step forward towards a disengagement of human beings of decision making processes regarding the current state of affairs. Indeed, the changes implied on ICT lies not only on the capabilities of communication in order to enhance human relationships, but also to regulate and to make homogeneous those grounds upon which decision making processes can be reached without compromising any subjectivity. In this regard, it is important to remark that one of the conditions regarding the bureaucratic organization, a form of social organization dominant in modern world, is precisely based on a pursuit of a ruling based on anomity and expertise in performing organizational actions. How it became a source for conducting human beings is explored in the next section.

GOVERNMENT IN THE PRESENT: A BRIEF INSIGHT

In Foucault account on the definitions of forms of government on contemporary society, there is a claim that the introduction of economy into politics constitutes a cornerstone for the development of government: "Government becomes concerned with the right manner of disposing things [..] so as to lead to an end which is "convenient" for each of the things that are governed. This implies a plurality of specific aims" (Foucault, 1991, p. 95).

These primary stages of government are considered by Foucault as a first step towards considering the population as the end of government. Population understood as a biological variable in which certain regularities are manifested such as mortality rates, diseases, etc. This allowed a definition of the objectives of a new form of government: "Interest at the level of consciousness of each individual who goes to make up the population, and interest considered as the interest of the population regardless if what the particular interests and aspirations may be of the individuals who compose it. This is the new target and the fundamental instrument of the government of population" (Foucault, 1991. p. 100).

It is important to note that the interest of government is towards the whole mass of undifferentiated individuals, but also to each individual in particular. Government is not only a search for common conditions for the enhancement of productive labour, but also to control the conditions of each individual. The rationality becomes this form of government is what Foucault calls as governmentality. It is understood to mean three things.

1. The ensemble formed by the institutions, procedures, analyses and reflections, the calculations and tactics that allow the exercise of this very specific albeit complex form of power, which has as its target population, as its principal form of knowledge politcal economy, and as its essential technical means apparatuses of security
2. The tendency which, over a long period and throughout the West, has steadily led towards the pre-eminence over all other forms of this type of power which may be termed government
3. The process, or rather, the result of the process, through which the state of justice of the Middle Ages, transformed into the administrative state during the fifteenth and sixteenth centuries, gradually becomes "governmentalized". (Foucault, 1991. pp.102-103)

From above, it becomes obvious that emphasis on the notion of population as the object of government ignores the ethical value of human being. It discards a capacity to criticize and engage in the definition of the norms governing the whole society. Political debate is replaced by technical knowledge about the management of population as a biological entity, hence a controllable variable (Gordon, 1991).

The apparatuses of security are not limited to a repressive side. It is much complex and includes other devices that control the conduct of the population. Health care and welfare policies are examples of such devices.

The control of behaviour became simultaneously more complex and stricter. The discourse about population is always a technical discourse in which statistics played a key role. Individuals showing their specific and peculiar conditions are subsumed under a homogeneus category (Gordon, 1991).

This approach to government as being focused in the commonest needs, but also, those whose satisfaction is at individual level, opened up a way to exercise the control of human beings without invading or becoming intrusive in the exercise of their freedom. In this regard, such a way of conceiving the government is very appropriate for a liberal viewpoint of society.

But, there is a feature that requires to be commented, the technical means based on security. Indeed, it has become paramount for this definition of government, a concern about the provision of security. In this sense, prediction and control become key factors to ensure that security is provided. Therefore, the issue at stake lies on a capability of ensuring that the conduction of society as a whole could be addressed in terms of ensuring a possible path of behaviour and mechanisms to control it.

From a liberal perspective, the emphasis would lie on orchestrating control without threatening individual's freedom and their individual pursuit of each individual's good. Therefore, it becomes critical to show any mechanism of control as a promoter or propellant of freedom. In this sense, there is not a community of autonomous individuals free to their own forms of life, but a set of individuals with a space for freedom defined through the intervention of social forms, such as the state, capable of orchestrating themselves through the pursuit of purposeful interests. In this sense, the path of behaviour is reinforced by considering freedom as the starting point and also the arrival of society as a whole when society is considered as a project. On the one hand, it is assumed that free individuals is the starting point to define and orchestrate the mechanisms of control, but on the other hand, the constitution of free individuals is the expected outcome from a successful government. The mechanisms of security will be identified as the way to ensure the exercise of freedom.

Therefore, it should become evident from above that liberalism constitutes a very sophisticated form of conducting behaviour by arguing that such a control is not possible. The disguise of a powerful force implying a form of social bonding by referring to freedom is paramount on the constitution of government as a technological device. Government in itself became a technology to exercise control by ensuring paths of behaviour without compromising individual freedom explictly, but imposing a form of life upon the grounds of defending freedom of choice. Such a peculiar condition of a government gaining its momentum by claiming that there is not an imposition or ordering of things is central to the constitution of government as a technology.

Consequently, governmentality is conceived as a historical consequence inscribed in the domain of liberalism. The exploration on other forms of social ruling or political ideology is not relevant for the argument developed in this paper. In this regard, it is important to remark that in the framework of liberalism and government, the ensemble of institutions becomes a hidden and not easily traceable form of power whose dominance stems from its non-indicable condition. The constitution of society as a whole as a dispositive of security to ensure freedom makes of government a technology for the control of society. It is now relevant to consider how ICT becomes a dispositive for the technology of government.

THE ROLE OF ICT IN THE TECHNOLOGY OF GOVERNMENT

In the previous section, the building of government as a technology to dispose of society upon the grounds of enhancing freedom by the provision of security showed us that such an account was basically driven by the historical outcome of the dominance of liberalism as the ideological framework in which modern society is grounded. In this regard, it could be argued that from the standpoint of a search of technological support for the governmentalization of society, the liberal standpoint is a very favorable framework upon which the incorporation of a technology to support government as technology could be supported.

In this sense, the provision of information and communication plays a key role in the process of reducing uncertainty and to allow the identification of possible courses of action for individuals and collectivities. Therefore, it should become evident that in order to exercise freedom of choice, there is a need of knowledge about choices available and the probabilities of success and risks associated with each one of them. But also, it will be helpful to gather data, information and gain knowledge about the regularities of the population as a whole and at individual level. In this regard, ICT provides a useful tool to grasp and deal with large amounts of data and to generate qualitative and quantitative information. The provision of statistics and a very complex set of tasks and procedures to data mining, prospective studies and forecasting plays a key role in the planning and public decision making procedures.

However, there is also a very relevant issue that arises with the presence of technological devices devoted to control information and communication. It refers to the capability of using technology as a anonymous device for the gathering of information and communication. The sudden and sustained battles on the legal arena and in the defense of rights regarding data access, information provision and the use of individual information to be used for other purposes different to those allowed at the moment of gathering them, are examples that unveil a way in which the use of ICT has become so dominant and relevant to exercise a close monitoring on the behaviour of individuals and collectivities.

In this sense, it should become evident that current use of ICT is particularly beneficial to the procedures of monitoring and controlling population without endangering their freedom. Let us now characterize the way in which this use of technology could enhance the way in which individuals exercise control over the institutions devoted to the government of society.

This side of the equation is rather blurred. Indeed, it seems that the exercise of control and monitoring of the state by its citizens is not currently an issue in which governments have placed their major concerns, efforts and resources. By contrast, it has been the emergence of a sort of a "digital citizenship" upon which the use and diffusion of ICT as a mechanism to exercise social control over the performance of the states and public offices has occurred. Such control held by citizens is some how a resemblance of the balance checking upon which the modern state was grounded. That is, the existence of independent political bodies such as national assemblies, parliaments and the administration of justice as separated and autonomous entities engaged in the self-contention of the state, in particular, executive power. However, an interesting issue at stake here is the individualization of the so called "political control" exercised through the use of ICT as a device for gathering information and diffusing information and knowledge. The potential political capability of influence upon the different sectors and issues sensible to public opinion is amplified to levels beyond any expectation in the past. The reconstitution of the public space and political debate due to the impact of ICT is still in progress and it will take different forms and reach different audiences depending on the way in which "digital citizenship" is conceived of and put into practice. In this sense, it could be argued that digital citizenship on its own implies and requires a very careful treatment as a object of study and reflection.

However, it is our contention that such a reconstitution of the public space and political debate is severely constrained by the presence of the government as a technology driven to control the individual in the framework of a dominance of liberalism. There is clearly an open question. What would be the role and contribution of e-government in a social setting driven by a different ideological framework? However, our intention is rather to address the consequences on the constitution of a public space based on the conditions brought forward by the incorporation of ICT in running public affairs under the aegis of liberalism.

THE ICT AND CITIZENSHIP: A PURSUIT OF A "NEW" SOCIETY

It is commonly argued that a new society is emerging from the use and incorporation of ICT in the public decision making structure and more importantly, in the provision of public services as it allows a configuration of the public space in which the definition of the rules of government is based, according to the principles of democratic ruling (Feenberg, 1992) (Cohen and Arato, 1992).

A very schematic account of liberalism would help us to understand the constitution of the public space already conceived as being historically driven by a liberal account of human being and society. Basically, there is a commitment to individual freedom and equality which acts as the boundaries for political authority. Closely connected to this conception are a notion of sovereignty (political authority) as a result of a contract between individuals. Second, a conception of universal human rights which are enshrined in the social contract and act as boundary concept to sovereignty. Finally, there is the notion of public and private sphere. The public sphere is regulated by the political authority and the private, in which individuals are free to pursue and revise the diverse conceptions of the good which they hold, is a space controlled by autonomous individuals (Owen, 1995 p.5).

How does ICT support these features? In the previous account we were concerned on the mechanisms through which the government developed a technology in which disposal of resources and conduction of human beings were based on the identification of regularities that are not tied to ideological concerns or specific forms of social life. In this regard, we argued about the relevance of security as the central issue to exercise individual and social control. The concern about security is closely related to the exercise of individual freedom. Therefore, access to data and information provides individuals with enough tools to reduce uncertainty about the whole environment in which individuals interact one with another. But, it is also important to consider a way in which citizens could draw and request the respect of boundaries imposed upon political authority. That is, to define the public sphere upon which political authority is exercised, and the ICT could become a tool to make issues visible and relevant to a large portion of society through mechanisms that are inexpensive and without any delay from the advocates of a given issue and the rest of society. In a sense, the public sphere is easily reached and influenced by those sectors of society using ICT. On the other hand, the metatopic condition of the public sphere is not any longer an attribute of it, but instead, it defines precisely what does public sphere means in a society where ICT is at the disposal of influential sectors of society. The public sphere does not require any longer active citizens and processes of organizing collectivities in order to enhance participation as it seems to be already provided by the self-organized structures of interaction built in the ICT as a tool for collective organization and decision making. Consequently, it could be argued that the exercise of citizenship has become more dependant on technical constraints or opportunities rather than being dependant on political will. Let us briefly address this issue as it could provide a way to understand the constitution of a new political subject.

The political struggles in the past had been related to processes of having access to media and mechanisms through which the diffusion and influence of a given account of reality could be finally draw in the participation of people in collective decision making processes. These processes were mainly driven by a sort of "mechanic solidarity" stemming not from rational debate and consensus, but instead from common interests assumed to exist previous to any social interaction. In current times, the widespread availability of media and communication resources to engage in dialogue and debate is making easier the access to those willing to do it and following the modern tradition of rational debate. Therefore, the separation between the political will and technical capabilities is becoming more relevant to the exercise of political action. It implies that even those with scarce interest on political dialogue and debate are capable of influencing and exercising social pressure by using technical capabilities. Of course, it also implies that political willingness is not enough to guarantee access and diffusion of a given political agenda. The incorporation of ICT is therefore generating a sort of a divide in terms of the exercise of citizenship that could draw two different and sometimes independent public spheres, although it could be more generally understood to be alternative public spheres whose interaction is very complex as it does not respond to any specific logic governing each one of those two spheres.

There is a public sphere constituted by the exercise of citizenship defined in terms of its involvement in the constitution of interests based on social interaction "face to face" and usually defined in terms of resolution of issues associated to a common place, and whose impact can be considered in local terms. The more abstract and general issues are relegated to processes of representation and it usually involves a constitution of political actors engaged in other spaces and modes of interaction. This is a sort of a local-based public sphere.

The other public sphere, which could be labelled as *"virtual public sphere"* is constituted by a meta-topical space in which interaction is mediated by technical devices and protocols of communication and interaction. In this sphere, the potential of political action is propelled by a capability of individuals and collectivities to make use of ICT. In this regard, citizenship can be exercised everywhere without being constrained by physical or spatial limits. Therefore, their capability of influencing and reaching other layers of political action in society is potentially higher than the local-based public sphere.

A very complex issue at stake refers to the constitution of the political will that suddenly is capable of reaching other layers of political action of society without any aggregation of interests, debate and collective construction of a political will, or the constitution of a social identity capable of performing its role in the definition of the political space of possibilities of society without putting sectional or individual interests as the first priority to be served by the whole political system.

The impact of individual strategies into the political system reaching higher levels of aggregation of interests implies that the way in which political authority is limited would be highly dependable on the exercise of individual's will. It should become evident that the public sphere is somehow "controlled" by those who have the knowledge and expertise in technologies associated with the dissemination of information and communication. In the case of civil society, the "virtual public sphere" would become the critical factor for the improvement of influence in the political system. The struggle is not any longer in the streets and massive demonstrations, but rather it becomes orchestrated by an unity constituted by a myriad of individuals without any need of building up identities, trust and solidarity beyond a post-materialist commitment for a betterment of society according to individual criteria before anything else.

From above, it becomes clear that both, the limits on the political authority and the constitution of a public and private sphere are reinforced by enhancing the capabilities of citizens to orchestrate political actions both as collective and as individuals to reach even the highest levels of public decision making. In addition, there is another feature of liberalism to consider. It is the idea of universal rights and the regulatory concepts of freedom and equality. In this regard, it could be considered that universal rights are easier to reach as the potential to request their fulfillment by the state is widespread among the citizens. However, a non equal access to ICT could endanger the whole system of demands and controls that is developing in the framework of a *"virtual public sphere"*. In this regard, the equality is not any longer a matter of a prescriptive rule. Instead, it becomes a casualty of a differential access to technology and knowledge about the use and impact of ICT. Indeed, it could become comprehensible that recent efforts has been orientated to the dissemination of ICT (labelled as "democratization of access to ICT") responds to a situation of unbalanced access to a tool from which political actions could become more relevant and decisive in the conduction of society as a whole. The relevance of such access is critical for the preservation of the social contract upon which liberalism grounds its legitimacy: the universality of rights. Hypothetically, it could be argued that the constitution of a "virtual public sphere" which contrasts and diminishes the value of traditional mechanisms of political articulation from localities to more abstract entities such as regions and the nation, could endanger the access to demand and control by those citizens without access to ICT. Therefore, citizens without access to IC would became "invisible" through the "screen" in which the state and citizens recognizes each other in such a new form of political interaction mediated by ICT.

The constitution of a "new" society apparently tests the technology of government as it developed historically under the dominance of liberalism. It could be argued that current state of affairs regarding the appearance of ICT in running public affairs, not only from the dimension of management of public offices but also in the constitution of a new political form of interaction among social actors, is probably showing some of the deficiencies of liberalism in relation to assume as the starting point for the constitution of a society as a whole, freedom and equity as given. However, on the other hand, the displacement of the condition for exclusion is moving from political arena towards technical competences then we could expect that the impact of ICT in running public affairs goes far beyond its consideration in the managerial milieu. The need of a more comprehensive and systemic perspective to grasp the implications of such displacement of technologies of government towards a more technological driven perspective is in order. It is the purpose of this final section to consider the relevance of a systemic perspective to grasp ICT as a more complex phenomena.

A SYSTEMIC PERSPECTIVE OF ICT AND TECHNOLOGIES OF GOVERNMENT

In the journey already taken, it has been disclosed that government became a technology under the aegis of liberalism and some of the consequences of incorporating ICT has been depicted following two standpoints to consider. First, the relevance of ICT for running public affairs from the perspective of the management of the state was sketched. Second, the impact of ICT was unveiled by referring to the characterization of liberalism as a normative framework of political interaction in which ICT fits properly.

It is now necessary to unveil a hidden inquisitive ground upon which this journey has been founded. The consequences of unveiling and explicitly carrying out the study of ICT and its impact in running public affairs and the constitution of a new citizenship is presented in the final chapter (see Ochoa and Petrizzo, 2009).

At the beginning of this chapter, it was argued that technology became a discourse so powerful that it finally became a dominant worldview. Such a dominance referred to the separation of subject and object of inquiry in the pursuit of an independence that could allow a better understanding of internal logics governing any phenomenon. In addressing the historical emergence of government as a technology and how this technology could make use of another technology (ICT) in order to improve its performance, it is becoming evident that the way in which the disclosure of their relationship was unveiled is possible because there is an standpoint which is not interested in unveiling the world as a dispositive to be used. What are the ground upon which this exploration has been performed?

First, it is evident that there is not an inquiry that is neutral. Any inquiry is founded over interests that cannot any longer be hidden behind a claim of pursuing knowledge for its own sake. Knowledge is a form of power that reveals its deepest seated interests when it is claimed that knowledge is neutral. Therefore, it is important to unveil the interests behind this exploration on government as technology and the use of ICT. The interest could be defined as a search for a historical constitution of the political subject under the aegis of a technological driven society in order to grasp the grounds upon which society orchestrates itself as a whole.

The wholeness of a phenomenon has been the hallmark of systemic studies and methodologies orientated to comprehend the world by requesting that a reductionistic account or analytical study of the world unveils a picture of the world that is not enough to comprehend the complexities of the world. Therefore, it has been assumed that systemic methodologies are different to scientific methods and approaches because they attempt to grasp something that is unreachable through scientific methods. However, in the historical development of systemic methodologies it is becoming clearer that systemic methodologies are also driven by an instrumental interest. Even more, it could be argued that systemic thinking has been regarded as the most advanced account driven by a technological will.

Looking at the evolution of systems thinking, it could be argued following Checkland's account on *Systems Thinking* (1981) and Jackson and Flood's *Creative Problem Solving* (1991), that it has been a concern of Systems Thinking to address the way in which information flows are related to organizational procedures when considering the organizational milieu as a object of systemic inquiry. In this regard, from cybernetics account of information flows and the soft systems approach to information management clearly provides a wide framework of inquiry in which procedures and the performance of the system as a whole is a critical factor for the success of the system as a whole. However, a large number of methodologies and approaches based on systems ideas have been conducted in order to enhance control and monitoring of activities. The use of information and rational decision making are central to some of those methodologies. In this regard, cybernetics has been a cornerstone upon which the use and enhancement of information allowed the recognition of information as a critical resource for the management of complex organization. Governmental offices and more important, governmental procedures had been enhanced from the knowledge and perspective embraced by cybernetics and other so called, hard systems approaches. However, it is our contention that it is required to assume a critical standpoint when considering the use of systems thinking when addressing complex issues such as the government of society.

In any case, systems thinking also claimed in the past to surpass such a deep-seated interest by requesting and involving itself in efforts to critically address issues in society and even to carry forward a criticism to the way in which boundaries are defined in terms of the object of study. The wide ideological range upon which systems thinking lies in terms of the epistemological and ontological grounds reveals at least a will of emancipating itself from a serious trap of becoming another instrument to intervene in the world. In this case, it becomes relevant to consider the critical standpoint provided by soft and critical systemic approaches to consider the meaning and relevance of e-government as a social phenomenon beyond the boundaries of the managerial efficiency and the technical interests. This critical standpoint will be a starting point to propel a reflection on technology capable of liberating human beings themselves from the trap of technology. In this regard, the whole account presented in this chapter is based on a critical interpretive systemic account of government and technology as developing a new space for the reflection and intervention from a systemic perspective

This chapter has been an effort to unveil how the notion of government in current society could be considered as a technology which is using other technologies in order to improve its performance without critically addressing the way in which we assume the conduction of human affairs in society from an instrumentalist perspective.

At the end, it seems that if we want to grasp a deeper meaning of e-government in current society, we should be aware of a challenge of overcoming the dominion of the assumption that knowledge is neutral and technology is only an instrument in our hands. On the contrary, the influence of technology as a way of looking at the world is putting us at the verge of becoming instruments of our most sophisticated instruments.

How can we deal with it? How does it poses a serious challenge in the future ahead?. These are just only two questions that demands answers from many different accounts of the government, citizenship, technology and knowledge. In this regard, the relevance of a systemic acount of these concepts should play a key role in unveiling ICT and government as a technology to be deeply connected issues in current society.

REFERENCES

Arendt, H. (1958). *The Human Condition*. Chicago: University of Chicago Press.

Checkland, P. (1981). *Systems Thinking and Systems Practice*. Chichester: Wiley.

Cohen, J., & Arato, A. (1992). *Civil Society and Political Theory*. Cambridge, MA: MIT Press.

Feenberg, A. (1992). Subversive Rationalization: technology, Power and Democracy. *Inquiry*, *35*(3/4).

Flood, R. L., & Jackson, M. C. (1991). *Creative Problem Solving: Total Systems Intervention*. Chichester: Wiley.

Foucault, M. (1991). Governmentality. In G. Burchell, C. Gordon & P. Miller (Eds.), *The Foucault Effect* (pp. 87-104). London: Harvester & Wheatsheaf.

Gordon, C. (1991). Governmental Rationality: An Introduction. In G. Burchell, C. Gordon & P. Miller (Eds.), *The Foucault Effect* (pp. 1-72). London: Harvester & Wheatsheaf.

Habermas, J. (1972). *Knowledge and Human Interests*. London: Polity Press.

Ochoa, A., & Petrizzo, M. (2009). *Information and Communication Technologies and Government: A challenge to citizenship?* In J.-R. Córdoba & A. Ochoa-Arias (Eds.), *Systems Thinking and E-Participation: ICT in the Governance of Society*. Hershey, PA: IGI Global.

Owen, D. (1995). *Nietzsche, Politics and Modernity*. Thousand Oaks, CA: Sage Publications.

Chapter 4
Patterns for E-Government Development

José-Rodrigo Córdoba-Pachón
University of London, UK

ABSTRACT

This chapter presents three different patterns to understand and manage e-government initiatives. The word pattern means a set of values, beliefs and actions that can be distinguished as informing e-government development. The patterns are derived from the literature on the information society and information systems practice. They aim to help people involved in e-government to make sense of their work and impacts and to facilitate communication. In the chapter we use the patterns to critically review an e-government initiative in Colombia called Gobierno en Línea (Online Government).

INTRODUCTION

The information society is unfolding worldwide. Its development started since the definition of the concept in the US (Gore, 1991) and Europe (CEC, 1997) and has been followed by global declarations to encourage the use of information and communication technologies (ICTs) to address social and economic problems (DOT Force, 2000). Several countries have embarked in formulating and implementing different initiatives and integrating them with other ideas related to knowledge-based societies (Japanese Government, 2000). These initiatives are now geared to facilitate the inclusion of citizens in government affairs. The term e-government has been defined to account for a variety of technological applications of service delivery which to some could bear the fruits of transformations towards having more participative, democratic and transparent societies (Banister, 2004; Marchionini, Samet, & Brandt, 2004; ODPM, 2003; Saxena, 2005).

From the conception of the information society at the policy level in some regions of the globe, it was envisioned that such type of society would yield benefits like (Information Society Commission, 1999):

DOI: 10.4018/978-1-60566-860-4.ch004

- Economic opportunities in the production, exchange and dissemination of electronic content, products and services
- Participation and access to education through electronic means
- Competitiveness due to intra and inter-organisational collaboration across geographical locations
- Citizen empowerment

The last of these benefits seems to be the current focus of policies under the banner of eGovernment. We now have several examples of policies and plans for eGovernment in Europe, the US, Scandinavian countries and the Far East. Their similarity makes us think that in parallel to this phenomenon of e-government globalisation, we are witnessing also the homogenisation of ways of thinking about it.

In developing a global information society, it is assumed that countries will find the best ways to accommodate to develop the above policies. Governments are called to assume either a neo-liberal or 'dirigiste' approach to the information society (Moore 1997; Mansell and Steinmueller 2000). The first one means establishing fair conditions for the emergence of products and services related to information. The second one is about prioritising certain areas of investment and regulating the implementation of plans and policies. In both governments are called to ensure that potentially disadvantaged groups are to be considered as beneficiaries of plans.

In both approaches, it is interesting to see that the information society (and thus e-government) have been seen as 'inevitable', and therefore the role of governments is to ensure their implementation. Somehow this type of visionary thinking has been adopted, and we're left wandering if there are other alternatives or challenges to consider (see introductory chapter of this book). For instance, how local differences to this 'inevitable vision' are to be considered or accommodated; how can people make the best of it; and how e-government could be used more strategically for our own purposes as individuals and societies.

It is necessary then to think about the ways in which governments conceive of and adopt e-government initiatives. This chapter enquires about the potential definition of *conceptual* elements to account for our thinking. We propose three (3) different patterns to consider the assumptions that enter into the formulation and implementation of e-government initiatives. For the purposes of the chapter, *pattern* means a series of assumptions, values and beliefs held by individuals which influence their action. The patterns are influenced by personal experience[1], conceptual and empirical research in the uptake of approaches for information and communications technologies (ICTs) planning (J. R. Córdoba, 2002; J. R. Córdoba & Midgley, 2006), and the use of systemic thinking to evaluate the implementation of ICT-related plans in the Colombian context (J. Córdoba & Robson, 2008; J. R. Córdoba & Robson, 2003).

The patterns are 1) idealist; 2) strategic; and 3) power-based. We use them to analyse the Colombian e-government initiative of "Gobierno en Línea" (Mincomunicaciones, 2007). Following our analysis, we conclude the chapter by drawing possibilities for systems methodology use under each pattern that could help those people involved or affected by them to improve e-government initiatives. Through the chapter we highlight the importance of having conceptual elements to help us assess and improve our thinking behind such initiatives. We do so by considering that, as being influenced by information systems and public policy practice, e-government knowledge imposes a sort of determined type of vision which is to be adopted. Out of this 'imposition' though, there is the opportunity to re-think about or shape our ways of conceiving e-government.

INFORMATION SYSTEMS IN THE INFORMATION SOCIETY

Whilst information systems practitioners argue worldwide if their area of knowledge is a discipline or not (Baskerville & Myers, 2002; Introna, 2003), what we have seen is that information systems pervade almost every aspect of our daily lives. The information society as a phenomenon only confirmed and reinforced what was happening in organisations since the early 1980s: The use of information systems and technologies to 'add value' to organisations, so they could become more competitive and serve their customers better (more efficiently) (Porter & Millar, 1985). It is now individuals who are supposed to gain value from the use of information systems (IS) and ICT in different realms of life.

Information systems research and practice have contributed with a series of *middle range* concepts to inform practice. One of the most prevalent is the link between corporate and information systems strategies. Alignment between online and offline operations (Henderson & Venkatraman, 1999; Ward & Griffiths, 2002) becomes the manifestation of how an organisation uses information systems and technologies to support their strategy. Those organisations which are more 'advanced' develop an information culture where the search for opportunities for competitive advantage is continuous, supported by appropriate structures and processes that facilitate learning (Reponen, 1998). Such structures will enable integration of data, and the soft generation of a culture of learning.

In order to start thinking about strategy, most organisations in the 1990s developed their thinking around a vision. A **vision** is a future state of affairs to be achieved. The vision could be made reality with the help of information and communication technologies (Hammer & Champy, 1995). Such technologies could even enhance the vision. Some organisations preferred to stick to a 'mission', a kind of declaration of their 'raison d'etre', their purpose, their business. In both of these cases or in the case of organisations simply defining plans to account for their next forecast in terms of sales or markets, a future state was defined which could then be used to define projects, activities and resources to be allocated and implemented.

Embracing the importance of a vision, the information society has already defined a future state for all of us, a state in which we use information in our every day lives, and in which there is access, opportunities and participation. It is interesting to see how policy documents take us to the year 2020 (if not 2050) and show us a state of affairs in which even corruption has disappeared from the face of the planet, also thanks to information systems (Vision-2019, 2005). The society we envisage gives opportunities for all. In it, economic opportunities arise without the need to have gone through industrialisation cycles.

Moreover, there are conditions that need to be in place so that this particular culture of everyday information develops. Such conditions include provision of communications infrastructure; appropriate regulatory environment (to facilitate generation of markets of products and services); and education to prepare the workforce and other groups in society to accede to information (CEC, 1997).

E-government seems to follow as an extension of this type of visionary thinking for the information society in the domain of public affairs. A good example is that of European policies, which first encouraged competitiveness of businesses and education; then social inclusion seems to have taken the stage (Commission of the European Communities (CEC), 2004). The government systems of the future are being envisioned as transforming service delivery and becoming **citizen-centred**. Recently the eEurope programme aims to bring 'online' people and provide elements for transparency and efficiency of public service deliveries. The vision of a citizen-centred Egovernment is that of having a 'one-stop shop' for citizens, so that people can get services that they need in every important event of their lives (European Commission, 2007).

As with the information society vision, the E-government one also requires some pre-conditions that see it as developing gradually from catalogue based to full interaction (Layne & Lee, 2001; Siau & Long, 2005; West, 2004). This brings a number of challenges in relation to making e-government co-exist (at least temporarily) with other forms of provision of information, and the 'old ways' of doing things. Temporality means that alignment is to be continuously assessed and developed.

Out of the inclusion of the management of government technological platforms, Dunleavy et al (2006) explore the development of several projects of online services implementation and conclude that the governance of information technology needs to be carefully looked at. Their conclusion is based on evidence that suggests that technology suppliers absorb key knowledge which should be transferred. Their findings echo those of Contini (2009) who argues that successful e-government implementations should avoid getting locked into what one ICT supplier (or technology) can deliver at a particular moment of time. Contini also adds that the regulatory and normative environment of institutions could inhibit widespread adoption of new systems and practices and therefore needs to be carefully managed. Thus, alignment of different elements (technological, organisational and regulatory) in e-government can be pursued but we need to consider how it needs to 'deviate' from continuous and linear progression to meet strategic requirements. This prompts us to consider how the information society and e-governments unfold and produce dynamically new configurations or forms of interaction among people and technology.

THE SOCIAL CONTEXT OF E-GOVERNMENT

Consideration of how the social context influences initiatives of the information society has been approached in several ways. Studies of the impact of initiatives show that there is not only one but several types of information society (R. Mansell, 2002; R. Mansell & Steinmueller, 2000; Webster, 2002). People have not only 'adapted' or localised information society initiatives. They have also shaped new forms of organisation and interaction between groups in society. The diversity of configurations makes it difficult to generalise, but it can be said that the configurations are evolving and dynamic.

In e-government, and as said in the introductory chapter of this book, new forms of developing and studying it are emerging. They are based on the *process* that e-government implementations generate along the way (Grimsley & Meehan, 2007; McLoughlin & Cornford, 2006) rather than only on what they should achieve (online transactions or one-stop portals). Alternative conceptions to e-government as a linear process yield insights into the nature of new forms of organisation and partnerships between governments and technology suppliers (Dunleavy, et al., 2006). They also show how different contexts of design, implementation and use are to be considered in e-government projects to improve chances of adequate success (Heeks, 2005, 2006). E-government projects have also been analysed as processes in which commitment of different actors go through different stages and how it can then contribute to ensure systems use (Pan, Pan, Newman, & Flynn, 2006; Tan & Pan, 2003). The attribution of 'public value' by citizens and users of e-government systems via perceptions of information, control and influence in the use of online services is also seen as a key constituent of systems use (Grimsley & Meehan, 2007).

These cases show how people, their perceptions and their organisations *shape up* technologies and at the same time, how technologies shape them in a mutual and continuous process of informing each other. In the information systems realm, this type of insight has also been identified and it is said that individuals' cognitive and social structures influence and are influenced by technology systems. With time these systems inscribe institutional properties (Bijker, Hughes, & Pinch, 1987; Loureiro, 2008; Orlikowski, 1992)

Dynamic configurations on e-government arise due to the interaction between technologies, institutions and people. To some they all can be considered 'actors' in a dynamic network (or groups of networks) of 'politics' whose interactions develop around the acceptance of a problem and the adoption of solutions (e.g. e-government system) (Heeks & Stanforth, 2007). Consideration, mapping and inclusion of stakeholders and their interests is a key issue to understand and be able to manage projects within a particular context. People's values and perceptions can provide useful meaning which influence e-government projects goals. These play against norms that must be followed or implemented according to government regulations or legal requirements (Contini, 2009).

The interplay of interests, values and norms together with the adoption of technologies generates tensions which then to be identified and managed. Studies on e-government evaluation aim to identify some critical moments, stages or junctures in projects in which the course of e-government projects (could) change (Heeks & Stanforth, 2007; McLoughlin & Cornford, 2006). Ideally, one as a manager of an e-government initiative should be able to identify (not only in retrospective) these moments, and provide questions or ways for people to positively influence change whilst this is possible (Grimsley & Meehan, 2007; Heeks & Stanforth, 2007). Questioning depends on what is 'accepted' in a context. We know of contexts where the accepted norm is first to implement, then to consult citizens (Tan & Pan, 2003).

What also emerges from these studies is the use of theories and approaches from several disciplines: Public policy, management, and information systems are some candidates. The cases cited here combine theories to devise frameworks. Such frameworks are devised to study the above dynamics of interest and action. From the use of frameworks, the idea is to draw some lessons to further orient action. As mentioned before, some lessons include for instance to adopt simple technology (also easy to use), and to avoid becoming interlocked in complex technological or regulatory arrangements (Contini, 2009). Others include to raise awareness on the implications of projects for diverse stakeholders and their interests; and to maintain some interpretive flexibility in design of systems so as to accommodate such variety (Heeks & Stanforth, 2007).

The study of the social context of e-government leads us to consider the degree of complexity of assemblages of policies, technologies, procedures and practices. It also requires us to match such complexity with appropriate understandings and principles. About the latter simplicity and flexibility seems to be appropriate elements of an initial e-government strategy to foster accommodation between actors, their interests and ways of doing things. What seem to be missing in the assumptions leading to 'assemble' concepts for studying such social context (possibly because of a dominant focus on the self-production of technological, regulatory or organisational aspects in existing studies) are two elements:

1. Dealing with the 'unexpected' in the midst of combining the old and the new on government affairs, so it is not only e-government systems (and their effective adoption) that needs to be the focus of study. Simplicity and flexibility could be helpful but not sufficient if we assume that the future stems from the present, and that the past influences the latter. We have been successful in limiting the scope of the study of e-government to what institutions are doing and what happens inside them (Heeks & Bailur, 2007). If we broaden the boundaries of e-government, we might then consider that institutional projects are having far from desirable consequences for the wider context. In such a wider context, people might be using other systems

and technologies and developing new forms of interaction which could be unknown to formal government institutions.

2. This type of unexpected and often undesirable challenge to the linear life of e-government projects (see introductory chapter in this book) could lead us to review our own role in change processes. How to provide room for ethical reflection by those individuals and groups involved in or affected by e-government projects becomes an issue to be addressed. Within the networks and assemblages that e-government initiatives generate, it should be possible to assess how people could act and do something about them so that they find appropriately their place (J. R. Córdoba, 2006; Heeks & Stanforth, 2007). This should be the result of what they think can be achieved to improve their relations and possibilities in different spaces for action. The review of our own role(s) in e-government projects can also lead us to think who we want to become as individuals and / or citizens. This is indeed an *ethical* question.

Therefore in dealing with the 'old' and the 'new', the desirable, the undesirable, and the ethical, agency needs to be rescued. To make room for it we now include the 'unexpected' in e-government.

THE UNEXPECTED

Focus on making retrospection about e-government discourses (e.g. interpreting them) is not only about deciphering assemblages or networks. Soon such network 'maps' become the 'old' within the dynamics of implementation of we seriously accept that e-government is a dynamic process taking place in a dynamic context. As much as we can know the 'map' of e-government, we need to deal with its terrain, what it continuously means for our own action in society.

In the realm of information systems, it is now being more widely acknowledged that system implementations are not linear, and that we encounter contradictions, abrupt changes and unintended results (Lyytinen & Newman, 2008; Robey & Boudreau, 1999; Sabherwal, Hirschheim, & Goles, 2001). Change could be abrupt as well as 'smooth' or gradual; a wider understanding of change could well complement what we see as currently happening in an e-government project. The challenges of transformation that e-government is encountering to develop a citizen-centred view as an ideal state of affairs (Layne & Lee, 2001) could become opportunities for more localised and ethically driven action.

Furthermore, the availability of information technologies and their varied use also confirms the emergence of the 'unexpected' in terms of technology use. Not only technologies like 'blogs', 'wikis', 'forums' and 'open source' tools demonstrate that people can self-organise and take co-ordination if they want to. Political movements have been organised through the internet; the internet was also an unexpected configuration that developed thanks to the mediation of key individuals and groups (Castells, 2001; Turner, 2006)

These emergent possibilities in both the technological and information systems domains show us that people can act among what seems to be the 'normal', the intended or the planned based on what is to be the case (for instance with a defined vision, an architecture or a communication channel). In e-government there should be possibilities to 'do otherwise' than is formulated, interpreted or (linearly) expected. Interpretive studies of the social context of e-government provide frameworks that help us formulate questions to help people gain a more profound understanding of what could be done to enhance e-government acceptance and to some extent to create some local and co-ordinated action to make use of e-government projects (Heeks & Stanforth, 2007). However, more needs to be said about what people can do beyond what is being established as a goal for their initiatives (e-government implementation). Further questions could be formulated about the purpose of e-government projects, who and what they are privileging (or excluding) and how else they could be developing better relations among themselves and with their governments. This sort of questioning could require us to even consider the possibility of radical change or departure from what an e-government initiative aims to achieve.

If we consider that it is possible to 'do otherwise' in e-government to account for the unexpected in the best possible way[2], the role of the **mediation** (also mentioned in the first chapter) gains importance in altering the course or trajectories of e-government projects configurations. This is the case when groups of people do not have to 'collide' or interact with each other, where the sort of interactions are not pre-determined by law or regulations (Contini, 2009). Influence in mediation could be then pursued further so that not only people shape up e-government projects, but even regulatory practices might be challenged via mediation. This opens up the possibility for e-government projects to become *empowering* initiatives in both participation, use and even political action[3]). Those individuals who are involved or affected by initiatives should be offered ways for them to understand how mediation operates within and beyond domains (technological, organisational) and how the 'old' and the 'new' could offer them a space for the reinvention of the present (Foucault, 1984b).

What we identify with this possibility is a sort of a normative stance, as with this sort of critical mediation we are rescuing a notion of 'governmentality' in which citizens can work out better possibilities of avoiding becoming 'normalised' or fully constrained in their action by government practices (Foucault, 1984a). The unexpected becomes then a way of making sense of the present whilst we use it to become what we want to become. Simplicity and flexibility in e-government projects management should then allow us to think 'otherwise' and 'beyond' whilst still being 'within' e-government arrays and assemblages. This can also apply to the use of technologies in this domain.

We now have possibilities to adopt three different ways of conceiving of e-government: a) One based on visionary thinking; b) another that is based on shaping up configurations between different domains to make e-government initiatives 'happen'; and c) a third one which is based on the possibility of re-thinking action in such initiatives within which 'doing otherwise' than expected should take place. With these perspectives we now propose three different *patterns* of action that can then guide further our understanding and management of initiatives. We name them as patterns, as they can embrace thinking and actions that are shared across organisations, individuals and groups[4]. Moreover, they embrace some core values that guide e-government implementation. In the next sections we define the patterns in terms of features of use, strengths and limitations.

AN IDEALIST PATTERN FOR E-GOVERNMENT

This pattern is influenced by writings about a visionary transformation of society with information and communications technologies (ICTs) improving quality of life as a main goal. It is assumed that electronically available information and control over it contribute to make individuals more autonomous, hence improving their lives (CEC, 1997; Negroponte, 1995). Information can be easily generated, stored and transmitted thanks to the convergence of technologies like computers, internet and telecommunications.

Improvement of quality of life with information yields a number of ideals: Equality can be achieved through access to information and with it to societal opportunities (DOT Force, 2000; Information Society Commission, 1999; ISC, 1996). Education and its quality are enhanced.

Knowledge to solve problems can be drawn from different locations; organisations can geographically distribute and streamline their operations; and citizens can participate more actively in their government affairs.

Information and communication technologies can help achieve these ideals but it is necessary to provide adequate (e.g. telecommunication) infrastructures, more service choices for users (presumably through competitive service markets), and government intervention policies to avoid exclusion of less privileged citizens (R. Mansell & Steinmueller, 2000).

Under this pattern of thinking, e-government is then about transforming relations so that citizens become closer to their governments, and in doing so democracy can be lived through more efficient and transparent, participative and accountable relations (West, 2004). This can then be achieved by putting 'every' public service online (McLoughlin & Cornford, 2006; Tan & Pan, 2003) so that their provision is streamlined and made efficient. This indeed is a radical departure from existing service provision (Tan & Pan, 2003). Subsequent service transformations could increase the coverage of services to different areas (i.e. tax payments, social security, etc) (Dunleavy, et al., 2006) and geographic locations (e.g. locally, regionally, nationally and trans-nationally). In all of these initiatives, the vision to be developed is going to happen inevitably.

The achievement of such vision on e-government requires a number of preconditions, one of the most important being the degree of 'readiness' which government institutions need to have. There is to be an appropriate democratic environment to facilitate communication with citizens. Connectivity among institutions and alignment between 'offline' and 'online' processes is to be achieved as part of a progression towards a 'one-stop shop' scenario. Education and involvement in, as well as assessment of the use of services by citizens are desirable conditions to ensure that systems become used. Inclusion needs to be guaranteed so that every citizen is brought to enjoy the benefits of being connected to their administrations.

Governments adopting an idealist pattern might like to define (or copy) a vision and the ideals that support them. We know that visions are also backed up with public policy discourses (for instance those of new public management) that make emphasis on citizen-centred government to describe services that are to become transparent and efficient. For instance, public tendering can be automated and accounted for to reflect transparency and efficiency. Citizens would like to see this so they feel the public good is secured. Moreover, they can be advised via the provision of information and the availability of choices for citizens (Grimsley & Meehan, 2007). This reflects a 'whole' picture of e-government.

A vision is not related to a particular organisation in general. Its adoption would require preparatory work which can be translated into formulating plans for infrastructure preparation (hardware, telecommunications), and for educational and regulatory policy formulation and implementation. It would also require defining some indicators of accomplishment, which are often translated into service use ones. Indicators could refer to the penetration of ICTs via internet usage, the number of services becoming online, number of computers and computer systems being interconnected within government institutions and the rate of service use by the public in general.

For developing countries the adoption of an idealist pattern of e-government could be taken as 'copying' what others have done; looking at best practices; and somehow fitting them in. This could include among other things, bureaucratizing public administrations (Rose, 2005), or working on improving their performance according to indicators including e-government readiness (United Nations, 2005). In terms of information systems, an idealist pattern would represent in this case a 'raw' translation of a context of design into a context of use (Heeks, 2005) via a system which could generate opportunities and tensions in its implementation.

In terms of strengths, adopting an idealist pattern can help organisations (including government) to define a 'north' or a vision for e-government, and encourage efforts to work towards it. Given the inevitability of technological (and with it social change) on which this pattern is based, attention is focused on how to achieve a vision rather than defining it. The availability of ideals through a vision can gather a number of different stakeholders around them and get resources committed for their achievement. It can also facilitate a quick start to the path of a citizen-centred e-government environment and its evaluation towards progress as based on efficiency, service penetration and use.

To sum up, an idealist pattern for e-government makes positive emphasis on:

- The definition of a vision
- The preparation of some conditions under which the vision is to be further enhanced. These conditions include infrastructure, education, interconnection (integration) and access
- The setting up of tasks or projects to make the vision reality, together with indicators of achievement or progress (for instance in terms of number of services online, cost reductions and efficiencies in transactions)

However and given its inevitable and uncritical adoption, the idealist pattern could abandon what is currently taking place in a context and replace it by a new set of ideals, or implement a number of initiatives similar to others that have worked *elsewhere* at any cost. Mere adoption of ideals can degenerate in simply copying and imitating, and then blaming technology or the institutional climate for any failures (Heeks, 2006), or accepting that there are other factors (i.e. economic conditions, corruption) which impact the achievement of e-government indicators (Bannister, 2007; Rose, 2005; United Nations, 2005:10) and which which would need special attention. To address the inclusion of the 'context' in which e-government initiatives can be developed, another pattern emerges: Strategic.

A STRATEGIC PATTERN FOR E-GOVERNMENT

The dominant degree of determinism with which a vision can be adopted leaves out any consideration of how people are already relating to each other in a particular context of change. Shaping up of change based on existing social relations is necessary and due. The people, their own values and their 'ideals' need to be part of e-government initiatives. Consultation about which type of services they want, as well as how much content they need instead of how much degree of automation to be provided (Bannister, 2007) is due.

Under a strategic pattern for e-government, it is assumed that outcomes of e-government projects are the result of interplays between a number of different actors who have diverse interests (Dutton, et al., 1996; R. Mansell & Steinmueller, 2000; Molina & Kinder, 2001). Such interests will reflect the types of information that people would like to manage in their relationships with each other (Brown & Duguid, 1999). Thus, change could be 'negotiated' and accommodated to concerns, values and perspectives of different groups (Checkland, 1981; Checkland & Poulter, 2006; Heeks, 2006). For negotiation, dialogue and participation of groups is essential in order to 'fit' ICTs and systems to the ways in which people already (want to) interact (Mumford, 1983). Dialogue, consultation and participation would also enable ICT designers and suppliers to suggest using e-government systems to improve the quality of life of those involved and affected by its changes, and ultimately the value that they assign to these systems (Grimsley & Meehan, 2007).

In the realm of e-government the strategic pattern can be seen more clearly in the adoption of hybrid methods and approaches to e-government design and implementation (Grimsley & Meehan, 2007; Heeks, 2006). It is considered that after all, e-government systems are information systems to support human activity and therefore need to be embedded into the activities that make *sense* to individuals (politically or rationally driven), their views and appreciations (Checkland, 1981). These appreciations can be mapped and aligned in relation to certain categories or principle whose management or adoption could contribute towards successful implementations of e-government systems (Contini, 2009; Grimsley & Meehan, 2007; Heeks & Stanforth, 2007; McLoughlin & Cornford, 2006). Questions to include such appreciations can be formulated and their answers considered in the implementation agendas of e-government systems.

The use of a strategic pattern allows the inclusion of a variety of groups involved and affected in initiatives at different levels to be mapped with their perspectives and managed (McLoughlin & Cornford, 2006; Molina & Kinder, 2001). Negotiation and accommodation of perspectives can also take place, and a number of considerations can be included to secure e-government implementation success. For instance, flexibility, simple use of technology and availability of a blue print of future impacts for groups can help people make sense of changes and put themselves in the picture (Heeks & Stanforth, 2007). Furthermore, ideals that support a vision (and the vision itself) can then also be re-defined according to what is considered appropriate for the context of e-government initiatives. There is the possibility to act upon and intervene on change.

To sum up, a strategic pattern for e-government will make emphasis on:

- Translating a vision into meaningful goals and activities for people, through the identification of their interests, issues and values
- Shaping up projects and outcomes according to these elements through dialogue, consultation and participation
- Preparing people to gradually assume and participate in the design of change

There is an assumption however that is embedded within this pattern in relation to participation. It is the assumption that groups will ultimately agree on what needs to be done, and will be committed to achieve it; in other words it is assumed that participation (or consultation) is missing and is the best vehicle to achieve change in a context. Differences in perceptions can be worked out by refining the description of groups which are mapped, involved or networked so that localised and meaningful action can be defined (Bijker, et al., 1987; Heeks & Stanforth, 2007). It is assumed that interpretations can be improved (or hybridised) so as to provide people with opportunities to share and agree on what to do to improve e-government initiatives (Heeks, 2006). However, if change is uncertain and subjected to the influence of different factors, people and their interests, how can it be fully defined and managed?

The study of how e-government changes are shaped needs to consider as Mansell and Silverstone (1997) and Hawkins (1995) point out, that often certain interests from certain groups are privileged at the expense of others, and that there is in reality limited scope for action. There is struggle or conflict (either implicitly or explicitly) when interests are to be accommodated, and therefore the existence of a number of constraints should be taken into account. Participative approaches to e-government need should take into account frailties and imbalances in participation generated by issues of power, politics, under representation and others that could be affecting the definition and projects' outcomes of initiatives (Heeks & Stanforth, 2007). Participation also needs ways of addressing such imbalances, and we think that this could be developed via empowerment within the old and the new in e-government.

A POWER-BASED PATTERN(S) FOR E-GOVERNMENT

Difficulties related to the accurate and linear implementation of a 'vision' (idealist pattern), and the shape of e-government initiatives via participation and dialogue ('strategic pattern) can lead us to consider a different pattern on e-government in which initiatives can be used for different purposes, some of which intended, some of which unintended. We have said that now the literature on information systems acknowledges the existence of theoretical and methodological approaches which study change as an intertwining of success and failure, benefits and harms (Lyytinen & Newman, 2008; Robey & Boudreau, 1999). Such approaches would enable us to contemplate and embrace apparently contradictory effects of systems (i.e. good and bad) in organisations and individuals. Embracing contradiction would also mean that effects of initiatives would include the opposite to those intended to be generated. Those involved and affected by initiatives can not only adopt but resist systems or use them for their own purposes (Bloomfield & Coombs, 1992; J. Córdoba & Robson, 2008; Doolin, 2004).

A pattern for e-government management can be then defined as based on embracing contradictions in the midst of complexity. If we embrace complexity in our future view of affairs (forward looking), we then include what we have called the 'unexpected'. A notion of power can help people to make sense of e-government projects with a view towards what could happen. Power exists not only as politics but in the relationships between individuals (Foucault, 1977). Through power, individuals influence other individuals' actions (Foucault, 1984a). Power has been acknowledged as an influencing factor in e-government configurations at different levels of action (local or regional) (Heeks & Stanforth, 2007). However and in our view power is not only about influencing but also acknowledging the influence of other people's actions on our own, so that our action is framed within existing relations between ourselves and others. Such a wider notion of power would then help us to interpret how actions (ours, other people's) could have unintended consequences at different levels (individual, institutional), and lead us to go beyond managing accommodations for successful project completion as the above evidence for a strategic pattern for e-government seems to suggest (Heeks, 2006).

Following Foucault's work, it can be said that power produces asymmetrical relations between people, and induces ways of knowing, acting and behaving (Foucault, 1977). Power needs to be managed so that individuals do not become 'normalised' by it or with no possibility to do otherwise than what is the 'norm' (Foucault, 1984a). The notion of power as proposed by Foucault encourages individuals to situate themselves as subjects of power within a complex, unpredictable and dynamic landscape of relations (Brocklesby & Cummings, 1996). In such relations, there are possibilities and constraints for future action. Foucault invites individuals to imagine themselves as different types of subjects (in relation to themselves and others) so that they can use their available freedom to become what they want (Foucault, 1984b). To do so they need to continuously assess what can be achieved in relation to power, in other words to map constraints and possibilities for action.

These possibilities and constraints have to do with who we have become in practice and what we can do about it, in other words with who we are as ethical subjects (J. R. Córdoba, 2006). In order to assess such possibilities and constraints, we need to develop self-reflection about what we have become and who we want to become in relation to ourselves and others; e-government projects can give us a good 'map' or 'terrain' of possibilities. Whilst people can still take part in designing, implementing and using e-government services according to pre-defined visions (e.g. citizen-centred government), or shaping such visions to their own values and concerns, we could still think of other ways of taking action to become the type of subjects that we want. Moreover, the availability of 'open' and 'flexible' information and communication technologies that has helped projects to get started (Contini, 2009) could also help us to imagine possibilities to promote electronically collective action with others and which could be beyond the 'norm' of what a project aims to achieve. This would mean for instance that as individuals, we could set up our own 'e-government portal' that mirrors an existing one but responds more appropriately to our own concerns. We know of cases in the region of Britanny in France where citizens created a 'counter-website' to the existing council one to show their discontent with their government. The other type of possibility is to continue being part of an e-government initiative, but use them to gain further benefits. Cordoba and Cegarra discuss in their chapter of this book how for instance small and medium enterprises (SME) could make use of the online government facilities to facilitate organisational learning and with it enhance enterprises competitive advantage.

The above are just two types of possibilities among many. Although some degree of local empowerment for groups is recognised in e-government project networks (Heeks & Stanforth, 2007), more needs to be studied and said about developing 'empowerment' or 'autonomy' practices in e-government, and using electronic resources to develop action according to individuals' own ethical agendas in relation to themselves, others and governments. In this regard, **mediation** could be key elements to bring together communities of different citizen groups or users to learn from each other, or to co-ordinate and develop action with the help of wikis, blogs, online repositories or discussion forums. Mediators could help making access to government services easier whilst enabling individuals to be more proactive in their own affairs. The flexibility and simplicity elements of e-government projects that have been interpreted under a strategic pattern could be taken further to become ways of enabling ethical development of individuals in different realms of society.

A power-based pattern can also lead governments to pay attention than is currently the case to the emergence of electronic objects that reflect the formation of communities around e-government (J. R. Córdoba, 2005). This and other manifestations could be considered by policy makers, designers or developers in order to explore a new terrain for developing relationships with other people (e.g. citizens). But this is not to say that governments are to absorb these elements and ultimately normalise them (for instance include them in an e-government portal). A relational view on power as the one that supports this pattern invites us also to consider that relationships and the use of technology within them are very dynamic elements, and what could be 'emergent' today could become the 'norm' tomorrow (Darier, 1999:18). For citizens, this means that they need to be continuously aware of the type of subjects they want to become, and ready to act in the face of situations of normalisation of their relations with governments.

To sum up, a power-based pattern for e-government will make emphasis on:

- Embracing both intended and unintended impacts of projects
- Enabling individuals to assess their possibilities and constraints for action in relation to who we have become in the light of power relations
- Use information technologies to take action to become the type of ethical individuals who we want to become, even if this implies going against the 'norm' in relationships with government and the supporting technology use
- Maintaining awareness on possible situations of 'normalisation' of relationships with governments and / or the use of systems and technologies

Although the above features seem adequate to develop ethically grounded relations through e-government initiatives (and thus addressing the lack of consideration of the social context and the unexpected of the other patterns), there is an inherent limitation in a power-based pattern: The degree of uncertainty and unpredictability with which individuals (i.e. citizens) could make use of power resources and possibilities given to them. Power could be confused with resistance and domination, and people might see it negatively when it comes to participate in government affairs or initiatives. This indeed is a situation of 'normalisation' in which their further thinking is to be constrained. Careful understanding of how power works in relation to individuals' freedom and how power is an inevitable element of society (Foucault, 1984a) is needed.

Having considered the above perspectives, what follows is a reflection of how they could be present and coexist within e-government initiatives. To explore this we refer to the case of a current Colombian e-government initiative called "Gobierno en Línea" (Mincomunicaciones, 2007; Sin, 2007). This exploration is based on an interview[5] with the initiative technical director of the programme under which this initiative is being develop (Sin, 2007), electronic material available on the internet and personal experience as a Colombian citizen and now e-government user.

A CASE STUDY: COLOMBIAN ONLINE GOVERNMENT INITIATIVE "GOBIERNO EN LÍNEA" (ONLINE GOVERNMENT), 2007

E-government strategy "Agenda de Conectividad" has been defined for 2007 as aiming to build "…a more efficient, transparent and participative state, which delivers better services to citizens and businesses" (Mincomunicaciones, 2007:3). Since its inception in 2000, has evolved through several stages. Currently is being focused on a programme (series of projects) under the name of 'online government' (Mincomunicaciones, 2007).

Projects include initiatives at the national and regional levels. For its definition and action plan, the online government programme has two key components: A definition of a vision and a series of architectures at the technological level that also define the ways citizens will operate with their governments. This vision statement provides elements of efficiency, transparency and participation. In this regard, it resembles other statements which incorporate the benefits of ICT to the development of democracy, for instance Heichlinger (2004).

Both the vision and architectures aim to transform government practices from being 'institution-centred' to become 'citizen-centred' (Mincomunicaciones, 2007). The architectures have in general two types of components: A 'front-end' (or tip of the iceberg), which comprises elements that the end-user is going to perceive and use; and a 'back-end' (or the iceberg itself), which includes the different elements to support the delivery of services. The aim of these architectures is to provide support for a shift in the conception and implementation of e-government, from being institution-centred to citizen centred. Projects are also being formulated to facilitate the inter-connection of government institutions systems and with regional administrations throughout the country. The list of current projects of the e-government initiative include (Mincomunicaciones, 2007):

- Information systems in public administration (i.e. portals)
- Online tendering, one-stop registrations and certifications
- Government intranet
- High speed communications network (RAVEC)
- Data centre and common services to access citizen's information
- Multimedia interactive centre for citizen's services
- Regional e-government systems

With projects to automate processes of tendering, registrations and certifications, the idea is to provide a more efficient and transparent environment for citizens so they see how systems reduce possibilities for error whilst providing faster response times (Mincomunicaciones, 2007; Sin, 2007). Corruption will be then avoided and those involved in the flow of transactions will become more accountable (Vision-2019, 2005). The 2007 strategy for online government has some principles stating multi-channel access and equitable participation. The objectives of the strategy are developing e-government services that respond to needs of citizens, and increasing the level of satisfaction when using systems. Transparency is to be achieved by providing a better degree of visibility of government decisions (i.e. through online services provision); creation of new channels to facilitate participation; and increase the use of citizens' feedback regarding their interactions with government.

At the level of strategies, the above principles are implemented via strategies that aim to to generate awareness by enhancing communication with different stakeholders, unifying the computational interface that citizens are to use for any service, giving incentives for the use of online services, and by integrating user's feedback in further systems and applications refinement processes (Mincomunicaciones, 2007; Sin, 2007).

As shown in Table 1, the use of patterns to enquire about the initiative shows that there is *interplay* of the different patterns, as there are elements of each of them. The above description of the Colombian initiative shows that a prevailing pattern is the idealist one. A vision has been set up (e.g. citizen-centred government) and a number of architectures have been planned to guide its fulfilment. Ideals are defined in relation to both government and technological aspects that support e-government implementations. Subsequent projects (now under the management of a centralised national office) aim to prepare the ground for an adequate environment (e.g. facilitating interconnection) as well as show how the vision can begin to be achieved (e.g. enabling transparent tendering and speeding up issuing of certificates and licences).

In relation to other patterns, the Colombian government has taken steps to facilitate inclusion of citizens and the provision of transparent and accountable services via principles of transparency and efficiency in services. This is to be achieved though by unifying strategies, for instance simplifying the service delivery through multiple channels. It is not clear though how this strategy

Table 1. Insights from the Colombian Initiative "Gobierno en Línea" (Online Government)

Pattern	Idealist	Strategic	Power-based
Insights	There is a vision to transform institution centred government into citizen-centred government, and a number of projects to take the vision to completion in a service portal. Technological infrastructures are being adequated. Integration between applications is being developed. Communication with government is supported through different channels Indicators are established for the provision / coverage of e-government services and their costing in initiatives.	Intended Consultation with different sector organisations about vision Different sectors (state, industry, the community) are considered in the definition of projects tailored to them. How will different groups be engaged, either through consultation or two-way communication in e-government projects or e-government services?	Action based on some principles (transparency, efficiency) and using available channels (e.g. portal) could be taken to offer services to support stakeholder. Autonomy can be extended in services (payments, certificates) Could foster community and citizen's action *within both* institution and citizen-centred developments, also making use of other technologies (electronic forums, blogs, alternative websites).

plays against citizen's views, and we know that in some other regions of the world less sophistication is preferred to more up to date and relevant content in e-government portals (Bannister, 2007). At a general level, there is concern with citizens' participation. At a more concrete level though, it is less clear how participation is to inform further design of e-government projects and how dialogue (rather than simply consultation) is to take place with different actors to shape up e-government services at local, regional and national level.

In relation to a possible consideration of a power-based pattern, it can be said that within existing policies there could be the foundations for citizen's own empowerment when using e-government services. Current principles bring possibilities for participation and engagement. For instance the principle that e-government is more than technology; or the principle that standardisation and quality of service will lead citizens to trust more in their governments. If they are applied to the notion of one-stop citizen centred e-government shop, some possible strategies that citizens could suggest within what currently happens are:

- Websites that help using such services
- Software facilities to encourage discussion, sharing of information, group-based decision making or service performance monitoring
- The extension of electronic procurement facilities for companies so they can for instance bid jointly for contracts or get advice for other types of processes are examples of what could happen within existing services

Beyond existing initiatives or projects, the potential availability of community-based portals, blogs or wikis could offer other opportunities for discussion or to exert accountability of services. The Colombian government could be using these to improve what they are doing. Citizens could use this to exert pressure on the 'official' e-government portal initiatives.

For the Colombian case in general, the intermixing of patterns also seems to suggest that they could occur sequentially and that an ideal pattern could be adopted initially. The subsequent adoption in the sequence strategic / power-based pattern(s), or the other way round is less clear. It can be suggested that an ideal pattern could be followed by a strategic pattern *and* a power-based pattern simultaneously or sequentially. If simultaneously, it can complement formal channels of participation. If sequentially, such channels can be re-designed and with them new ideals, forms of engagement and behaviour in relation to citizenship can be developed.

IMPLICATIONS FOR SYSTEMS METHODOLOGY USE

Assuming that those involved and affected by e-governments want to pursue action under each of the above patterns explained, with an adequate degree of their strengths and limitations, the concluding section of this chapter provides some guidance on how systems methodology use can help them. This section complements the initial thoughts offered in the introductory chapter of this book.

To develop an idealist pattern for e-government, systemic thinking considers that organisations are part of a wider system (society), and that therefore any planning about an organisation's future should be developed with a wider view of how society is to be envisaged (Ackoff, 1981). A number of methodologies including interactive planning (Ackoff, 1981) and soft systems methodology SSM (Checkland & Poulter, 2006) can help people to envisage a new state of affairs, and to explore the type of support required by technology to offer services or features to individuals. Creative thinking can also be encouraged to consider perspectives in relation to develop a vision and providing benefits to a wider variety of stakeholders (De Bono, 2004). Visions could be validated in terms of their implications for stakeholders, and how decisions implied in a vision have consequences in terms of inclusion, exclusion and marginalisation of people and issues (Midgley, 2000; Ulrich, 1983).

To develop a strategic pattern for e-government, systemic thinking can provide methodological tools to explore the context of e-government use by surfacing people's concerns about a current situation and express them as a whole system of human activity (Checkland, 1981). Systemic thinking also can help people to engage in meaningful dialogue to seek alternatives for improvement. Soft systems methodologies like the above mentioned could help people structuring debate with stakeholders concerned about the messiness of a situation (e.g. an e-government project). Other methodologies like strategic assumptions surface testing (SAST) (Mason & Mitroff, 1981) could also help people to explore implications of options (for instance implementing an e-government service online). The role(s) of information systems and technologies could also be discussed as supporting the addressing of people's concerns,, and suggestions for improvement of the situation as well as for designing information systems could be elicited (J. Córdoba & Midgley, 2008).

To develop a power-based pattern for e-government and in relation to systems methodology use, a complementary approach between power-ethics analysis needs adopted given that systems methodologies do not explicitly address issues of power. We have begin exploring how such a complementary approach would sit together with systems methodologies (J. R. Córdoba, 2006). Our reflections from practice using the ideas from this approach in practice (J. Córdoba & Robson, 2008) lead us to consider the importance of analysing power formations and reflecting on how action (for instance developing alternative e-government services or technologies) can be informed by ethical reflection about who we want to be as individuals.

These implications give us a sense of hope that we can do something about e-government by reflecting on different ways in which we believe it has been developed, and that we can continue using systemic thinking and systems methodology use to engage people (including ourselves) in improving relationships with others and our governments. The door is now open to continue developing ways of working for this purpose. How to continue developing inclusive, reflective and ultimately systemic reflection is an area that we would like to continue taking into consideration for further research, development and policy making efforts on e-government.

CONCLUSION

In this chapter, an attempt has been made to define three patterns to understand e-government in the context of the information society. These patterns are 1) idealist; 2) strategic; and 3) power based. Each pattern has been described with its particular strengths and weaknesses. These patterns have been used to analyse a particular e-government initiative; their future use could be extended to inform analysis and improvement of other initiatives. We conclude by suggesting how best to use each of these patterns in e-government practice.

The idealist pattern can be helpful to gather different organisations, individuals and other stakeholders related to e-government initiatives to define a vision, and the means (i.e. organisational, technological) to pursue it. It can also be used to 'copy' foreign visions and practices without considering their effectiveness in the particular context (local, regional) where e-government initiatives are to be implemented.

The strategic pattern would enable shaping of e-government initiatives by different groups involved and affected by these initiatives. Accommodation among different interests could be promoted; mediation could address the gaps between norms and practice, between technology and law, between people and e-government solutions. However, as it is often the case accommodation can be the result of imbalances and frailties of participation in the context where e-government initiatives are to be implemented. Shaping up of e-government initiatives should also consider further analysis of the context in which such initiatives are being developed. The chapter of Córdoba and Orr in this book provides an approach to help exploring such context.

A third pattern was defined to address the shortcomings of the other two. Based on a relational view of power, this pattern would facilitate the exploration of opportunities and constraints for action given the old and the new in terms of government and e-government practices. It could also be used to empower individuals and give them opportunities for technology use so they can take advantages of e-government principles, initiatives and projects according to what they consider beneficial, right or wrong for them. Again, exploring opportunities and constraints for action could be enhanced by a deeper degree of analysis of what happens in the context where e-government initiatives are being developed.

The use of systems-based methodologies can support the development of each of these patterns and help people involved to make better sense of what is happening with and because of e-government initiatives. Methodologies can also help guiding enquiry about suggestions to improve the current situation. One area in which we would like to use these ideas is in the evaluation of e-government initiatives, in which different aspects (not only financial ones) are to be considered (see for example the chapter of Kromidha and Córdoba in this book). In this and other areas we see great potential to take the above conceptualisation of e-government development to practice and to open up new areas of enquiry in e-government research.

ACKNOWLEDGMENT

We would like to thank Dr. Hugo Sin from the Colombian Ministry of Communications and technical director of the "Gobierno en Línea" initiative for his valuable support during our visit in March 2007.

REFERENCES

Ackoff, R. (1981). *Creating the Corporate Future: Plan or to be Planned for*. New York: John Wiley and Sons.

Banister, F. (2004). *Deep E-government*. Paper presented at the European Group of Public Administration (EGPA) Annual Conference, Ljubljana.

Bannister, F. (2007). The curse of the benchmark: an assessment of the validity and value of e-government comparisons. *International Review of Administrative Sciences*, *73*(2), 171–188. doi:10.1177/0020852307077959

Baskerville, R., & Myers, M. (2002). Information systems as a reference discipline. *MIS Quarterly*, *26*(1), 1–14. doi:10.2307/4132338

Bijker, E., Hughes, T., & Pinch, T. (Eds.). (1987). *The Social Construction of Technological Systems: New Directions in the Sociology and History of Technology*. Cambridge, MA: MIT Press.

Bloomfield, B., & Coombs, R. (1992). Information technology, control and power: The centralization and decentralization debate revisited. *Journal of Management Studies*, *29*(4), 459–484. doi:10.1111/j.1467-6486.1992.tb00674.x

Brocklesby, J., & Cummings, S. (1996). Foucault plays Habermas: An alternative philosophical underpinning for critical systems thinking. *The Journal of the Operational Research Society*, *47*(6), 741–754.

Brown, J., & Duguid, P. (1999). *The Social Life of Information*. Harvard, MA: Harvard University Press.

Castells, M. (2001). *The Internet Galaxy: Reflections on Internet, Business and Society*. Oxford, UK: Oxford University Press.

CEC. (1997). *Green paper on the convergence of the telecommunications, media and information technology sectors, and the implications for regulation: Towards an information society approach*. Brussels: Commission of the European Communities (CEC).

Checkland, P. (1981). *Systems Thinking, Systems Practice*. London: John Wiley and Sons.

Checkland, P., & Poulter, J. (2006). *Learning for Action: A Short Definitive Account of Soft Systems Methodology and Its Use for Practitioners, Teachers and Students*. Chichester: John Wiley.

Commission of the European Communities (CEC). (2004). *Framework to reinforce the exchange of good practices in e-government: A contribution to eEurope 2005*. Retrieved April 27, 2004, from http://europa.eu.int/information_society/programmes/egov_rd/doc/good_practices_framework_baseline.pdf

Contini, F. (2009). ICT, assemblages and institutional contexts: understanding multiple development paths. In F. Contini & G. F. Lanzara (Eds.), *ICT and Innovation in the Public Sector* (pp. 244-271). London: Palgrave.

Córdoba, J., & Midgley, G. (2008). Beyond organisational agendas: Using boundary critique to facilitate the inclusion of societal concerns in information systems planning. *European Journal of Information Systems*, *17*, 125–142. doi:10.1057/ejis.2008.4

Córdoba, J., & Robson, W. (2008). Power and Ethics in IS Evaluation. In D. Jemielniak & J. Kociatkiewicz (Eds.), *Management Practices in High Tech Environments* (Vol. forthcoming). Hershey, PA: Idea Group.

Córdoba, J. R. (2002). *A Critical Systems Thinking Approach for the Planning of Information Technology in the Information Society.* Unpublished Doctoral dissertation, University of Hull, Hull, UK.

Córdoba, J. R. (2005). Communities and evaluation of e-government services. In S. Clarke & E. Coakes (Eds.), *Encyclopaedia of Communities of Practice in Information and Knowledge Management* (pp. 32-34). Hershey, PA: Idea Group Publishing.

Córdoba, J. R. (2006). Using Foucault to analyse ethics in the practice of problem structuring methods. *The Journal of the Operational Research Society*, *57*(9), 1027–1034. doi:10.1057/palgrave.jors.2602080

Córdoba, J. R., & Midgley, G. (2006). Broadening the boundaries: An application of critical systems thinking to IS planning in Colombia. *The Journal of the Operational Research Society*, *57*(9), 1064–1080. doi:10.1057/palgrave.jors.2602081

Córdoba, J. R., & Robson, W. D. (2003). Making the evaluation of information systems insightful: Understanding the role of power-ethics strategies. *Electronic Journal of Information Systems Evaluation*, *6*(2), 55–64.

Darier, E. (1999). Foucault and the environment: An introduction. In E. Darier (Ed.), *Discourses of the Environment* (pp. 1-33). Oxford, UK: Blackwell.

De Bono, E. (2004). *How to Have a Beautiful Mind*. London: Vermillion.

Doolin, B. (2004). Power and resistance in the implementation of a medical management information system. *Information Systems Journal, 14*(4), 343–362. doi:10.1111/j.1365-2575.2004.00176.x

Dunleavy, P., Margetts, H., Bastow, S., & Tinkler, J. (2006). *Digital Era Governance: IT Corporations, the State, and E-Government*. Oxford, UK: Oxford University Press.

Dutton, W., Blumler, J., Garnham, N., Mansell, R., Cornford, J., & Peltu, M. (1996). The politics of information and communication policy: The information superhighway. In W. Dutton & M. Peltu (Eds.), *Information and Communication Technologies: Visions and Realities*. (pp. 387-405). Oxford, UK: Oxford University Press.

European Commission. (2007). *Information Society and Government: Linking European Policies*. Brussels: European Commission.

Force, D. O. T. (2000). *Okinawa Charter on Global Information Society, Declaration of the G8*. Retrieved May 2008, from http://www.mofa.go.jp/policy/economy/summit/2000/documents/charter.html

Foucault, M. (1977). *The History of Sexuality Volume One: The Will to Knowledge* (Vol. 1). London: Penguin.

Foucault, M. (1984a). The ethics of the concern of the self as a practice of freedom (R. e. a. Hurley, Trans.). In P. Rabinow (Ed.), *Michel Foucault: Ethics Subjectivity and Truth: Essential Works of Foucault 1954-1984* (pp. 281-301). London: Penguin.

Foucault, M. (1984b). What is enlightenment? (C. Porter, Trans.). In P. Rabinow (Ed.), *The Foucault Reader: An Introduction to Foucault's Thought* (pp. 32-50). London: Penguin.

Gore, A. (1991, July 1991). A national vision. *Byte,* 188-189.

Grimsley, M., & Meehan, A. (2007). e-Government information systems: evaluation led design for public value and client trust. *European Journal of Information Systems, 16*(2), 134–148. doi:10.1057/palgrave.ejis.3000674

Hammer, M., & Champy, J. (1995). *Re-engineering the Corporation: A Manifesto for Business Revolution*. London: Nicholas Brealey.

Hawkins, R. (1995). Enhancing the user role in the development of technical standards for telecommunications. *Technology Analysis and Strategic Management, 1*, 21–40. doi:10.1080/09537329508524193

Heeks, R. (2005). e-government as a carrier of context. *Journal of Public Policy, 25*(1), 51–74. doi:10.1017/S0143814X05000206

Heeks, R. (2006). *Implementing and Managing Egovernment: An International Text*. London: Sage.

Heeks, R., & Bailur, S. (2007). Analyzing e-government research: perspectives, philosophies, theories, methods and practice. *Government Information Quarterly, 24*, 243–265. doi:10.1016/j.giq.2006.06.005

Heeks, R., & Stanforth, C. (2007). Understanding e-Government project trajectories from an actor-network perspective. *European Journal of Information Systems, 16*(2), 165–177. doi:10.1057/palgrave.ejis.3000676

Heichlinger, A. (2004). *eGovernment in Europe's Regions: Approaches and Progress in IST Strategy, Organisation and Services, and the Role of Regional Actors*. Maastricht (Netherlands): European Institute of Public Administration.

Henderson, J., & Venkatraman, N. (1999). Strategic alignment: Leveraging information technology for transforming organisations. *IBM Systems Journal, 38*(2&3), 472–484.

Information Society Commission. (1999). *Information society Ireland: Discussion document on IT access for all.* Dublin: Information Society Commission.

Introna, L. D. (2003). Disciplining information systems: Truth and its regimes. *European Journal of Information Systems, 12*, 235–240. doi:10.1057/palgrave.ejis.3000465

ISC. (1996). *Information Society Ireland: Strategy for Action*. Dublin: Information Society Commission Ireland.

Japanese Government (2000). *First meeting of the digital opportunity task force (dotforce)*. Okinawa: G8.

Layne, K., & Lee, J. (2001). Developing fully functional E-government: A four stage model. *Government Information Quarterly, 18*, 122–136. doi:10.1016/S0740-624X(01)00066-1

Loureiro, C. (2008). A theoretical framework for a structuration model of social issues in software development in information systems. *Systems Research and Behavioral Science, 25*(1), 99–109. doi:10.1002/sres.868

Lyytinen, K., & Newman, M. (2008). Explaining information systems change: a punctuated socio-technical change model. *European Journal of Information Systems, 17*(6), 589–613. doi:10.1057/ejis.2008.50

Mansell, R. (Ed.). (2002). *Inside The Communication Revolution: Evolving Patterns of Social and Technical Interaction* (1st ed.). Oxford, UK: Oxford University Press.

Mansell, R., & Silverstone, R. (1997). Introduction. In R. Mansell & R. Silverstone (Eds.), *Communication by design. The politics of information and communication technologies.* (pp. 1-14). Oxford, UK: Oxford University Press.

Mansell, R., & Steinmueller, W. (2000). *Mobilizing the Information Society: Strategies for Growth and Opportunity*. Oxford, UK: Oxford University Press.

Marchionini, G., Samet, H., & Brandt, L. (2004). Digital government. *Communications of the ACM, 46*(1), 25–27.

Mason, R., & Mitroff, I. (1981). *Challenging Strategic Planning Assumptions: Theory, Cases and Techniques*. New York: John Wiley and Sons.

McLoughlin, I., & Cornford, J. (2006). Transformational change in the local state? Enacting e-government in English local authorities. *Journal of Management & Organization, 12*(3), 195–208.

Midgley, G. (2000). *Systemic Intervention: Philosophy, Methodology and Practice*. New York: Kluwer Academic.

Mincomunicaciones (2007). *Agenda de Conectividad - Gobierno en Línea y Plan de Acción 2007* (Presentation). Bogotá: Ministerio de Comunicaciones.

Molina, A., & Kinder, T. (2001). National systems of innovations, industrial clusters and constituency-building in Scotland's electronics industry. *International Journal of Entrepreneurship and Innovation Management, 1*(2), 241–275. doi:10.1504/IJEIM.2001.000456

Mumford, E. (1983). *Designing Human Systems for New Technology: The ETHICS Method*. Manchester: Manchester Business School.

Negroponte, N. (1995). *Being Digital*. Boston, MA: MIT Press.

ODPM. (2003). *Local e-Government. Process Evaluation of the Implementation of Electronic Local Government in England*. London: Office of the Deputy Prime Minister (ODPM).

Orlikowski, W. (1992). The duality of technology: Rethinking the concept of technology in organisations. *Organization Science*, *3*(3), 398–427. doi:10.1287/orsc.3.3.398

Pan, G., Pan, S., Newman, M., & Flynn, D. (2006). Escalation and de-escalation of commitment: A commitment transformation analysis of an e-government project. *Information Systems Journal*, *16*(1), 3–21. doi:10.1111/j.1365-2575.2006.00209.x

Porter, M., & Millar, V. (1985). How information gives you competitive advantage. *Harvard Business Review*, *63*(4), 149–160.

Reponen, T. (1998). The role of learning in information systems planning and implementation. In R. D. Galliers & W. R. J. Baets (Eds.), *Information Technology and Organisational Transformation* (pp. 133-149). Chichester: John Wiley and Sons.

Robey, D., & Boudreau, M. (1999). Accounting for the contradictory organizational consequences of information technology: Theoretical directions and methodological implications. *Information Systems Research*, *10*(2), 167–185. doi:10.1287/isre.10.2.167

Rose, R. (2005). The internet and governance in a global context. *Journal of Public Policy*, *25*(1), 1–3. doi:10.1017/S0143814X05000267

Sabherwal, R., Hirschheim, R., & Goles, T. (2001). The dynamics of alignment: Insights from a punctuated equilibrium model. *Organization Science*, *12*(2), 179–197. doi:10.1287/orsc.12.2.179.10113

Saxena, K. B. C. (2005). Towards excellence in e-governance. *International Journal of Public Sector Management*, *18*(6), 498–513. doi:10.1108/09513550510616733

Siau, K., & Long, Y. (2005). Synthesizing e-government stage models - a metasynthesis based on metaethnography approach. *Industrial Management & Data Systems*, *105*(4), 443–458. doi:10.1108/02635570510592352

Sin, H. (2007). *Personal Interview, Gobierno en Línea*. Bogotá, Colombia.

Tan, C., & Pan, S. (2003). Managing e-transformation in the public sector: An e-government study of the Inland Revenue Authority of Singapore (IRAS). *European Journal of Information Systems*, *12*, 269–281. doi:10.1057/palgrave.ejis.3000479

Turner, F. (2006). *From Counterculture to Cyberculture: Stewart Brand, the Whole Earth Network, and the Rise of Digital Utopianism*. Chicago: University of Chicago Press.

Ulrich, W. (1983). *Critical Heuristics of Social Planning: A New Approach to Practical Philosophy.* Berne, Switzerland: Haupt.

United Nations. (2005). *Global E-Government Readiness Report: From E-Government to E-Inclusion*. New York: United Nations.

Vision-2019 (2005). *2019 Visión Colombia II Centenario (Colombia's vision 2019 executive summary report) - Propuesta para Discusión - Resumen Ejecutivo.* Retrieved October 2007, 2007, from http://www.dnp.gov.co/paginas_detalle.aspx?idp=366

Ward, J., & Griffiths, P. (2002). *Strategic Planning for Information Systems*. Chichester, UK: John Wiley and Sons.

Webster, F. (2002). *Theories of the Information Society (2nd Ed.)*. London: Routledge.

West, D. (2004). E-government and the transformation of service delivery and attitudes. *Public Administration Review*, *64*(1), 15–27. doi:10.1111/j.1540-6210.2004.00343.x

ENDNOTES

1 The experience includes participation in a forum of policy makers, politicians, educators and information technology managers during 1997 to formulate policies for the adoption of the information society in Colombia.

2 Of course we are not able to account for all the unexpected phenomena that surround us, but at least we can explore some consequences of our actions. Systemic thinking can help us in doing so.

3 Petrizzo and Palm refer to 'active citizenship' in their chapter of this book. We sympathise with their argument as well as with Heeks and Stanford (2007)'s notion of 'power to'. But as will be argued in the chapter, the notion of power needs to be extended to account for power as influencing our action as well.

4 These could be considered as intermediate level constructs (Lyytinen and Newman 2008)

5 Questions of the interview were focused to elicit the following elements in the government strategy: 1) An overview of the development of the e-government initiative. 2) Aspects / strategies that have been included to facilitate dialogue with different stakeholders. 3) Aspects / strategies that have been conceived of to facilitate the autonomy of service users or their empowerment in understanding and using e-government services.

Chapter 5
Shifting Discourses on E-Government:
From Piecemeal Planning to Boundaries and Traditions

José-Rodrigo Córdoba-Pachón
University of London, UK

Kevin Orr
University of Hull, UK

ABSTRACT

This chapter develops a synergy of ideas from systems-thinking and public policy analysis to help make better sense of why e-government initiatives unfold in particular ways. Specifically, the chapter emphasizes the importance of critical reflection about systems boundaries and local government traditions. These two notions complement each other and could help practitioners to understand the dynamics of contexts in which e-government initiatives develop. The aim of this synergy is to suggest an alternative for what we see as piecemeal planning and implementation approaches to e-government.

INTRODUCTION

More often than not, information systems practitioners and policy makers commonly assume that they need to plan and implement e-government by focusing on different levels of analysis, service and action. This leads both of them to define *piecemeal* plans which are partial and partitioned. Such plans only make sense at a particular level (regional, local or communitarian) but leave out any critical analysis of the dynamics of such levels. To date piecemeal planning privileges dealing with issues on e e-government like information security and confidentiality. But tomorrow things could change. Will we ever stop planning for the next issue? Can we better ground our understanding about what needs to be attended?

E-government has being served by policies, projects and plans that aim to pervade every aspect of citizen's life. We have plans in areas like infrastructure, telecommunications, services and information protection to name a few. Moreover, a vision of a citizen-centered e-government has been put forward. It requires the fulfilling of certain preconditions in societies so that we prepare the

DOI: 10.4018/978-1-60566-860-4.ch005

ground for the achievement of such a big ideal. In all these, there is scope for action, but there are continuous tensions which lead us to suggest that boundaries of e-government are far from clear.

In this chapter we argue that despite this apparently harmless style of piecemeal planning in e-government, we need to consider e-government as a 'whole', not only a collection of pieces, projects, regulations or procedures. This becomes more evident at the local government level, how this is being influenced by different aspects (international, national, as well as cultural and historical issues), and generating challenges for analysis, policy making and management. Local governments have to deal with relations, people and technologies at the same time as they aim to maintain the life of their communities around. They have to fight against the globalization of spaces that technologies like the Internet have brought to their geographical regions (Castells, 2001). If we decide to do 'otherwise' than accept globalization and its discourses in the context of local governments, we are left to our own devices to define ways to understand and conceptualize e-government at this level. This also requires us to critically reflect on the meaning of 'local'.

Our work suggests that using the verb form of e-government – talking about 'e-governing' – may be appropriate. An appreciation of the contested nature of how initiatives in this arena are construed and constructed casts light on the flux and ambiguity of the term e-government. We aim to move towards a more dynamic slant on this, including towards traditions of governing and e-governing in which the goals and values shift according to what is held as important to different situated actors. E-government is a concept which comes alive through its local translation by actors, read through the lens of myriad traditions of governing. This interpretation of e-governing steers our attention to the ideational basis of policy making and local practices. Given this degree of dynamism it is necessary to be critical and able to reflect on what happens within such practices.

To support reflection on e-government, we make use of the ideas of 1) systems boundary critique and 2) local government traditions, also as a way of redressing what we see as negative consequences of piecemeal planning. In our view, this style of planning has contributed to generate islands of e-government and non-e-government developments which to many authors are difficult to bring together (Bannister, 2007; Contini, 2009; Dunleavy, Margetts, Bastow, & Tinkler, 2006). To bridge gaps between technologies, government practices and citizens' perceptions, further consideration of purposes, values and ways of doing things embedded in local contexts is needed.

Systems boundary critique can help us surfacing the consequences and impacts of policies and plans. It can show us the terrain of possibilities for action. Applied in a local context of analysis and intervention, critique on systems boundaries can help us identify purposes and values that we use to privilege our decisions and thus to help us become critical of e-government projects or initiatives. Nevertheless, this is only part of what we see as an alternative conception of e-government planning. Boundary critique should be followed by or complemented with a deeper degree of analysis to reveal why and how people in such a context come to privilege values and issues at the expense of others. To do so we propose the concept of traditions in local government (Orr & Vince, 2009); the use of the concept in practice will complement and be complemented by boundary critique. Therefore, this chapter will present the main tenet of both notions and will finish by offering policy makers, information systems practitioners and other stakeholders some reflections resulting from combining them.

The chapter begins by contextualizing the emergence of alternative notions for e-government analysis by highlighting what we see as a traditional (piecemeal) model of planning in the context of the information society. We then present a set of criteria to help us put forward the two notions we have mentioned. Our intention is to shift our thinking and attention from what we see as a dominant stance on e-government to another one which helps us be critical of the former. We hope that our ideas will open up further discussions and inspire new forms of enquiry.

PIECEMEAL PLANNING IN THE INFORMATION SOCIETY

Strategy-oriented planning seems to beget what we call piecemeal planning. It has now made its appearance in the public sector. Ideas from strategic (corporate) management are now being used commonly. Concepts of value-based, evidence-based and the like embed the assumption that planning is about defining the best ways of serving users; that they need to see value for money and they also need to continuously perceive that the public good is being protected if not enhanced (Her Majesty's Stationery Office, 2000; UK Department of Health, 2002). Efficiencies, cost reductions and transformations are to be achieved. Whilst this could be seen as a potentially beneficial way of thinking, awareness on its shortcoming and possible negative consequences for organizations, for instance the alienation of employees and other stakeholders and the lack of discussion about means and ends employed (Alvesson & Willmott, 2003) needs to be further developed.

Unfortunately this has been left out of discussion in the public sector at least in countries like the UK. Strategic thinking has materialized in the definition of projects and programs, as they become the focus of investment and decision making. As Hodgson and Cicmil comment (2006), the notion of *a project* has become paramount in modern society, because a project becomes a hub where skills and knowledge converge; this is where people can find opportunities to improve not only a situation but also themselves as societal individuals; it is where society sees its progress being defined. Furthermore, careers and lifestyles are being affected by the emergence of project management as a scientific discipline. Discussion on how these projects are being defined, what their focus is and how they aim to bring benefits to public in general is just now being brought to the attention of governments, in particular at the local level.

Despite many governments claiming that project ideas have been out in public consultation exercises, and that local and cultural differences are being taken into account to build integrated solutions for the benefit of citizens, it is strikingly surprising that there are many similar projects to automate tax payments, issue certificates, renew licenses and the like in almost every country (Dunleavy, et al., 2006), let alone projects to automate local provision of services. As if all citizens thought in the same ways regardless of their locality.

In a way, and following Heeks and Bailur (2007), it can also be said that up to date research on e-government has also contributed to privilege piecemeal planning. A persistent aim of reported research is the generation of models that explain *how* e-government unfolds and *how* it is accepted or rejected without asking deeper questions about *why* this happens. Strategic planning as a way to define and deliver e-government projects has found an ally on this type of e-government research. As Heeks and Bailur contend, it is necessary to be more rigorous and critical of e-government. It is necessary to unearth theory that informs studies, to draw theoretical insights from existing e-government research and to give relevant insights to practitioners as to how to improve their current situation.

Piecemeal styles and their influence on e-government can be also seen when looking at many local government websites[1]. E-government websites show us how different areas of government being dealt with (education, housing, the environment, crime, etc). The quest for many e-government project managers becomes that of enriching the portfolio of online services, often to the detriment of citizens who find these websites increasingly complex to navigate or meaningless (Bannister, 2007). With piecemeal styles for e-government, Government becomes a competitive process to produce efficiency and cost reductions in service delivery. Acceptance models implicitly assume and do not contest these drivers or that the main driver is benchmarking based.

Together with concepts of 'new' public management, governance and the like, piecemeal planning and delivering of public services becomes a way (if not the way) of seeing the world. It defines a series of transformations to be achieved. To the pervasiveness of this style of planning, there has been a number of authors advocating an alternative way which is more holistic and focused on societal development rather than growth (Ackoff, 1981; Jackson, 2003; Mason & Mitroff, 1981; Midgley, 2000), emerged locally (Ochoa-Arias, 2006). This is about considering planning as a learning process in which organizations are part of bigger systems whose needs are to be looked after so that resulting plans create a better future for many stakeholders (not only organizations).

An alternative approach to piecemeal planning needs to shift the focus of thinking and action from performance to long term, from growth to development, from piecemeal objectives / initiatives to whole benefits to organizations and society, and from concern with competitiveness in the delivery of value to a redefinition of it. We see that focus needs to be regained in relation to ethical implications of plans and policies implementation. In relation to e-government and taking it into a local level of analysis or intervention, an alternative approach would then consider a deeper degree of critique that attempts to unearth the emergence of e-government as a project, and challenges its basic assumptions in terms of the value that it provides, who is to benefit and why. It also considers the potential emergence of negative consequences for different groups whose local involvement, influence and future can be brought up to discussion.

The 'holism' sought by Ackoff, Midgely and Jackson above mentioned involves not just identifying and looking at larger systems 'outside' the organization. It also entails recognizing the complexity of the organizing spaces in which local actors operate. One way of doing this is by exploring the ideational basis of local complexity. A second implication of our work is that in relation to Bannister's (2007) contribution, policy makers might need to appreciate the contested nature of evaluative criteria for e-government projects and services, and recognize that the disputes surrounding 'value' are embedded in differences surrounding 'values'. In this regard, our work aims to facilitate systemic identification of issues of concern that impinge on e-government as well as the tensions between different forces whose origins are necessary to unearth and make problematic.

ALTERNATIVE APPROACHES TO PLANNING: SOME CRITERIA

In order to justify more fully the need for alternative ways to enquiry on e-government within a local context, we see it necessary to develop a set of criteria to assess how they could better contribute in the context of the information society. A detailed description of these criteria can be found elsewhere (J. Córdoba & Midgley, 2008). We now summarize the main tenets of each criteria.

a. Dealing with marginalization.

As described by Madon et al (2004), marginalization refers to the continuous and dynamic problem of access of people to electronic information services or content. It is continuous because 'catching up' is a never ending process. New services, new software, hardware and possibilities put those who initially had access to old versions or technologies back to square one. Those with no initial access can be beneficiaries of a government program or initiative. There is a need to identify marginalization as a phenomenon that relates people to certain issues of inclusion and exclusion. Lack of limited resources in relation to who to benefit requires thinking about consequences of leaving some people out, including consequences that access gap will be widened.

If we take this to the domain of electronic government (provision of government services with the help of information and communication technologies) what we see is that tackling exclusion is a goal to be achieved in world regions like Europe (European Commission, 2007)). In parallel to mainstream developments of enabling electronically mediated communication in certain government affairs (e.g. voting, paying taxes, issuing certificates), the impacts of these developments is assessed in sectors like low income socio-economic groups. It is necessary to consider that both mainstream and marginalization are part of the same system or situation of concern. Therefore the assumptions and consequences of defining initiatives with a certain group of individuals in mind should be subjected to critical reflection. Origins, manifestations, dynamics and implications need to be identified and debated upon appropriately.

b. Developing critically informed planning

Following from the above, lack of consideration about the importance of marginalization lead us to consider the need to be critical about the effects of piecemeal planning. Questioning it can be subsumed to the assumption that plans have to be produced for the sake of organizations and society's survival, leading us to accept that if plans have worked elsewhere we inevitably need to catch up. Piecemeal planning makes knowledge unquestionable. Its degree of sophistication, its scientific rationale and expected outcomes (returns on investment or achieving of social goals) is seen as the result of analytically sound and value neutral approaches.

The results of piecemeal planning in e-government see it going through different and 'natural' stages (Layne & Lee, 2001) (see also introductory chapter of this book) towards a state of affairs where online interaction and accountability is the norm. Challenges are to be overcome rather than paid too much attention to. Planners and policy makers become the champions of transformational efforts and detached of what currently happens (the present). The future is ideal, and with it is difficult to criticize visions that support it. An alternative approach to planning should then rescue the so-called neutrality of planning instruments and outcomes. It should offer a space for people to question the rationales behind them, and assess the impacts for different groups (those formulating plans, those being affected by or involved by their implementation).

c. Including opposing views in debate

As mentioned before, marginalization can be considered a dynamic process. Those with access to public services today might not be those with access tomorrow. Conversely, those individuals and groups being marginalized today can become a privileged one tomorrow. Analytical planning sticks to a pre-defined set of goals; in case of any contingency there might be variation in the priorities and resources given to each goal. New goals are added to existing portfolios. It is up to program or project managers or policy makers to find the right place for their efforts within the existing constellation and survive within it.

Moreover, the 'side-effects' of plans are not considered or embraced. Just now we have started to acknowledge the importance of sudden change, unpredicted outcomes and inclusion of the 'undesired' into our understandings of what happens when systems and technologies are deployed (Robey & Boudreau, 1999). This comes about because within the analytic (and linearly directed) style of planning, have the opportunity to rebel themselves and do otherwise than is

Table 1. Criteria to develop alternative approaches to planning

Criteria	Description
1. Dealing with marginalisation	Enabling continuous assessment of who is (to) benefited and marginalized from the process, as well as what is to be considered relevant as the content of plans
2. Critically informed planning	Enabling questioning the neutrality of planning tools, sources of expertise, and effects of planning outcomes in groups of people
3. Including opposing views in debate	Embracing both the intended and unintended potential and actual impacts of plans and policies.

expected, often taking control of systems and technologies (Doolin, 2004). An alternative approach to planning should be able to relate to the dynamics of situations, so that continuous assessment of consequences of decisions is undertaken, and the unintended or undesirable is included or considered as part of the system of concern. Our work on traditions suggests how complex these local dynamics might be. For electronic government services, their design and deployment should be continuously assessed, in particular because new technologies are not only being used by governments but by citizens and other societal stakeholders. Consequences of their use could be (in the eyes of governments) (un) desirable. Continuous awareness should be developed in relation to what is intended, and what is achieved in e-government initiatives.

Table 1 summarizes the criteria explained above.

We now present the main tenets of both boundary critique and traditions, and in our presentation we also relate their ideas to the fulfillment of the above criteria.

THE NOTION OF SYSTEMS BOUNDARY CRITIQUE

In the words of Churchman (1968, 1971), a systems boundary is a social and inter-subjective construction that defines the knowledge and the people that we consider relevant in a situation to guide our decision making activities. In other words a boundary defines what/ who is to be inside our system (situation) of concern, and what/who could be still be considered important but not relevant. For Churchman the value of this concept resides not in enabling the identification of an 'all comprehensive' boundary, something that is impossible to achieve in practice. Rather, it is to enable critical reflection through 'pushing out' what we identify as the boundaries that guide our decision making. By doing so, even our assumption of who is a decision maker in a situation can change. Churchman suggests a process of dialectics through which we subject to debate our assumptions with their 'enemies' (people with different perspectives, often analytical planners) (Churchman, 1979). If they survive dialectical debate, we can then genuinely pursue societal improvements. The situation changes dynamically with the adoption of certain boundaries which in practice will lead us to 'push them out' again later on.

Our work on traditions overlaps with Churchman's concept in relation to the idea of inter-subjectivity. In other words traditions are sets of ideas, values and assumptions that are capable of being held on common. Our elaboration of one set of constructed traditions (fourteen in all) points to the plural nature of what is held in common by different, shifting coalitions in different spaces, at different times. The history of governing is littered with doomed policies that failed to recognise the extent to which alternative ideas may be 'relevant' and their carriers 'important' enough to impact critically on the success or failure of implementation. Our work builds on Churchman's idea in that it suggests more than a duality at any given time, and rather a 'melange' of competing, interacting and inter-subjective voices, some in the foreground and some in the background.

In the work of Churchman there was a link between boundaries, assumptions and human values which Ulrich (1983) clarifies and extends methodologically. For Ulrich, the setting up of certain boundaries is intimately linked to human values because our assumptions about a system are backed up with values we want to pursue (for instance we aim to benefit the elderly with a policy because we care for them as we do for our parents). Conversely, values influence the setting up of boundaries (e.g. with our policy we decide to target the elderly and not the youngsters). Ulrich (1983) provides a sets of questions that enable us to ensure that debates on setting of boundaries can be structured so that boundaries adopted are collectively thought of and agreed. The methodology is called critical systems heuristics (CSH) and has been applied in processes of policy evaluation, information systems design and citizen empowerment. In line with CSH's main tenets, our approach suggests the futility of imagining that there can be a commonly agreed and collectively understood boundary that is not constantly being contested, thus leading us to consider the dynamic nature of enquiry.

From a slightly different perspective on the notion of a systems boundary (perhaps more contextual), Midgley (2000) asks the following question: What happens when two groups of people set two different types of boundaries around the same situation? The situation can be best illustrated by Figure 1.

The figure shows two types of boundaries in relation to a system of concern. The inside boundary can be called primary and represents what one group define as relevant issues and people to inform decision making. The outside (secondary boundary) represents what another group of people want to privilege. Between the boundaries there is a set of marginal elements (issues, people) which represents the focus of conflict. Conflict is seen as endemic to society and it has deeper roots in how society has developed. Moreover, it is not something that we could solve in the short term.

Figure 1.

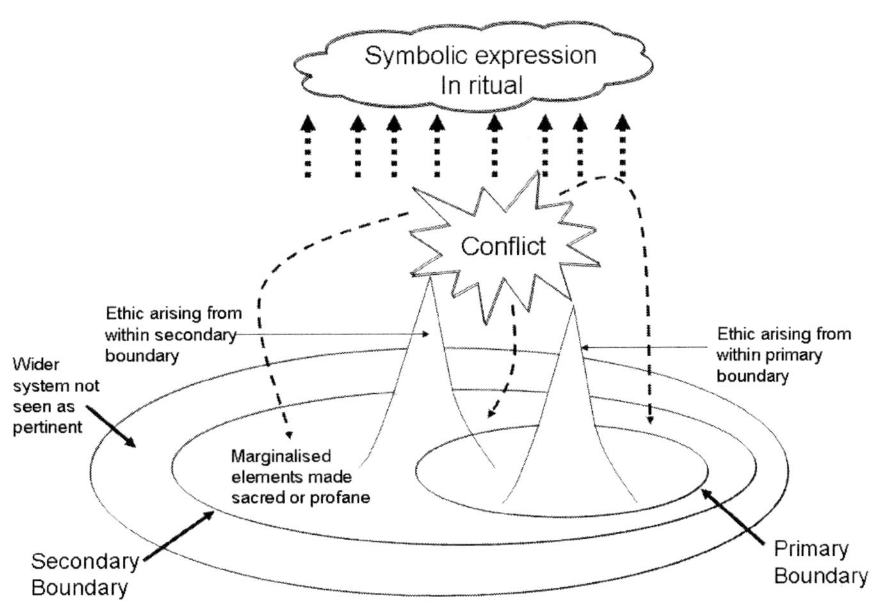

What happens with conflict is that it is stabilized by the imposition of a status on the marginalised elements (Midgley, 2000). Making them 'profane' privileges the primary boundary and its associated ethics. Conversely, making marginalised people and issues 'sacred' gives value to the secondary (wider) boundary. The terms 'sacred' and 'profane' indicate the valued or devalued status of these elements. There is rarely a consensus about whether marginalised issues or people should be seen as sacred or profane, but one of these interpretations generally comes to dominate – especially when it is institutionalised in social ritual so as to temporarily stabilize conflict. Midgley (2000) claims that this understanding of marginalisation can usefully inform intervention: patterns of marginalisation can be identified and addressed as part of the intervention process, enabling us to reflect and act on situations where we see this happening. Questions about the consequences of making marginal elements sacred or profane can be asked.

In practice, we find that Midgley's use of boundary critique can help us surface a number of issues that people have so as to identify potential situations of marginalization, and then to reflect on the consequences of adopting one boundary at the expense of other (s) (J. R. Córdoba & Midgley, 2006). Following from this, a useful way of combining both Ulrich and Midgley's ideas on boundary critique is to surface a number of issues, identify potential marginalisations, debate about them and then test validate the degree of comprehensiveness of the boundary being adopted using Ulrich's critical systems heuristics (CSH) questions. We have previously used this combination in processes of information systems planning (J. Córdoba & Midgley, 2008). On reflection, it can be said that the notion of systems boundary critique helps us to open up debate to a variety of planning approaches, interpretations and ways of doing things. It can also help us to promote dialogue and reflection on societal improvements.

However, the notion of boundary critique has limitations, one of them being that it can help us visualize but not deal with the origins and effects of power relations (Midgley, 1997) which influence the definition of possibilities for improvement; in practice we find that power relations define what constitutes a good plan for improvement. Although we have provided a complementary way to conduct power analysis (J. R. Córdoba, 2006), there is still a long way to go to be able to fully explain the dynamics of contexts in which boundaries emerge and are dealt with so that certain improvements are defined possibly at the expense of others. We then move to present more specifically the ideas of local government traditions to progress in relation to the issue identified. They will help us to promote critical reflection on boundaries and will help us extend the scope of such reflection to account for values, beliefs and actions of those people involved or affected by e-government initiatives.

The notion of traditions accounts—in principle for local phenomena. At the concluding section of the chapter we will highlight possibilities and constraints of this type of analysis.

THE NOTION OF LOCAL GOVERNMENT TRADITIONS

Orr and Vince (2009) have articulated a number of inter-related local government traditions. These traditions represent a process of deliberate and active construction. The traditions are suggestive of a set of connected, inherited and inter-subjective (capable of being held in common) 'beliefs and habits' which have been created, sustained and modified by actors and have developed within particular groups across time.

In relation to the notion of systems boundary critique, each tradition can offer us a sense of what might be held to be sacred, and profane, within each tradition. Traditions are:

Traditions of Politics:

(1) Localism
(2) Democracy
(3) Party Politics
(4) Governance
(5) Mayoralty

Traditions of Organization:

(6) Professional
(7) Regulatory
(8) Management
(9) Consumerism
(10) Partnership

Traditions of Critique:

(11) Crisis
(12) Fragmentation
(13) Centralisation
(14) Modernisation

The above grouping of traditions - as 'Traditions of Politics', 'Traditions of Organization', and 'Traditions of Critique' - suggests thematic commonalities across a range of traditions. 'Politics' is a group of traditions which highlight aspects of the political nature of local government, such as its democratic functions, its relationships with localities, the importance of political parties and the role of elected leaders. These traditions imply a greater external relation than some of the other traditions, as they are related to notions of how councils govern. 'Organization' draws together different facets of the structures, design and processes of local authorities, including the influence of professional experts, the impact of management, and the way in which councils organise their activities in ways which to greater or lesser extents support partnership working or relationships with 'consumers'. This grouping underlines ways in which councils seek to make relationships and establish networks in inter-dependent ways. The third category is 'Critique' and emphasises the ways in which different ideas and discourses can be mobilised to argue for or against certain courses of reform, to support certain changes and to resist others. This grouping emphasises the way in which the normative purposes of local government are contentious and contested and that the mode of critique is well developed.

Of course, other arrangements of the traditions are feasible – complex and diverse narratives are open to complex and diverse interpretations (Orr and Vince 2009) - but this pattern enables some important thematic connections to be made. Nor are these traditions either fixed or, in a positivist sense, discoverable. As Bevir and Rhodes suggest, traditions are '...contingent, produced by the actions of individuals. The carriers of a tradition bring it to life' (2003, p.33).

As we explore in this chapter, traditions may help us to understand how initiatives such as e-government come to be seen, related to, and shaped by situated actors operating within and across these traditions and how and why boundaries influencing action on e-government emerge. These traditions have been constructed with intra-organizational practices and assumptions in mind. However, in this chapter we also suggest the potential to shift the focus of work on traditions to look 'outwards', to think about how these traditions might suggest the prioritization of some groups of people, and interests, over others. We will discuss this further at the end of the chapter.

The carriers of a tradition bring it to life. Actors may be largely unaware of how some traditions play through their context but struck by the 'solidity' of others. The traditions will provide dissonance and consonance in different ways for different actors according to values, expectations, preferences, interests, history and so on. Traditions offer narratives, images, values, concerns and priorities which can be drawn upon by actors in these settings, and are constitutive of their choices and actions. These kinds of interactions with traditions sustain organizational life and in a process of becoming shape the 'complex and diverse narratives' about particular changes, such as e-government, and their alternatives. They suggest the ideational basis of 'boundaries' as well as their political nature.

Traditions are power-laden contexts which reflect different interests and orders and which actors have actively created, disrupted or sustained in processes of struggle, challenge, and reproduction. In this way we contend there is a strong and purposeful dynamic involved in the construction and deployment of the ideational resources within these traditions: the construction of 'traditional' narratives are attempts to impact upon the process and outcomes of change.

In this regad, an important implication of our approach is that rather than search for essential truths about public management and the implementation of initiatives such as e-government, learning can occur through embracing situated actors' constitutive stories about how they see and experience the processes which surround working with (or against) such initiatives. Our interest is in the ways in which traditions provide a range of cognitive filters and ideational resources, and constraints, for making sense of such initiatives in ways that include closing down, or opening up, certain courses of action, and of how, within these dynamics, e-governing includes and excludes different groups within communities. We point to a reflexive dynamic which is contingently and differentially experienced and acted upon.

LEARNING FROM EACH NOTION

Overall, it can be said that the use of both notions (boundary critique and local government traditions) highlights the dynamic and contested arenas in which processes of e-governing are introduced. They can alert us to the different ways on which they are experienced and represented, and how e-government can come to be seen and understood and valued (or made profane) in different contexts.

The notions provide lenses through which such e-government initiatives may be viewed. However, the word 'initiative' here is itself problematic in that while for some it may capture an appropriately positive, strategic and forward-looking concept, for others e-government may lack these very qualities though they may be implied by the term. Official discourses of e-government can be considered normative expressions which privilege certain kinds of values and priorities over others. In turn and in differential ways they are reinforcing and challenging of values and assumptions held by others.

Looking at boundaries and traditions alerts us to just how dynamic, complex and potentially exclusive are the processes of organizing, managing and resourcing action in relation to something like e-government. Appreciating that there are multiple traditions and their associated boundaries heightens the sense that local government is not a unified homogeneous organizational entity into which e-government can thrive as an 'initiative' or 'project' in neatly predictable ways, but is rather a mélange of voices, interests and assumptions about how to organize, prioritize and mobilize action in which have implications for which 'programs' are supported, which resisted and which transformed.

Such implications can be identified and addressed with these two concepts, and in what follows we show how by doing that the use of traditions can learn from the use of boundaries, and vice –versa.

Traditions and Boundaries

The use of traditions points to one way of developing a nuanced appreciation of the complexities of government processes, one which goes beyond crude dualities of in-groups and 'out-groups', the elite and the non-elite, the e-literate and the e-illiterate, and suggests instead buzzing, interacting coalitions and oppositions which arise in relation to the introduction of new ideas and practices. Traditions can be used to support the exchange of strategic knowledge about the richness of local government organizations and the implications that this disorderliness has for policy processes, action and change. They can help build knowledge about which traditions support each other and which exist in opposition, in ways that can inform the strategic choices of leaders and managers in relation to reforms such as e-government. Such a process of knowledge construction can be visualized and debated upon with a boundary like form of representation.

Our work suggests that critically informed planning can benefit from using the lens of traditions to explore the ideational contexts in which processes of e-governing will play out. This implies the need for attention to the sociology of ideas as well as the classic planning notions of socio-economic groups, demography, and topography among others. The realm of ideas does not displace material factors; rather it goes some way to explaining how material advantage and disadvantage is protected and reinforced through different policy initiatives. How do groups construct a sense of the sacred? How do they render alternatives profane? These are important questions for policy makers to understand. With both traditions and boundaries, we could have the means to make these questions more visible and relevant for e-government

Each of the traditions suggests a different articulation of 'public value' (Moore, 1995), which then can be represented and shown so that individuals can reflect on how they see their own notion of public value displacing or being displaced by other values. Each of the traditions emphasizes the legitimacy of certain goals and objectives over others. They have distinctive implications for the processes of governance adopted as well as for the outcomes to be delivered. In parallel ways the different traditions of organizing give a sense of the range of forms, routines, processes and role boundaries that may be preferred in different organizational settings. They can be used to engage practitioners with the idea that different traditions inform political and managerial practices and processes in councils. This line of enquiry encourages and enables researchers and practitioners to have a reflexive view of local circumstances and provides resources that can help diagnose the 'softer' contextual variables in these organizations.

Boundaries and Traditions

Putting the use of boundaries together with that of traditions offers us the insight that the maintenance, reproduction and disruption of traditions have ethical implications for the maintenance, reproduction and disruption of ideas such as e-government. The assumptions which are sacred and profane within different traditions have implications for how different groups come to be seen, what aspects of community life are thought of as virtuous or problematic and what decisions are made about organizing in relation to the needs and circumstances of different people – in other words how different people come to be marginalized, or not. This indeed can help us to visualize consequences of decisions, reflect on their origins and act accordingly. Insights from debate can support work that analyses more general policy development in local government settings and the impact that different traditional interests and ideas have upon the process, and to cast light on the messiness of implementation.

Furthermore, rather than simple boundaries between the 'organization' and 'society' or 'the environment' or between different groups, by introducing the notion of traditions we see a mélange of complex voices and interests, forming and reforming dynamically in contested spaces. In this regard, our desire to represent boundaries needs to consider the fluidity and dynamism of traditions and the possibility that individuals not only privilege one tradition over another but both at the same time. This leads us to consider the possibility of shifting our boundary critique focus on visualization and debate to that of questioning.

As we see it, our learning from one notion to another has just now begun and could continue in looking at the realities of local government and how it can be better approach in a more holistic and critical way.

CONCLUSION

This chapter has attempted to provide a synergy of ideas of systems thinking and public policy analysis by combining the use of the notion of boundary critique with that of local government traditions. This is the result of what we see as a reaction to 'piecemeal planning', a way of developing societies which in our view has brought more fragmentation and marginalization. We hope that readers will appreciate the value that this combination brings to enrich the exploration of e-government and the development of alternative approaches to planning to replace a dominant perspective on this area (piecemeal planning). For ourselves we see a number of relevant possibilities to continue this work in practice and in providing better ways for people to interpret their own situation in relation to e-government and act accordingly.

REFERENCES

Ackoff, R. (1981). *Creating the Corporate Future: Plan or to be Planned for*. New York: John Wiley and Sons.

Alvesson, M., & Willmott, H. (2003). Introduction. In M. Alvesson & H. Willmott (Eds.), *Studying Management Critically* (pp. 1-22). London: Sage.

Bannister, F. (2007). The curse of the benchmark: an assesment of the validity and value of e-government comparisons. *International Review of Administrative Sciences, 73*(2), 171–188. doi:10.1177/0020852307077959

Beck, E., Madon, S., & Sahay, S. (2004). On the margins of the Information Society: A comparative study of mediation. *The Information Society, 20*, 279–290. doi:10.1080/01972240490481009

Castells, M. (2001). *The Internet Galaxy: Reflections on Internet, Business and Society*. Oxford, UK: Oxford University Press.

Churchman, C. W. (1968). *The Systems Approach*. New York: Delacorte Press.

Churchman, C. W. (1971). *The Design of Inquiring Systems*. New York: Basic Books.

Churchman, C. W. (1979). *The Systems Approach and its Enemies*. New York: Basic Books.

Contini, F. (2009). ICT, assemblages and institutional contexts: understanding multiple development paths. In F. Contini & G. F. Lanzara (Eds.), *ICT and Innovation in the Public Sector* (pp. 244-271). London: Palgrave.

Córdoba, J., & Midgley, G. (2008). Beyond organisational agendas: Using boundary critique to facilitate the inclusion of societal concerns in information systems planning. *European Journal of Information Systems, 17*, 125–142. doi:10.1057/ejis.2008.4

Córdoba, J. R. (2006). Using Foucault to analyse ethics in the practice of problem structuring methods. *The Journal of the Operational Research Society, 57*(9), 1027–1034. doi:10.1057/palgrave.jors.2602080

Córdoba, J. R., & Midgley, G. (2006). Broadening the boundaries: An application of critical systems thinking to IS planning in Colombia. *The Journal of the Operational Research Society, 57*(9), 1064–1080. doi:10.1057/palgrave.jors.2602081

Doolin, B. (2004). Power and resistance in the implementation of a medical management information system. *Information Systems Journal, 14*(4), 343–362. doi:10.1111/j.1365-2575.2004.00176.x

Dunleavy, P., Margetts, H., Bastow, S., & Tinkler, J. (2006). *Digital Era Governance: IT Corporations, the State, and E-Government*. Oxford, UK: Oxford University Press.

European Commission. (2007). *Information Society and Government: Linking European Policies*. Brussels: European Commission.

Heeks, R., & Bailur, S. (2007). Analyzing e-government research: perspectives, philosophies, theories, methods and practice. *Government Information Quarterly, 24*, 243–265. doi:10.1016/j.giq.2006.06.005

Her Majesty's Stationery Office. (2000). *Statutory Instrument No 896: The Local Government (best value) Performance Indicators Order*. London: Crown.

Hodgson, D., & Cicmil, S. (2006). Making projects critical: An introduction. In D. Hodgson & S. Cicmil (Eds.), *Making Projects Critical* (pp. 1-25). London: Palgrave.

Jackson, M. C. (2003). *Creative Holism: Systems Thinking for Managers*. Chichester, UK: John Wiley and Sons.

Layne, K., & Lee, J. (2001). Developing fully functional E-government: A four stage model. *Government Information Quarterly, 18*, 122–136. doi:10.1016/S0740-624X(01)00066-1

Mason, R., & Mitroff, I. (1981). *Challenging Strategic Planning Assumptions: Theory, Cases and Techniques*. New York: John Wiley and Sons.

Midgley, G. (1997). Mixing methods: Developing systemic intervention. In J. Mingers & A. Gill (Eds.), *Multimethodology: The Theory and Practice of Combining Management Science Methodologies*. (pp. 249-290). Chichester, UK: John Wiley and Sons.

Midgley, G. (2000). *Systemic Intervention: Philosophy, Methodology and Practice*. New York: Kluwer Academic.

Moore, M. (1995). *Creating Public Value: Strategic Management in Government*. Cambridge, MA: Harvard University Press.

Ochoa-Arias, A. (Ed.). (2006). *Aprendiendo en Torno al Desarrollo Endógeno (1st Ed.)*. Mérida, Venezuela: Universidad de Los Andes (Venezuela) and Fundacite-Mérida.

Orr, K., & Vince, R. (2009). (forthcoming). Traditions of Local Government. *Public Administration, 87*. doi:10.1111/j.1467-9299.2009.01770.x

Robey, D., & Boudreau, M. (1999). Accounting for the contradictory organizational consequences of information technology: Theoretical directions and methodological implications. *Information Systems Research, 10*(2), 167–185. doi:10.1287/isre.10.2.167

UK Department of Health. (2002). *Getting the best from Best Value: experience of applying Best Value in social care*. London: Department of Health (Social Services Inspectorate).

Ulrich, W. (1983). *Critical Heuristics of Social Planning: A New Approach to Practical Philosophy.* Berne, Switzerland: Haupt.

ENDNOTE

1 In the UK councils websites for instance, Different areas are presented: Education, housing, the environment, businesses, etc. In each there is then a policy, program or set of initiatives with a number of indicators to be reported against by the end of it. Websites of many councils look alike! Perhaps this is part of a central government directive or set of policies, in which case planning style and outcomes are becoming homogeneous, or that there is a generalized lack of design flair or civic imagination on e-government websites.

Section 2
Mediation in the Context of E-Participation

Chapter 6
E-System for Public Health in India:
Towards an Architectural Framework Incorporating Illiteracy and Linguistic Diversity

Rajneesh Chowdhury
PricewaterhouseCoopers, India

Deepankar Medhi
University of Missouri–Kansas City, USA

ABSTRACT

Public health stands for the study and practice of those activities and initiatives that result in the prevention and reduction of incidences of illnesses and diseases in the population. The application of Information and Communication Technologies (ICT) can considerably facilitate Public Health project initiatives. In spite of the huge benefits of using ICT in Public Health, it can also pose considerable challenges in certain populations, pertaining to the access and comprehension of information shared through modern technology stemming from a range of issues such as illiteracy, demographic and linguistic diversity, differing economic strata of people, and differing priorities. In this chapter, after presenting a discussion on the issues faced by the public and relevant systems thinking approaches that may enable addressing the same, we propose a visionary architectural framework for ICT in Public Health through the eye of systems-thinking. We have called this framework e-System for Public Health (ePH). The understanding draws heavily from the Indian context as the country presents an interesting array of the challenges that we have mentioned above.

INTRODUCTION

Public health stands for the *study and practice of those activities and initiatives that result in the prevention and reduction of incidences of illnesses and diseases in the population*. The application of Information and Communication Technologies (ICT) can considerably facilitate Public Health project initiatives. In spite of the huge benefits of using ICT in Public Health, it can also pose

DOI: 10.4018/978-1-60566-860-4.ch006

considerable challenges in certain populations, pertaining to the access and comprehension of information shared through modern technology stemming from a range of issues like illiteracy, demographic and linguistic diversity, differing economic strata of people, and differing priorities. ICT may still be relied on to overcome such challenges, for the delivery of Public Health information in a variety of formats, forms, and languages in a range of tools as user-interfaces for access of the general population, if fostered and supported by the public sector.

When complexity is high and the range of interrelationships between various elements is varied and diverse, it is important that any Public Health initiative be able to capture different co-existing factors that influence success, consider the inter-relationships between different elements in the environment, and understand how, because of the inter-relationships, the system as a whole transforms and evolves. In order to appreciate this, we have resorted to taking a systems perspective in evolving the architectural framework.

After presenting a discussion of the systems thinking approaches and the issues faced in public health, we propose an architectural framework for ICT in Public Health through the eye of systems thinking with a futuristic vision. Hereon, we will call this e-System for Public Health (ePH). The understanding draws heavily from the Indian context as the country presents an interesting array of the challenges that we have mentioned above. Furthermore, rather than being drawn from existing ICT, our proposed approach is visionary and forward-thinking in terms of what we want to see in future ICT in order to enable the ePH. The country is also on the verge of significant change in terms of ICT in public services and related e-governance initiatives. The framework that is proposed will not only be relevant to India, but learning from this can also be inferred for other countries with a similar environment.

We will first elaborate what we understand by Public Health and discuss the challenges in terms of diversities that are posed by India as a country. We will then elaborate why, as a result of this, systems thinking and approaches can lend a perspective to work towards an architectural framework for ePH in India. We then consider how this can be woven into the e-governance framework of the country, and finally, touch upon some of the challenges of the ePH that need to be considered during its implementation.

WHAT IS PUBLIC HEALTH?

We have defined Public health as *the study and practice of those activities and initiatives that result in the prevention and reduction of illnesses and diseases in the population.* What accounts for rectification of diseases when there is an outbreak, does not represent Public Health; rather, the process of designing and implementing initiatives that will result in the prevention of such incidences in the first place will account for Public Health. These may include educating the public and raising their awareness of health and disease specific issues, and implementing health and hygiene initiatives. Hence, Public Health does not mean treatment. Public Health can therefore be classified as being proactive rather than being reactive, and being preventive rather than being curative. Dasgupta (2005) notes:

Public health services are architecturally distinct from medical services. They have as a key goal to reduce a population's exposure to disease, for example through assuring food safety and other health regulations; vector control; monitoring waste disposal and water systems; and health education to improve personal health behaviors and build citizen demand for better public health outcomes (p.1).

Public health normally consists of community-wide health and welfare initiatives that are ideally facilitated by the government, and in certain cases supported by Non Government Organizations (NGOs), international bodies and self-help groups.

CONSIDERING THE CASE IN FOCUS

It is understood that Public Health and ePH initiatives exist in developed countries with a certain degree of maturity, given the ready availability of funds and the relative homogeneity of the population. Our interest is in evolving an ePH architectural framework for a geography that may pose challenges of a very different kind in terms of its socio-economic inequality and cultural diversity. Such a situation will present its own uniqueness. The country that is considered here is India, due to the uniqueness it presents in its diversity.

India has a population of over one billion people according to the Census of India (2001). The population is spread over an enormous landmass spanning over 3,287,263 square kilometers that includes mountains, deserts, swamps, rainforests, valleys, and plains, nurturing a rich and equally diverse eco-system. The massive population harbors a huge array of languages and dialects, and it follows different faiths. While the number of languages used in India can be identified to be over one-thousand, twenty-two of these languages are recognized by the Indian Constitution. The majority of the languages stem from two families: Indo-Asian and Dravidian. The former is spoken by about 70% of the population and the latter is spoken by about 22% of the population. The rest of the languages stem from families like Dardic, Tibeto-Burman and Austro-Asiatic families (Languages of India, 2008). The primary religions to be found in the country are Hinduism, Islam, Christianity, Zoroastrianism, Judaism, Buddhism, and Sikhism. A majority of the population follows Hinduism, followed by Islam and Christianity. Furthermore, Hinduism has numerous sects and sub-faiths that yield different levels of influences and beliefs.

This diversity of India is no less reflected in its economic scenario. An aggressive growth rate, rampant industrialization, unprecedented technological innovation, and heightened Foreign Direct Investments (FDI) have created a class of people that has immediately and directly benefited from the situation and who lead a lifestyle that is extravagant, elitist, and indulgent. There is another class of people that has been completely left out of the benefits of economic prosperity, who live a life of severe deprivation. Whereas India's 40 wealthiest men are worth more than $351billion according to Forbes (Sappenfield & Chopra, 2007), the United Nations Human Development Report states that about 34% of the population in India still continues to live below $1 a day (HDR, 2007-08).

It is worth noting that the current health scenario in India is quite dismal, contributed significantly by the lack of any coordinated public health initiative. This statement can be substantiated by information that we have collated from Garai & Shadrach (2006) who share the following unwelcomed information for India:

- Lack of adequate medical attention results in one maternal death every five minutes
- An existent infant mortality rate of 70 per 1000 live births
- The highest number of cases of HIV/AIDS outside of South Africa, with 5.1 million people infected
- Incidences of non-communicable diseases on the rise with 25 million cardiovascular patients, 2.4 million cases of cancer, and 25 million diabetics
- It is estimated that 24.7 million deaths are caused every year due to water contamination, poor sanitation and poor hygiene
- A serious lack of proper sanitation and toilets in rural areas

Demographic, economic, and linguistic diversity, coupled with a lack of coordinated efforts, give rise to a situation where the population has varying degrees of understanding of health-related issues due to varying literacy levels, and varying kinds of access to public health facilities. This may give rise to a situation where the access and use of any ICT enablement in Public Health gets restricted only to the literate and privileged sections of the society. Therefore, our interest is in working towards an ePH architectural framework that will benefit the underprivileged sections of the society. This can also be used to directly inform the e-governance agenda of the Indian government.

REQUIREMENTS FOR EPH DESIGN IN CURRENT CASE

An ePH architectural framework can be considered that would enable holistic information availability of a defined range of health-related topics aided by user-friendly interfaces, enabled by effective ICT. An ePH approach will need to deal with the scale of health related information that needs to be shared and the various media by which this needs to be provided. The common ways by which ICT supported Public Health initiatives have been implemented or conceptualized in India include the following (Garai & Shadrach, 2006):

- Disease specific websites
- Online training to paramedics & nurses
- E-learning packages like interactive web-based portals and CD-ROMS
- Knowledge gateways for personalized counseling services
- Virtual knowledge communities for sharing of information and experiences
- E-consultation of sensitive conditions like HIV/AIDS
- Online repository of latest medical journals and periodicals

However, it is crucial to note here that where the level of complexity and diversity is exceedingly high, there will be a requirement that the ePH architectural framework has the following qualities:

Inclusive

The ePH needs to be inclusive so that it caters to the requirements of the underprivileged sections of the society. For instance, a significant population of India remains non-literate yet they must still have access to Public Health information. A traditional ICT approach is not suitable for this population. Rather, there may be a requirement to develop a text-free User Interface (UI) for this population (Medhi, et. al. 2005; Medhi & Kuriyan 2007). Even within non-literate users, there is a cultural difference between different religious communities in understanding information. It was found that while non-literates among Hindus expect time to flow from left to right, the non-literate Muslims expect this in the other direction, although they do not have any knowledge of Arabic script (Donner et. al. 2008). In other words, the ePH needs to be outlined in such a way that it is able to take into consideration various requirements and sensitivities in the design and implementation phases, so that the system is inclusive in nature.

Dialectic

In a climate that is diverse, it is very common to have diametrically opposing viewpoints and apparently opposing ideologies about how the ePH needs to be developed. We may take the example of sex education in schools in India. Earlier in 2008, the health ministry proposed to introduce compulsory sex education for schoolchildren to increase awareness of sexual health and reduce incidences of sexually transmitted infections. However, this attracted tremendous resistance from a section of parents, who believed that sex education would infringe upon the morality of their children. The debate today is struggling to achieve a balance between rationality and morality in the country. The ePH is required to be designed considering such diametrically opposing ideas and approaches rather than trying to discount the same. Public participation should be dialectic and be able to arrive at a consensus, if not agreement, in order to have the ePH that is able to address the differing needs and belief systems of the society. Dialectic debate is the process in which opposing ideas are put forward in front of each other by their proponents and dialogued about to arrive at an agreement. The value and contribution of dialectics in the evolution of human thought has been extensively documented in social and management theory (Stump, 1989; McTaggart, 1964; Eemeren, 2003).

Adaptive

Technology is changing rapidly and so are requirements of and expectations from technology. What is therefore required is an ePH that has the ability to be adaptive to change and innovation. This can only be achieved by avoiding an ePH architecture that is rigid and straightjacketed. Architecting ePH needs to be able to make provisions of a system that is able to appreciate and address the variety in the environment. This should be designed in the line of a complex adaptive system that can change and shape itself according to the demands of the environment, so that control and order become emergent, rather than pre-determined (Dooley, 1997).

Evolving

Adaptiveness is the prelude to evolution. An ePH should be able to adapt to differing requirements in a continual manner, but at the end of a particular time span, the resultant character of the ePH should be qualitatively different from and superior to what it was when it started. This is important because pubic health requirements can themselves change over a period of time. For instance, in India, plague was an issue of severe public concern in the 1990s. However, the concern gradually moved to HIV/AIDS a decade later. Currently, there are other forms of illnesses such as Avian Flu and Severe Acute Respiratory Syndrome (SARS) that are capturing the attention of the public and the policy makers. The way the ePH will have to address these differing health requirements needs to evolve as the quality, quantity, sensitivity, and modes of delivery of the information (pertaining to the different demands) will also change.

Robust

It is important that in spite of the ePH being inclusive, dialectic, adaptive, and evolving, it needs to be robust with the ability to withstand the test of changing times and demands. The challenges that this will give rise to are the ability to be robust yet flexible, to be system defined yet adaptive, and to be specification-centric yet evolving. However, robustness in this case is understood not from negating its counterpoints, be it inclusiveness, dialectic, adaptiveness or evolutionary, but by being able to derive its strengths from these negations.

People Sensitive

Any ePH must be people sensitive; this means that it must be able to absorb the dynamics involved in people-machine interactions within the existing socio-political framework. With new technological provisions, an ePH may seem like a liberator. However, the environment within which this architecture needs to be managed may have acculturated and ingrained logics of working that may be averse to the use of ICT as well. Conventional planners and practitioners may be highly conservative about the way they have been doing a particular job for quite some time, and they may not want to bring about changes in their working patterns. Hence, the framework should also be able to absorb such resistances. Change management can appear to be a challenge, and innovation acceptance, a threat to established practices. The analysis and understanding of the human element is of utmost importance in considering the successful introduction of new ICT systems. Unfortunately, planners tend to pay more attention to the technology element rather than the people element.

From the above discussion, we arrive at the understanding that an ePH for a diverse population with challenges around language, literacy, poverty and deprivation, needs an approach that is able to encompass these factors and look at the process of working towards this framework in a holistic and systemic manner. This holistic and systemic approach will not only be required in the conceptual stage of working towards the ePH architectural framework, but also in its planning and implementation. Together with this, a range of different approaches will be required to take into consideration the pluralism of elements that exist in interplay within the system and how they can be integrated to make further application-sense for the target population. Considering these factors, we have expressed a bend towards the Critical Systems Thinking perspective that we will elaborate on in the following section.

CRITICAL SYSTEMS THINKING AND APPROACHES

Descending from the Greek verb *sunistanai* the word 'systems' originally means "to cause to stand together" (Senge, et. al., 1994). Systems thinking involves appreciation of the collective character of elements, enabling a macro view beyond micro sub-systems and activities, advocating the idea that the whole is more than the sum of its parts. Chowdhury, et. al. (2007) note that "The central idea behind systems thinking is that organizations are constituted of sub-systems, or elements, that are in interrelationships with one another, and that exist within a boundary. What are important are not the elements per se, but the interrelationships between them, because it is the nature of the interrelationships that give character to the system" (p.10). Systems thinking therefore, lends the idea that in order to understand the complexity of social systems, we need to assess how different sub-systems are organized within a particular system and attempt to gauge how this can result or is resulting in a larger system that is qualitatively different from the sum of its parts. Therefore, elements are to be looked at as a whole from the beginning, rather than looking at them in isolation; therefore, there is a direction to gauge for emergent patterns, complex relationships, and unobvious interactions.

Critical Systems Thinking (CST) is a specific area within systems thinking that inspires the inclusion of diversity and pluralism in approaching and understanding situations with human betterment as its core concern. This involves the ability to take into consideration different, and sometimes differing, worldviews, challenging preconceptions and boundaries that people create in their thought processes and the ability to be empathetic to situations. Churchman (1971) also refers to this as the "sweeping in" of ideas and factors in the intervention process.

According to Midgley (1996), there are three fundamental commitments of CST:

- **Critical awareness:** Examining and re-examining taken-for-granted assumptions, along with the conditions that give rise to them
- **Methodological pluralism:** Using a variety of research (or intervention) methods in a theoretically coherent manner, becoming aware of their strengths and weaknesses, to address a corresponding variety of issues
- **Emancipation:** Ensuring that research is focused on "improvement", defined temporarily and locally, taking issues of power (which may affect the definition) into account (p.11)

CST can therefore be considered as a step ahead within systems thinking where holism is regarded as a characteristic to be derived from a critical understanding of the stakeholders who form part of the system and who take a proactive approach in improving their situation within the system.

Apart from systems thinking being a perspective, it has also inspired a range of approaches, methodologies, and techniques that have been extensively used in real world situations to define and resolve complex problems and arrive at viable solutions and alternatives. Methodologies and techniques, inspired by systems thinking, have undergone significant changes over the years (Clarke, 2001) since the 1950s to the present time. These changes stem from formulating and applying techniques and methods from a positivism-oriented hard systems approach, to where the thrust had been on an interpretivism-oriented soft systems approach, to where the interest has leaned towards a postmodernist-oriented critical systems approach[1].

There has been a significant amount of work done in the field of demonstrating which methodologies and techniques are applicable to which situation (Jackson & Keys, 1984; Jackson, 1990; & Jackson, 1995). Jackson (1995) argued that depending on the system (or context) and the people complexity of the situation under consideration, different approaches and methodologies may need to be adopted. He concluded his understanding in a framework that he named the System of Systems Methodologies (SOSM), which is depicted in Figure 1.

In the SOSM, the x-axis indicates people complexity. A *Unitary* context implies that there is a pre-existing agreement on the end result and there need not be any requirement of a detailed discussion about the same, or any intervention for systems design. Participants have similar values, belief systems, and interests. They share a common purpose and are involved in decision making to reach a particular defined objective. A *Pluralist* context would indicate there is some difference between the participants on what an end result of any intervention or systems design may be; however, these differences can be overcome by engaging the participants in a collaborative exercise, wherein the differences can be explored and a resolution arrived at. There is normally no hidden intension or agenda of the participants; whatever

Figure 1. SOSM (adapted from Jackson & Keys, 1984, Jackson, 1990 & Jackson, 1995)

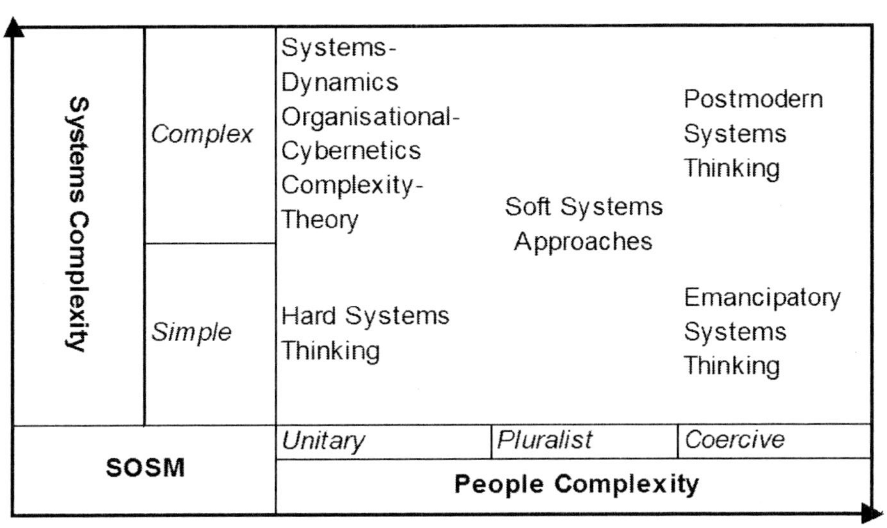

difference of opinion exists, they are highlighted with the desire of reaching a consensus. A *Coercive* context would indicate a situation where participants have different agenda and intentions that are influenced by different worldviews. Hence, such a situation is intensely political and outcomes are determined by a wielding of power and related influences. Reaching a consensus or agreement in such a situation is challenging and involves highly critical thinking to be able to infer unstated truths and be able to look beyond the obvious.

The y-axis indicates system complexity. A *Simple* context implies that elements in the environment have a straightforward relationship. Constituent elements that have identifiable interlinkages and deviances in one element result in easily inferable deviances in the following element in the value-chain. A simple system tends to remain stable over time and relatively unaffected by how each sub-system operates. A *Complex* environment, on the other hand, indicates the existence of a wide range of elements and sub-systems organized and in interaction in such a way that its relationship is not easily identifiable and comprehensible. Deviances in the system may be a result of a deviance in functioning or in the relationship between a wide range of elements, or the reasons may not be traceable at all. Complex systems do not remain intact over time, but they adapt and evolve because of their own action and the action of the larger context in which they operate. It is important to note that environmental complexity may not be dependent on the number of elements or people involved. An arrangement where there are few elements or people may be organized in a complex manner, therefore resulting in a complex environment. On the other hand, there may be a large number of elements or people organized in a very systematic process-based manner, therefore resulting in a simple environment.

Jackson (2003) notes that combining the two axes in the SOSM-framework leads to six ideal-types that can be described as "stating logical extremes that can be used to construct abstract models of general realities" (p.19). These are simple-unitary, simple-pluralist, simple-coercive, complex-unitary, complex-pluralist, and complex-coercive. Specific methodologies and techniques may be associated with each of these ideal typical situations by understanding which systems approach may be appropriate for which situation. For instance, hard systems thinking and approaches are deductive and formalistic; they carry the opinion that situations can be approached by esoteric mathematical and unitary methods. These approaches became popular in the post-war period that was focused on rapid rebuilding and reorganizing structures that could deal with limited resources to be deployed within a specific space and time. These approaches worked for the situation, and they therefore emerged during this period (Jackson, 2003). However, as time passed, higher complexities started emerging in societal and organizational systems, for which, dealing with these hard systems approaches was not only insufficient, but it was also inappropriate. This resulted in the need for approaches and methodologies that could take into consideration a range of different factors and people-issues during an intervention. These approaches would help in understanding the deeper structural relations in real world problem situations to enable organizations to adapt and evolve with internal and external contingent factors on them. The approaches and methodologies that emerged because of this were labeled later under the banner of 'soft systems approaches'. The advent of soft systems thinking has a great deal to contribute as we move towards the right hand side of the horizontal axis of SOSM. Soft systems approaches enabled the realization that worldviews are not always the same and differences are inevitable in the dynamic situations that organizations find themselves in. Methodologies influenced by soft systems thinking offered the possibility to facilitate divergent worldviews into agreement in an atmosphere of constructive debate and discussion. These approaches encourage the accommodation of crucial factors like values, beliefs, and human feelings. Moving further towards the right in the horizontal axis of the SOSM-framework, there is an indication of the coerciveness in human relationships, including situations of hidden or negative power dynamics, imposition of ideologies, and the idea of might having the right. The interest of the interventionist in such situations is to appreciate and unearth different dynamics in play and explore how, because of their intervention, there may be an improvement in the situation. To deal with such situations, methodologies that encourage an emancipatory[2] and postmodernist outlook can be of assistance to the systems practitioner. These enable looking at the same situation from different perspectives, appreciating opposing agenda, and creating a platform where it can be recognized that there is no one best way to do things. These approaches indicate an interest toward exploring those hidden agenda that are inherent in the design and deployment of complex systems.

CRITICAL SYSTEMS THINKING AND EPH

We believe that for the ePH to be successful in the Indian context, it needs to be positioned within a multi-contextual situation considering social, political, economic, demographic, and cultural factors. Therefore, architecting an ePH framework will entail that the above elements are considered, not in isolation, but in interaction with one another. Any ICT infrastructure will need to be able to capture and address factors that are over and above those that are directly lifestyle- and illness-related, towards those that are inherently related to the multi-dimensional factors mentioned above. As Midgley (2006) observes, "the whole concept of public health is founded on the insight that health and illness have causes or conditions that go beyond the biology and behaviour of the individual human being" (p.466). Architecting effective ICT for public health will need to adopt approaches that are able to recognize interrelationships between elements stabilize difference of opinions and emancipate, to whatever extent possible, the voices and opinions of those sections of society that are easy to be ignored or overlooked. Therefore, it is important to approach the architectural framework design of an ePH for the Indian context from the standpoint of Critical Systems Thinking. In the context of Public Health, Leischow & Milstein (2006) note:

Equally important (to considering directly health-related aspects) is an emphasis on relating different types of structures that shape our lives, including the biological systems of our bodies, the organizational systems in which we work, and the political systems with which we govern public affairs. Although there is no single operational method for identifying and interpreting these relationships, there is, in fact, a common architectural orientation recognizable as a systems approach: it is a paradigm or perspective that considers connections among different components, plans for the implications of their interaction, and requires transdisciplinary thinking as well as active engagement of those who have a stake in the outcome to govern the course of change (p.403).

A range of systems approaches and methodologies have been used in various ways in Public Health, research planning for healthcare, and intervention and policy formulation for healthcare. The methodologies and techniques used include System Dynamics, Critical System Heuristics, Soft System Methodology, Strategic Assumption Surfacing and Testing, and Interactive Planning and Strategic Choice (Cohen & Midgley, 1994; Sudhir et. al., 1997; Maani & Cavana, 2000; Fahey et. al., 2004; Chowdhury, 2006; Chowdhury, Clarke & Butler, 2007; Chowdhury & Nobbs, 2008)[3].

Considering our discussion on the socio-economic diversity of India and the factors that interplay therein, we have placed the Indian context under consideration in the complex-coercive space of the SOSM (System of Systems Methodology)-framework. This directs us towards considering a postmodernist approach in the architectural framework design for ePH that will appreciate the diversity the system will deal with. A postmodernist systems approach will encourage appreciating the diversity of factors that interplay beyond the surface level and consider how an attempt can be made to reach out to the benefits of the subaltern and deprived sections of the society. In this light, we focus on proposing a visionary architectural framework, rather than being tied by what is possible within the current ICT systems such as the ones proposed as of date.

EPH AND E-GOVERNANCE

Kanungo (2004) defines E-governance as "the application of information and communication technologies to transform the efficiency, effectiveness, transparency and accountability of informational and transactional exchanges within government, between government and government agencies of national, state, municipal and local levels, citizen and businesses, and to empower citizens through access and use of information". E-governance can contribute immensely for both the government and citizenry. Das & Chandrashekhar (2006) talk of how e-governance can contribute towards governance, public services, and management. They highlight the advantages in the following classifications:

Governance

- *Transparency*
- *People's participation*
- *Promotion of a democratic society*

Public services

- *Efficient, cost-effective and responsive governance*
- *Convenient services to citizens and businesses*
- *Greater citizen access to public information*
- *Accountability in delivery of services to citizens*

Management

- *Simplicity, efficiency, and accountability*
- *Managing voluminous information and data effectively*
- *Information services*
- *Swift and secure communication (p.4)*

The Government of India introduced the National e-Governance Program (NeGP) in 2006 with the vision to make "All Government services accessible to the common man in his locality, throughout his life through One-stop-shops (Common Service Centers) ensuring efficiency, transparency, and reliability at affordable costs to meet the basic needs of the common man" (NeGP, 2006). However, we have not come across any conceptual deliberation on how this vision can be translated into action in the realm of Public Health. We therefore believe that the ePH we propose can serve as an important step towards crystallizing a conceptual integration of the Public Health dimension in the country's e-Governance framework.

In the next section, we discuss the architectural dimensions for our envisaged ePH framework, considering a range of interplaying factors. A Critical Systems Thinking (CST) perspective has resulted in our consideration of a diversity of variables, directly affecting those factors that can result in the betterment of the situation of the disempowered and underprivileged.

EPH: SYSTEMS DIMENSION AND CATEGORIZATION

With the above background and in order to present an architectural framework for the ePH system, we are now ready to propose our thoughts on an architectural framework for ePH. In order to do that, we first identify the following important primary dimensions:

Population Factor

While the ePH system is meant for all citizens, there are certain aspects about the Indian population that must be taken into consideration. Consider that the top 5% (Farrell & Beinhocker, 2007) of the population (the middle class) is highly educated and has either access to computers easily or own a computer with an Internet connection. This group already has the knowledge and resources to know what to look for in terms of understanding about a particular public health need. In fact, this group is well versed in the English language and can use any popular search engine such as Google or Yahoo to search for information about any public health issue. Thus, in our systems consideration, we assume that this population group is *not* the primary target for the ePH in India. In other words, the target population group for this ePH system is the lower middle to lower strata of the society. In the context of India, this group can be further divided into different categories by considering the following factors: language, demography, and literacy.

Mode of Access to Information

There are various modes of providing information about public health to citizens: TV, print medium (e.g., newspapers), billboards, the Internet, and Mobile Phones. India uses billboards to make its citizens aware of a particular public health concern (e.g. HIV-AIDS). Furthermore, TV and the print medium (newspapers) have also been commonly used over the years for the same purpose. Two of these media (print and billboards) assume that the target group is literate. Three of these media (TV, print, and billboards) do not fall under our consideration. In other words, we focus primarily on the following two modes for access to ePH: the Internet and Mobile phones.

Internet access from desktop computers is now available in many places, including in rural areas where this access is available at community centers through the effort of the National Informatics Centre. Besides such Internet access, the mobile phone is an important component of the ePH system for India. There are four primary reasons for this: (1) the newer mobile phones have a number of capabilities such as a multi-media display; (2) due to infrastructure limitations of wired phone connectivity, the mobile phones have made inroads to the remotest parts of the country, even in places where the availability of electricity is limited. In many rural areas in India, the wireless base stations for mobile phone services are routinely powered by diesel-powered generators; (3) From a communication technology point of view, the infrastructural side of providing mobile phone services is moving toward the Internet protocol (IP) technology, to be in common with the wired Internet; (4) Web access is already available on many mobile phones, albeit for a fee.

Public Health Priorities and Types of Information

While the basic architectural framework should not depend on a particular public health priority, for the purpose of illustration, we must prioritize public health information in terms of dissemination of information in two directions. One would be for routinely known needs such as AIDS, malaria, and so on, and the other for newly emerging issues on a short or near term basis; for example, the impact of Avian Flu, Swine Flu, Severe Acute Respiratory Syndrome (SARS), Chikungunya, Dengue, Coronary Heart Disease, and Diabetes.

Content Type

Any information about a particular public health issue may be available in four basic modes for on-line availability: text, picture, audio, and video. 'Text' means a web page that may include some graphics as necessary. 'Picture' means depicting a particular piece of information pictorially for ease of understanding; this type is meant to serve the need of non-literate or low-literate people who would benefit from having this feature. 'Audio' refers to the ability to listen to public health information, be it via the Internet or a mobile phone. 'Video' refers to a video clip to explain a particular public health issue. In general, 'audio' and 'video' are important for non- and low-literate users as they can listen or watch to understand a phenomenon.

The above four dimensions are inter-related for the ePH consideration. For instance, a particular public health priority may be available primarily in one of the content types, but not all. That is, from an architectural point of view, we do not want to imply that it is necessary to have all information contents available in all modes. Development of these will depend on cost as well as needs. Secondly, when we consider the population factor and the mode of access, we can see that it is necessary to cater to different needs. For example, for a non-literate user, the assumption that the information that is available through a classical text-based User Interface (UI) on a computer is viable is neither practical nor sufficient. Rather, for a non-literate user, there is a strong need for a *text-free* UI. Such text-free UIs can be pictorial in nature so that the user can click on a button to understand what type of information it refers to; not only that, the button, when the mouse is moved over, plays a small audio to explain what the button is about. In the case of a mobile phone, an audio service may be provided so that the user can directly talk to, which in turn, the service understands and interprets the information. Note that unknown to the user, the feature may generate a call to an automated system or an actual person, who can then offer feedback to the user on the required information either as an audio or a video clip or a text message.

We next focus on the users themselves. In India, there are many different languages and demographic groups, further stratified by economic factors. For instance, users in a particular language may be more non-literate than users in another language. Secondly, even among literate users, a significant portion of the population does not know or understand English. Currently, India has deployed computers in many village areas – with its English-based interface, but only a small fraction of the literate population can take advantage of simple features such as web-browsing and searching; thus, such computers are rarely used. While Hindi is a national language, only small fractions of literate users understand Hindi to the fullest; i.e., the problem is not simplified if the user-interface is in Hindi. What this means is that providing a text-based UI in just a single language does not satisfy the need of the overall population. In fact, it is critical to develop text-based UIs in different languages. With the advent of Unicode (Unicode), characters in fifteen Indian scripts are already available. In fact, this set encompasses many Indian languages. For example, the Devnagari script is used in many Indic languages in the northern part of India; Bengali script in Unicode can be used in Bishnupriya Manipuri and Meitei languages since the script is the same; furthermore, Bengali script, as defined in Unicode, includes a few extra characters, that are needed for the Assamese language—this means the Assamese language is also covered by Unicode. Thus, it is now a matter of developing language-based text-based UIs. Our architectural framework recognizes the need for text-based UIs in different languages in order to reach a larger population group.

In summary, by considering different dimensions, we envision the following broad categories from a systems approach perspective that must be considered in the architectural framework:

- **ePH content:** Either routine contents such as AIDS or Malaria, or non-routine contents such as when needed in the case of an epidemic (e.g. avian flu, swine flu)

- **Information type:** Text, picture, audio, video
- **Language factor:** A variety of languages
- **Literacy factor:** Non-literate, low-literate users
- **User's access mode:** Text-based or text-free UI, or mobile phone

The next question is how to consider these categories in the architectural framework of the ePH system. Although the information availability mode is via both the Internet and Mobile phones, we must address the literacy factor to recognize that the user's access mode must cater to three different modes: text-based UIs, text-free UIs, and mobile phones. Text-based UIs and text-free UIs are both required for Internet-based access to cater to literate, low literate, and non-literate users who need access to the ePH system. Moreover, the text-based UI must be available in different languages. Regarding mobile phones, we consider it as a single mode or simply, as a third User Interface. This is because traditionally, a telephone service does not require a user to be literate (because language or written texts are non-issues). However, at a much deeper implementation level, it would need to have different features for interactive sessions for the users of different literacy groups. Consider when a mobile phone is turned off; it is not uncommon for the caller in India to hear a message in three languages about the mobile phone being turned off: English, Hindi, and the local language. Finally, we believe that the demographic factor does not need to be highlighted separately; this can be addressed through a combination of the language and the literacy factors.

ARCHITECTURAL FRAMEWORK

Taking a holistic systems approach, along with the critical dimensions for India outlined above, we now propose a conceptual architectural framework for the ePH system. Our architectural framework for the ePH system has four primary functional components:

- User-Interface
- Information base
- Request Interpreter
- Content translator/broker

A pictorial view of their functional relationship for the proposed architectural framework is shown in Figure 2.

We now elaborate on each of these components.

User Interface

The *User-Interface* is the entry point for a user to the ePH system. From an actual physical device point of view, this breaks r down further to a computer for Internet access or a mobile phone. In the case of a computer, we may have both text-based UIs (albeit for multiple languages) and text-free UIs as discussed earlier.

Information Base

The *Information Base* is the storage of public health information that would be of interest to users. They would be available in a variety of media: text, pictures, audio, and video. Furthermore, information may be contained at different levels of content depth, and in different languages. It should be noted, that the Information Base is not meant to be a single entity. While from a functional point of view, we refer to the Information Base as one, it can be broken down into multiple different entities based on media types such as text, picture, audio, and video, and/or based on languages.

Request Interpreter

The *Request Interpreter* can serve a number of different roles. For instance, a user may request information about AIDS to be provided in her

Figure 2. ePH architectural framework

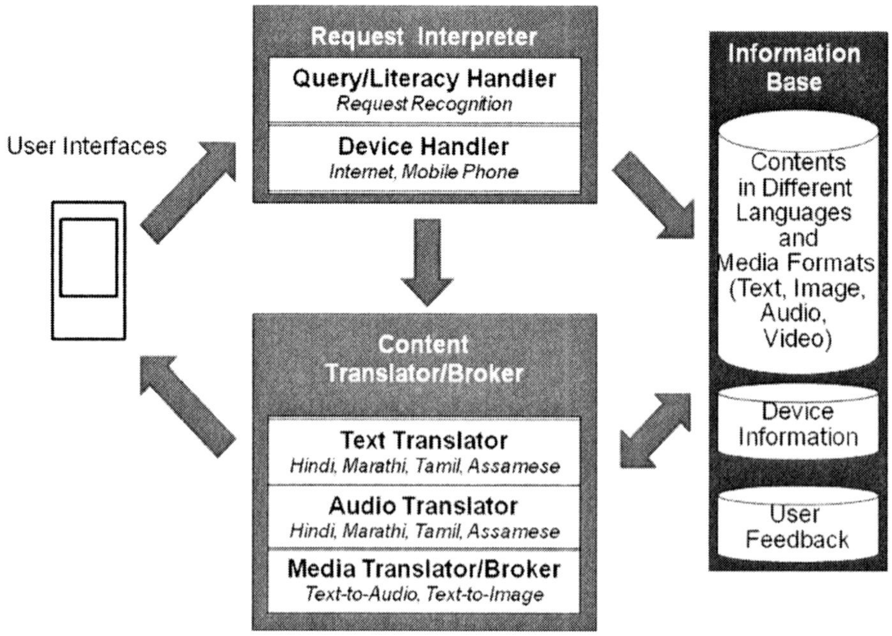

native language. In this case, the Request Interpreter contacts the Information Base for a copy to be provided in that language. Another type of request may be that a user indicates (through the User Interface) being non-literate but provides his preference of native language. The Request Interpreter then contacts the Information Base to provide the information to the user either in a pictorial form, or an audio, or a video clip in the user's native language.

Content Translator/Broker

The *Content Translator/Broker*, in its simplest form, works as a simple forwarding function when the Information Base already has the content that the user requested in the desired mode. Its role, in the more complex form, is multi-faceted. For example, if the request of a user cannot be served by the Information Base in the preferred type requested by the user, an alternate form may be presented. Consider a user requesting information about malaria in the Bishnupriya-Manipuri language in audio form. Suppose that the Information Base currently does not have any information about it in this format, but has a description about malaria in textual format in the Bishnupriya-Manipuri language. In this case, the Information base sends the response first to the Content Translator/Broker, that in turn does a text-to-audio translation in real-time and plays it out to the user. For another request, the Content Translator might serve the role of providing the response in an alternate format. For example, the user requests a video format; however, there is no content available in a video format, but it has it available in an audio format. In this case, the Content Translator/Broker plays the role of a broker to provide the content in a different format. Over time, with the advent in technology, a Content Translator/Broker would be able to translate from one language to another language in real-time, both for textual contents and also for audio clips.

The most important aspect of the proposed architectural framework is that it presents a futuristic vision that takes into consideration the requirement of an ideal ePH system for a country of the size of India with its diversity from many different directions. On the one hand, it is important to consider all the various factors, while on the other hand, the simplicity of our architectural framework is that it focuses on the important goal of serving the diverse end users without unduly complicating the framework. We believe that from a functional point of view these four major components would satisfy the need.

IMPLEMENTATION CHALLENGES

Having discussed the ePH architectural framework, we now discuss certain implementation challenges that can arise.

Consider first the Request Interpreter. This is not to be misconstrued as a basic web-server or a web-client. In fact, as of now, there is no known Request Interpreter that can satisfy the need of this architecture. The current search engines such as Google or Yahoo are text-based search engines; secondly, they cannot meet the user preference coming from heterogeneous devices; thirdly, they cannot serve as a resolver. For instance, a Request Interpreter must also serve the role of a resolver to different forms and/or locations of information in the Information Base. The concept of a resolver has been around on the Internet since its nascent days. An example of a resolver is when a user requests to go to a web site (such as http://www.google.com) that first identifies its domain (www.google.com); the domain is first resolved to a valid IP address to which the request is routed. A more sophisticated resolver is the Handle system (Handle). In our case, we envision the proposed resolver to work at an application level with a much broader scope as its role is to resolve requests based on multi-lingual texts as well as audio requests with multiple modes of operations. Consider a voice-command based search request through the phone system that is used in many parts of the world. For example, in such a system, if a user calls a phone number for flight information, a voice-prompt based interactive system identifies what the user is interested in knowing (the specific flight number, arrival or departure, and so on). We believe this mode of operation is extremely important to provide with the text-free User Interface for non-literate users so that they can use the microphone on a computer or a mobile phone to speak and request any particular information. Thus, such requests will be directed to the Request Interpreter that may involve a resolver to identify where to direct the request. Note that audio over the Internet is now possible since voice over Internet (VoIP) technology is now a standardized mechanism allowing users to make voice communications over the Internet.

In essence, the Request Interpreter is an abstraction for certain types of actions. For a specific medium, different search mechanisms or a request resolver may be developed or customized for different types of users. Rather than being a complex monolithic system, the Request Interpreter function may be implemented using multiple different units for different purposes by providing functions for different types of searches and capabilities for the resolver function through different computing units. It is important to recognize that unlike a web-browser, the Request Interpreter does not reside at the user's end. Rather, the user's end starts with the User Interface (UI). The UI is provided with certain intelligence that is more than a web browser, with the capability to direct the request to the right Request Interpreter.

Consider next the Content Translator/Broker. This is another function that is not currently available as we are envisioning it and it also does not need to be served through a monolithic system. Instead, different requests can be routed through a different Translator/Broker in order to meet the needs of the ePH system in terms of translation from one medium to another or being able to offer alternate solutions. In fact, going from the architectural view to an actual computing system design would lead to this function splitting into multiple sub-functions to minimize software system implementation and computing system management complexity. For instance, a particular computing system may handle all the requests related to the text-to-audio translator. Thus, when the brokering function determines that this translation needs to be invoked, then the request is forwarded to the appropriate translator. Not only that, such translator systems may be organized separately for different languages. It must be noted, that we are not advocating that every such translator be developed from scratch; rather, there are existing off-the-shelf text-to-audio translators for some of the languages that can be used in this translation function. Another translation function is when a specific text needs to be translated from one language to another. Such translations may potentially change the semantic meaning. Since we are dealing with Public Health, it is important to ensure that such translations do not give information that can lead to misunderstandings. Therefore, consistency in vocabulary that is aligned with the unified medical language systems (UMLS) is important since UMLS provides facilities for natural language processing. Along with this, a domain model that captures key concepts would be important to consider (Oldfield 2002). Recently, there has been some work on controlled vocabulary in Public Health (Niedźwiedzka et.al., 2008).

The Information Base is another function that serves as the abstraction for different content bases. In practical implementations, this function may be divided based on languages and different media types. It is, however, worthwhile to develop a good naming scheme for different content identifiers, along with associated attributes so that it is helpful both for the Request Interpreter in identifying correct documents (be it text, audio, or video) and for the Translator/Broker to further resolve it. The document object identifier (DOI) is a well-developed concept that has already been used in practice (DOI). The DOI system is currently used for identifying documents, including persistent identification. Either the DOI may be extended for the purpose of the ePH system or a similar identifier system would need to be developed.

Consider next the User Interface; in particular, consider text-free User Interface for low-literate and non-literate users. There has recently been some development in this regard (Medhi, Pitti & Toyama, 2005; Medhi, Sagar, & Toyama, 2006; Medhi & Kuriyan 2007; Sherwani et. al. 2007, Medhi & Toyama 2007a; Medhi & Toyama 2007b). We believe that such text-free UI plays a very critical role in the ePH system, especially since a sizeable audience requiring Public Health in India is from this demographic group. In general, most non-literate users, given that they are from the lowest income group, would not be able to afford a computer. On the other hand, due to efforts by governmental organizations such as the National Information Centres (NICs) and non-governmental organizations (NGOs), the non-literate user group is expected to have access to computers at certain locations ("community centers"), where the users can go to use the facility. It is important that such computers be customized to cater to the language (sometimes multi-lingual) needs of the local community as well as with the capability for text-free UIs with audio capabilities. In some instances, such devices may be customized so that some user preference/settings may be set up beforehand; in this regard, some functional components from the Request Interpreter may be off-loaded to the user end.

It is crucial to note at this point that the system is proposed with security robustness and foolproof access to ensure authenticity of information. This factor is important not only to enable high quality reliable information, but also to ensure that the system does not fall into the hands of miscreant elements of the society promoting their own agenda. This thought is important given the fact that our approach is influenced by postmodernism that propagates the existence of multiple interpretations; it should not pave a way for multiple interpretations of health related and medical facts. Postmodernism influences the approach and model, not the content. For this to happen, a multi-level governance mechanism needs to be introduced into the system so that information goes through several levels of rigorous reviews before it is deemed publishable.

We can see from the above discussion that there are many components that would need to be built to fit into our proposed architectural framework. Certainly, not all can be accomplished in a short period of time. A systemic approach is needed to reach the end goal. Earlier in this chapter, we discussed that our approach to the ePH system with critical systems thinking (CST) put our framework under the postmodern systems thinking within the scope of the System Of Systems Methodologies (SOSM). In addition, the proposed architectural framework is also partly influenced by the network-based distributed computing[4] paradigm. Rather than taking a software process approach to distributed computing, we take a systems approach to identify different functional components in a distributed fashion. It can be said that our approach then is a hybrid of an N-tier architecture (Manuel and AlGhamdhi 2003) and a peer-to-peer architecture[5] (Oram 2001) when we consider both the functional and sub-functional components of our proposed framework.

It is important to point out that neither the SOSM nor the distributed computing addresses the actual development of the overall computing system to reach the end goal. Given that many components are not going to be ready in short time intervals (both from a logistics and cost point of view), we propose that the actual system development should be a spiral-like approach. The spiral model is a well-known model for software development and enhancement (Boehm 1988). In our case, the spiral model also encompasses design decisions and enhancement. Rather, we argue that the end goal is not fixed; it will evolve with time as new issues in Public Health need to be addressed. Thus, the design phase from the architectural framework must be extensible with a spiral-like fashion, as opposed to a design for a purely closed loop system.

We end this section by commenting on the actual communication technology for Internet connectivity. Depending on different parts of the country, the network connectivity may not be the same; in fact, certain areas may have very low bandwidth network connectivity. Thus, in such cases, both the User Interface end as well as the Request Interpreter would need to have bandwidth-aware intelligence so that the response provided is not bandwidth consuming. This would require customization functions as appropriate.

USER-FEEDBACK

We presented our architectural framework with a goal to communicate a vision. This framework also incorporates user-feedback that is collected as needed for further processing (shown within the Information Base in Figure-2). We also discuss how we envision e-governance in a collaborative setting.

A system development of the scale of ePH in India must have a feature built in for user-feedback for its different interest groups. On one end, is the end user who either may not get the information she wants or the content is not of an adequate nature. Therefore, the end user UI must also have the ability to provide user feedback. For example, for non-literate users, this can be done by asking them to reply to a series of questions after certain information is provided to the user. Their responses to the questions, although to the user they appear to be going to the same system, would in fact go to a functional component that handles user feedback. So that a user is not "penalized" for not giving a high approval rating, this phase must have the ability to randomize the request where the responses are stored in a secure way so that the feedback evaluator does not know the actual user who provided the feedback and the information the user provided is available only through the user's anonymity. Such feedback should then go back to the system designer (another reason for our proposal to consider a spiral model) so that enhancements to the ePH can be done. Since the specifics of the user-feedback handling is more of a software system implementation level issue, this is only shown as a component within the Information Base in the architectural framework as the collection point of the feedback information. Certainly, additional interfacing by external entities to this component is needed for further processing

Besides the end user who provides feedback, the overall system must also allow other constituents to provide feedback on a variety of things. For example, non-government organizations (NGOs) are going to be an important user group of the ePH system and their needs and requirements may be quite different from that of an average end user. In other words, people's participation through user-feedback is an important component in e-governance.

Such user-feedback capabilities go to the heart of e-governance. For example, electronic feedback can be processed and summarized to understand new needs or changes to the currently developed system. These can then be taken into account quickly to make enhancements.

Unlike traditional governance, contents for the information base can also be contributed by the citizens of the country through another User Interface. This will help speed up information availability to the citizens, especially when it may be extremely time consuming to provide every piece of information in every language and media type. In addition, it is certainly possible that such contents are not reliable information. On the other hand, however, what we have seen with services such as Wikipedia on the Internet is that such contents often converge to highly reliable information. Such a system may be supplemented with a scoring system, where knowledgeable citizens can also give scores on the validity of the content. Thus, user-provided information may be classified as "not approved" to a point in time, when through the scoring system, it becomes recognized that this information has validity. Thus, in the e-governance context, we see contents can be broadly separated into three groups: user-provided but lacking an adequate score, user-provided that has an acceptable score, and approved information (either generated by public health officials or moved into this group through the scoring system).

The application of the ePH we have modeled in this chapter may offer the public authorities a framework for effective integration of ICT enablement in Public Health. This may in-turn facilitate the access of its services for the diverse population of the country. The ePH architectural framework we have evolved can be used as an integral part of the overall e-governance agenda of the Government of India. Integration of this kind of a framework would enable efficiency, effectiveness, clarity, transparency, and accountability of Public Health services within the e-governance framework.

REFLECTING ON THE EPH

The ePH that we have proposed in this chapter is an attempt to address the Public Health information requirements for the diverse population of India. We realized at the outset that given the uniqueness of the Indian context and the challenges that we have discussed associated with this context, the ePH will have to be designed in such a way that it is able to accommodate the contextual variety and complexity, rather than underplaying and simplifying the same. Taking a critical systems approach, we have chosen to adopt the postmodernist style that led us to consider a range of design variables and see how these can be coherently understood within a conceptual framework. We also considered a range of challenges that can emerge for the framework – a style that again emerges from postmodernism and critical thinking that encourages self-critique and paves the way for challenging the basis of one's own thinking. This lends the idea that there is no absolute truth, but truth is what we create as humans out of our own impressions of the situations around us. We have considered our own ePH framework as an impression, open to debate and questioning. Our attempt has been to propose a framework that is likely to be able to accommodate changing demands and situations as they emerge in the model of a complex adaptive system, taking cues from a critical systems approach. This has been considered with the proposed integration of feedback loops, spiral design, and a distributed computing paradigm. In so doing, we have also made an attempt to bring together various design parameters, including involvement of diverse stakeholders both in the planning and implementation stages. A critical systems perspective has enabled us to approach the ePH architectural framework in the light of the above considerations.

CONCLUSION

In this chapter, we have attempted to evolve an architectural framework for the ePH, relevant to the Indian context, with a postmodern systems thinking perspective. We first elaborated on what we understand Public Health to be and then discussed the challenges in terms of diversities that are posed by India as a country. We then elaborated, why as a result of this, systems thinking and approaches can lend a perspective to work towards an architectural framework for the ePH in India and considered how this can be woven into the e-governance framework of the country. Finally, we touched upon some of the challenges of the ePH that need to be considered during the implementation of the ePH.

Although we have taken India as the case-in-focus, we believe that this framework can be replicated and adapted to other countries with similar situations or with a similar stage of economic development, where challenges posed by illiteracy, ignorance, and lack of access may be of a similar nature.

It needs to be considered that given the fact that this framework has been evolved in a context where there is considerable complexity in the absence of any such prior framework at the national level, it is likely to face certain challenges in its implementation (some of which we have already tried to articulate in this chapter). Along with this, it is probable that there will be certain critiques of this framework itself. Our approach does have an affiliation with the postmodernist point of view, but at the same time, we have taken just two variables in the systems design – language and literacy. This approach needs to be widened with the inclusion of further variables in the systems design. Language and illiteracy were the two major variables we identified for this stage of a zero-based design for the ePH. We therefore propose this framework to our informed readers as a work-in-progress, open to new ideas and further deliberation on new variables and newer paradigms of developments. Our interest is to further explore how we can evolve a methodology for implementation of this framework. We equally invite deliberation from our audience for ideas on how we can best leverage our understanding of Critical Systems Thinking, in general, and the use of postmodern systems thinking, in particular, in this exercise, along with its integration with the distributed computing paradigm.

ACKNOWLEDGMENT

The authors would like to thank the following for reading a draft of this chapter and for providing valuable comments: Uttam K. Borthakur, Yugyung Lee, and Howard E. Thompson, and an anonymous reviewer. The authors also appreciate Yugyung Lee's help in improving the presentation of Figure-2. Finally, we thank Nancy Lorenz for her careful proofreading of this chapter.

REFERENCES

Attiya, H., & Welch, J. (2004). *Distributed Computing: Fundamentals, Simulations, and Advanced Topics (2nd Ed.)*. Hoboken, N.J.: Wiley-Interscience.

Boehm, B. (1988). A Spiral Model of Software Development and Enhancement. In *IEEE Computer,* 21(5), 61-72.

Census of India. (2001). New Delhi: Office of the Registrar General & Office Commissioner of India.

Chowdhury, R. (2005). *The Information Messiah: Issues & Debates within the NHS*. Working Papers in Social Sciences and Policy No. 15. Hull, UK: University of Hull.

Chowdhury, R., Clarke, S., & Butler, R. (2007). Healthcare IT Project Failure: A Systems Perspective. In *Journal of Cases in Information Technology* (pp.1-15). Hershey, PA: Idea Group Inc.

Chowdhury, R., & Nobbs, A. (2008). Critical Insights into NHS Information systems Deployment. In D. Jemielniak & J. Kociatkiewicz (Eds.), *Management Practices in High Tech Environments* (pp.245-264). Hershey, PA: Idea Group Inc.

Churchman, C. W. (1971). *The Design of Inquiring Systems, Basic Concepts of Systems and Organizations*. New York: Basic Books.

Clarke, S. (2001). *Information Systems, Strategic Management: An integrated approach*. London: Routledge.

Cohen, C., & Midgley, G. (1994). *The North Humberside Diversion From Custody Project for Mentally Disordered Offenders: Research Report.* Hull, England: Centre for Systems Studies.

Das, S. R., & Chandrashekhar, R. (2006). *Capacity Building for E-Governance in India*. UNDP Asia-Pacific Development Information Programme. Retrieved July 21st, 2008 from, http://www.apdip.net/projects/e-government/capblg/casestudies/India-Chandrashekhar.pdf

Dasgupta, M. (2005). *Public Health in India: An Overview.* World Bank Policy Research Working Paper 3787, New Delhi. Retrieved September 13th 2006, from http://www-wds.worldbank.org/servlet/WDSContentServer/WDSP/IB/2005/12/02/000016406_20051202162921/Rendered/PDF/wps3787.pdf

Document Object Identifier System. (n.d.). Retrieved from http://www.doi.org/

Donner, J., Gandhi, R., Javid, P., Medhi, I., Ratan, A., Toyama, K., & Veeraraghavan, R. (2008). Stages of Design in Technology for Global Development. *IEEE Computer*, *41*(6), 34–41.

Dooley, K. (1997). A complex adaptive systems model of organizational change. *Nonlinear Dynamics Psychology and Life Sciences*, *1*(1), 69–97. doi:10.1023/A:1022375910940

Eemeren, F. H. v. (2003). Anyone who has a view: theoretical contributions to the study of argumentation. In *Argumentation library* (Vol. 8, pp. 92). Dordrecht, The Netherlands: Kluwer Academic.

Fahey, D. K., Carson, E. R., Cramp, D. G., & Muir Gray, J. A. (2004). Applying systems modelling to public health. *Systems Research and Behavioral Science*, *21*, 635–649. doi:10.1002/sres.602

Farrell, D., & Beinhocker, E. (2007). Next Big Spenders: India's Middle Class. In *McKinsey Global Institute Business Week*. Retrieved on January 5, 2008, from http://www.mckinsey.com/mgi/mginews/bigspenders.asp

Garai, A., & Shadrach, B. (2006). Processes and Appropriation of ICT in Human Development. In *Rural India: Bridging the Research and Practice Gaps*. Retrieved on August 15, 2008, from http://www.dgroups.org/groups/oneworld/OneWorldSA/docs/TICTEIV_pdf.pdf

Handle System Architecture. (n.d.). Corporation for National Research Initiatives. Retrieved from http://www.handle.net/overviews/architecture.html

Human Development Report (2007/2008). *United Nations Development Program*. New York.

Jackson, M. C. (1995). Beyond the Fads: Systems Thinking for Managers. *Systems Research, 12*(1), 25–42. doi:10.1002/sres.3850120106

Jackson, M. C. (2003). *Systems Thinking: Creative Holism for Managers*. Chichester, UK: Wiley.

Jackson, M. C., & Keys, P. (1984). Towards a System of Systems Methodologies. *The Journal of the Operational Research Society, 35*(6), 473–486.

Kanungo, V. (2004). *Citizen Centric e-Governance in India: Strategies for Today, Vision for Future*. New Delhi: Society for Promotion of E-Governance. Retrieved June 22, 2008, from http://en.wikipedia.org/wiki/Languages_of_India

Leischow, S. J., & Milstein, B. (2006). Systems Thinking and Modeling for Public Health Practice. In *MPH. American Journal of Public Health, 96*(3), 403–405. doi:10.2105/AJPH.2005.082842

Maani, K. E., & Cavana, R. Y. (2000). *Systems Thinking and Modelling: Understanding Change and Complexity*. Auckland, New Zealand: Prentice Hall.

Manuel, P. D., & AlGhamdi, J. (2003). A data-centric design for n-tier architecture. *Information Sciences, 150*, 195–206. doi:10.1016/S0020-0255(02)00377-8

McTaggart, J. M. E. (1964). *A commentary on Hegel's logic*. New York: Russell & Russell.

Medhi, I., & Kuriyan, R. R. (2007). Text-Free UI: Prospects for Social Inclusion. In *Proceedings of International Conference on Social Implications of Computers in Developing countries*, Brazil.

Medhi, I., Pitti, B., & Toyama, K. (2005). Text-Free UI for Employment Search. In *Proceedings of Asian Applied Computing Conference*, Nepal.

Medhi, I., Sagar, A., & Toyama, K. (2006). Text-Free User Interfaces for Illiterate and Semi-Literate Users. In *Proceedings of International Conference on Information and Communication Technologies and Development*, Berkeley, PA.

Medhi, I., & Toyama, K. (2007a). Full-Context Videos for First-Time, Illiterate Users. In *Proceedings of ALT. CHI forum at ACM CHI '07*, San Jose, USA.

Medhi, I., & Toyama, K. (2007b). Full-Context Videos for First-Time, Non-Literate PC Users. In *Proc. IEEE/ACM International Conference on Information and Communication Technologies and Development*, Bangalore, India.

Midgley, G. (1996). What Is This Thing Called CST? In R.L. Flood & N.R.A. Romm (Eds.), *Critical Systems Thinking: Current Research and Practice*. New York: Plenum Publishing Corporation.

National E-governance Plan (NeGP). (2006). *Framework India*. New Delhi: Ministry of Communication & Information Technology. Retrieved on June 24, 2008, from http://www.mit.gov.in/download/eu/spsingh.pdf

Niedźwiedzka, B., Czabanowska, K., & Śmietana, R. (2008). Controlled Vocabulary in Public Health: An Overview of the Achievements to Date. *Journal of Public Health*, accepted. Retrieved November 25, 2008, from http://www.springerlink.com/content/31-5862w47215-m058/

Oldfield, P. (2002). *Domain Modelling*. Retrieved November 25, 2008, from http://www.aptprocess.com/whitepapers/DomainModelling.pdf

Oram, A. (2001). *Peer-to-Peer: Harnessing the Benefits of a Disruptive Technology*. Sebastopol, CA: O'Reilly & Associates.

Sappenfield, M., & Chopra, A. (2007). India's superrich get even richer. *The Christian Science Monitor*. Retrieved October 11, 2008, from http://www.csmonitor.com/2007/1218/p01s04-wosc.html

Senge, P. M., et al. (1994). *The Fifth Discipline Fieldbook: strategies and tools for building a learning organization*. London: Nicholas Brealey Publishing.

Sherwani, J., Ali, N., Mirza, S., Fatma, A., Memon, Y., Karim, M., et al. (2007). HealthLine: Speech-based Access to Health Information by Low-literate Users. In *Proceedings of IEEE/ACM International Conference on Information and Communication Technologies and Development*, Bangalore, India.

Stump, E. (1989). *Dialectic and its place in the development of medieval logic*. Ithaca, NY: Cornell University Press.

Sudhir, V., Srinivasan, G., & Muraleedharan, V. R. (1997). Planning for sustainable solid waste management in urban India. *System Dynamics Review*, *13*, 223–246. doi:10.1002/(SICI)1099-1727(199723)13:3<223::AID-SDR127>3.0.CO;2-Q

Unicode Consortium. (n.d.). Retrieved from http://www.unicode.org/

Unified Medical Language System. (n.d.). Retrieved from http://www.nlm.nih.gov/research/umls/

ENDNOTES

1 Positivism is a school of thought according to which, knowledge can only be derived from sense experiences, and the methods to be followed for the same are to be rational, formulae driven, and relate to what is inferable. The methodologies inspired by positivism include hard systems thinking, system dynamics and organizational cybernetics, which are driven by obvious objectivity and observatory knowledge.

Interpretivism is a school of thought according to which, knowledge can be derived only by delving into the thoughts and feelings of the observed, and identifying the unobvious and what is beyond the surface level. Systems methodologies inspired by interpretivism are soft systems approaches that make provisions of taking different human perspectives into consideration in the intervention process.

Postmodernism is a school of thought according to which, there can be numerous interpretations of knowledge, which can be arrived at with various approaches and perspectives; no form of knowledge can be considered as more valid than a different form of knowledge.

2 Emancipatory approaches are those that make human emancipation as the basis of the intervention process with the belief that the end result of any systems exercise should be the betterment of human beings.

3 For a detailed narration of these methodologies and approaches, refer to Jackson (2003).

4 Distributed computing refers to a system with more than one processor and/or storage that is distributed over multiple computers for a certain task (Attiya & Welch 2004).

5 An N-tier (typically, a three-tier) architecture in a distributed software system is a client-server architecture that includes a

middle-tier that processes a client's request to forward to the appropriate server (for example, a database server). In a peer-to-peer architecture, an entity can serve both as a client and a server at the same time, i.e., it is a server to others while it can take turns to be a client to request from others.

Chapter 7
E- Government Systems Architecture:
Contextual and Conceptual Level

Ioannis Savvas
Agricultural University of Athens, Greece

Elias Pimenidis
University of East London, UK

Alexander B. Sideridis
Agricultural University of Athens, Greece

ABSTRACT

Based on the inclusion of several types of stakeholders for e-government services, the authors propose an architecture of multiple levels that ensures adaptability to new technological and organizational demands for these services; thus the authors emphasize the possibility of representing public services as electronic services which can be continuously improved.

INTRODUCTION

A common problem in the selection of appropriate e-government systems and the relevant software development is the consideration of the whole set of stakeholders and their requirements. This is the most critical phase in establishing a problem's environment or in representing a "real world domain". We usually refer to it as the contextual level (why questions). It provides context information. In Cap Gemini's Integrated Architecture Framework (IAF) (CapGemini, 2006) for systems design, four levels of abstraction are recognized: contextual, conceptual, logical and physical. The first, contextual, is for answering the "why" question providing context information and key principles that support the value proposition for the architecture to be developed. The conceptual level addresses the "what" aspect of architecture design. It defines the services that are required and what is required from each service. The logical level derives "how" the customer needs can be realized, showing how components interrelate and where components 'implement' services. The last one, physical, level addresses the "with what" aspects of architecture design and defines standards, products (catalogues), guidelines, etc. for further development and implementation.

DOI: 10.4018/978-1-60566-860-4.ch007

In this chapter the authors propose a methodology that identifies all stakeholders in e-government domain. Through Public Administration (PA) theory and practice, stakeholder categories can be identified. These can provide a full set of principles which will govern a new architecture. Principles are guiding statements about fundamental beliefs, truths, rules and qualities that guide objectives and the decision making process. Architecture is linked to business needs through these principles. The principles / (non-functional) requirements are deduced through publications, formal documents, expressed opinions and experience. Coherent analysis of these high level non-functional requirements can lead to more specific ones. Even if the final result cannot be as detailed as an implementation engineer would like it to be, it provides a very satisfactory context allowing, requirements control. Instantiation of stakeholders' categories can lead to individual needs within the limits of each category.

As the efforts for having a software meeting the needs of the whole set of stakeholders begun from the enterprises themselves, findings from such efforts can be transferred to the e-government field as best practices. However, an evident and explicitly stated difference between PA and the private sector appears to be forgotten. While PA can no longer be a "bureaucratic monster", but a structured citizen and business oriented service, it cannot set aside its fundamental principles though.

In the following a modelling of PA's function is presented. The modelling refers mainly to PA of the "normative" countries but it could be expanded to the operational model countries as well. This modelling provides the conceptual level of the architectural design for e-government systems. The different groupings of styles of PA found in Europe are discussed at a later point in the chapter.

In this work's functional model, PA's function is represented in an independent way. Goals and limitations, quantitative objectives and law restrictions are being made obvious allowing tradeoffs and negotiations between trends. It highlights areas where controversies can deploy and allows instantiations for real life argumentation.

Finally towards the end of the chapter the authors provide application examples of this approach tested in the reality of the Greek PA system.

CONTEXTUAL LEVEL

In Savvas et al. (2007a), PA's stakeholders were identified and their strategic relationships in the socioeconomic environment, at national and supranational level were defined. Stakeholders were defined on both sides of public service provision, supply and demand.

- The demand side includes citizens (also as employees) and businesses. Judicial power (administrative courts) and Legislative power can also be classified here. Parliament receives PA services in law making process and it is interested in the application of the laws it provides. Courts are control mechanisms regarding public service provision. They are interested in the application of their decisions concerning administrative acts and they support administrative processes providing jurisprudence.
- The supply side includes the indivisible of governance. Government national and supranational (EU case). When we are referring to a certain service though, final provision is being made from one Public Organization (PO). The demand side then might includes other POs too.

Especially for the case study of the Greek PA a first set of stakeholder requirements has been presented. In this case, stakeholders are not only national but supranational as well, as Greece is part of the E.U. Stakeholders belong to the direct and the indirect environment of PA and have been defined as: Government, the EU, citizens/businesses, public organizations, public servants, the Law courts and country's Parliament.

Various papers and documents were used to determine goals of each stakeholder category and in thus identify and elicit their needs. These papers and documents used to form the first set of requirements are provided below in form of categories. In each category a set of indicative documents are given. The purpose of the use of these documents is to identify the basic principles. Thus if there is an official document explicitly expressing the principles of a stakeholder category this could be sufficient:

1. Strategic papers and reports as well as fact sheets of every member state.
2. E-Government Communications, (e.g. COM (2003)567, 26 Sept 2003), Working Papers (e.g. on eGovernment Beyond 2005"), White Papers (e.g. on European Governance", COM (2001)428, 25/7/2001), The new European strategy for Information Society i2010.
3. Projects (e.g. e-Government Economics Project (eGEP): e-Government Unit, DG Information Society, European Commission, 4/2006), Surveys (e.g. "Online Availability of Public Services: How Is Europe Progressing?" Web Based Survey on Electronic Public Services Report of the 6th Measurement, Capgemini, June 2006), Studies (e.g. "Study for problem's management of citizens and businesses". The "KAFKA" Plan, April 2006, Greek IS Observatory, "National Study on New technologies and IS held on Greek and foreigners", 2005, EDET S.A)
4. Organizational reports regarding performance of public organizations and thus expressing personal experiences of higher executive officers and their requirements for a proper and sound function of the organization they govern/administer. These reports can be found in public organizations' sites like the site of the ministry of interior (www.gspa.gr)
5. Documents from the website (www.adedy.gr) of the major labour union for public servants in Greece.
6. Articles of high ranked judges and presidents of the highest courts
7. Mission of the Parliament and statements of the President of the Parliament

Then sets of needs per stakeholder are formed correspondingly as follows:

1. Development, prosperity, equity, justice, freedom, democracy– transparency -participation form *group A*. These goals are for the functional definition of PA consequences. These goals are to be achieved with Efficiency-effectiveness-economy.
2. Additionally to the above, Pan-European sustainable- innovative services. Transparency belongs in *group J*. Inclusive services in *group C*. "Inclusive" concerns the external communication between PA and citizens/businesses, which has to include the whole of the population. Thus the approach should be *multi-channel:* (sites, fax, mobiles, unified administrative offices, call centres, m– government, infokiosks). Accessibility, multi language, security form *group D*.
3. Public money exploitation-equity, Less time, less trouble belong in *group A*. Sufficient information in *group J*. Protection (Personal data protection, Identification – authentication, Identity management, Data and network security, Battle against e-crime) form *group E*.
4. Interoperability (technical-organizational form *group G*-semantic form *group F*). These groups comprise European Interoperability Framework's principles, like Accessibility, Multilanguage, Security,

Privacy, Subsidiarity, Usage of open standards, Assessment of the advantages of open source software, Usage of Multilateral solutions, Collaboration culture, right data and their usage form *group H.*

5. Lower work load, less routine tasks in *group I*– education – training
6. Courts care for the execution by the PA of their decisions and for the soundness of acts issued so as to be relieved from the burden of appeals.
7. *Group B* refers to principles of administrative law like reasoning, fairness, trust, proportionality etc. Additionally, parliament uses PA for the preparation of laws and for getting feedback from their application

These needs concern, in the case of e-government, technology resources that would in turn pose specific requirements on human and financial resources

The above can be further explained in the following notes:

Efficiency and effectiveness based on van Dooren's (2006) work are expressed as ratios: Efficiency = input/output and Effectiveness = output/effect (1), and effect/consequence (2). Output is the administrative act, effect is the service and consequence refers to the first measurable specification of goals of the constitution or politics. Further analysis is provided in next section's PA's functional model.

Citizens' and businesses' needs for faster services are concurrent to governments' requirements for efficient administration. Time is included as parameter in the ratio of efficiency, both in person-hours and in terms of communication resources.

The needs for less discomfort and less bureaucracy are also included in the above ratio and concern the number of internal entities required and the swiftness and effectiveness of communication.

An additional issue is for the new administrative systems to earn the trust of the citizens and their legal consolidation. The three main categories of legal issues being, (a) legal validity of e-government, (b) e-government trust and security and (c) legal remedies of e-government.

The ability of the citizens to understand administrative language and practice or the ability of PA to address the external environment in a language that is understandable, in order to collect/provide useful and sound information is a further point of consideration. This affects the semantic interoperability between PA and external entities.

Public servants as stakeholders might pose additional points for consideration, such as (a) resistance to change (e.g., introduction of technological solutions), due to fear of loosing vested interests and (b) lack of knowledge sharing because of fear of loosing competitive advantage for promotion.

In all the above points there is an evident conclusion that stakeholder needs may often be contradictory to each other. In such cases a decision as to which is the prevailing stakeholder must be made. Negotiation between stakeholders could be discussed through various theories in the field and it is beyond the scope of this chapter. In the case discussed here, this decision-making is ad-hoc, depending on the service, that is the legal and the operational practices. Thus national government might draw back in favour of an EU decision if the Greek Constitution will allow it. There are cases, though, where strong cultural characteristics are present in public servants' and citizens' requirements and partially in those of public organizations. In such situations the order of prevalence may be reversed.

CONCEPTUAL LEVEL

The conceptual level addresses the "what" aspect of architecture design. It defines the services that are required and what is required from each service. The architecture is intended to provide services to the operational environment. To determine these supporting services normally an analysis of business processes is performed. In this case a generic solution, applicable to several widely different types of services is sought after.

E-government often coincides with the reorganization of processes and methods followed in public service provision. E-government systems though have to suit/facilitate a PA's rationale. This work addresses the function of PA at top level reusable mode. The approach adopted here is a goal oriented one, placing the administrative act at the centre of PA's function. To this effect, a method is proposed which in addition can spot problematic areas in public service provision and suggest actions and the introduction of appropriate e-government systems aiming to remedy the situation.

In its daily function, PA repeats a sequence of activities in order to achieve its mission, to serve citizens and businesses. Even if for the majority of the citizens terms PA and bureaucracy bring forth negative associations (Wilson, 1993) this has nothing to do with the real meaning of these terms. From Weber's definition (1946) to more recent trends (Peters and Savoie, 1998) and introduction of "New Public Management" theories (Boston et. al., 1991), bureaucracy's function is a prerequisite for the success of governmental programmes, effective and good governance. This activity is not always in accordance to contemporary or short term rationale, but in line with "public weal" or in other words in a sustainable and constitutionally sound way.

PA aims at achieving goals like development, prosperity, equity, transparency, justice, freedom, democracy. To achieve these goals PA provides certain *services*. In order to provide services PA issues administrative acts. The issuing of acts is the core activity of PA; it is always a State activity and concerns e-government. PA services could be divided into material (e.g., construction of a road) and non-material (e.g. Change of the marital status in a personal details database).

In this work, material activities are considered up to the issuing of the act that commands them, as the following activities are of no interest to the e-government domain. Contrary to that, non-material activities are fully in-scope with e-government as they comprise an additional circle of communication – information in order to implement a transactional service. Material services could be outsourced more readily.

A European PA Classification

There are several criteria for classifying PA profiles across Europe. Napoleonic and non, civil and common law, continental and Anglo-Saxon etc.

In a previous work (Savvas et al., 2007b) the authors classified the PAs of the fifteen older member states of the EU on the basis of vision and objectives for e-government, in two dominant trends: a) the empowerment of democracy through an open, transparent and participatory society (social state model) and b) improvement on monetary performance through cuts on state expenses or on returns based on raising competitiveness and on increasing of the number jobs offered by businesses. Citizens obtain additional gains through tax reduction. (Market driven model)

A different classification by Billiets et al. (2006) distinguishes between "normative" and "operational" models:

1. A normative model is characterized by increased interest for legal formality. A distinctive type of law, public law, governs the functioning of the state as well as the relations between public entities and civil society. Such PAs are rule oriented mechanisms
2. "Operational" administrative systems are result-oriented mechanisms. Legal tools are not ignored, but quantitative methods based on the use of performance indicators, strategic and operational planning, cost-benefit analysis and other similar techniques, mostly borrowed from business management, are the backbone of administrative working methodologies.

In both the above mentioned approaches Southern European countries like Greece, Spain, Italy, Portugal, and those of continental Europe like Belgium, France, Luxembourg, Germany, Austria are grouped together as to the first of the dominant trends identified. The model of PA's function discussed in this work adopts an independent point of view. Goals and limitations, quantitative objectives and legal restrictions are being made obvious allowing tradeoffs and negotiations across different trends. It identifies areas where controversies can be deployed and allows instantiations for real life argumentation.

The above concurs with Sabucedo and Rifon's (2006) point of view that in all democracies across the world the use of documentary evidence to support PA operations is a common feature.

A Public Administration's Functional Model

To represent PA's operation the authors use an input-output model:

The above modified input/output model for public sector function results from van Dooren's work (2006), which in turn was based on Pollitt and Bouckaert (2004).

Within society there are socioeconomic issues that create needs to citizens and businesses. Such needs are usually translated by politicians to policy objectives. In response to these objectives, inputs are assigned in the form of resources which by certain processes produce/provide outputs. Under the influence of the environment these outputs produce outcomes (effects and consequences). These respond to existing socioeconomic issues or prevent new ones from arising. PA in its broader sense is where governance as formed by politicians is exercised and embraces the whole cycle as pictured in figure 1. Public Organizations' and public servants' tasks are usually kept to efficiency matters.

The ratio of input over output expresses a measure of efficiency (horizontal axis). The outputs are expected to have outcomes on society. These outcomes can be intermediate (usually short-term, effects) or final outcomes (usually long-term, consequences). The final outcomes in particular, are influenced by the environment on which the organization or the program has a limited or no impact. The ratios of output over effect and effect over consequence are two effectiveness measures (vertical axis to the right).

Based on the above input-output model (figure 1) and setting service as effect, administrative act as output and finally as consequence of an administrative action, the long term effect of which is going to be aligned with the aggregation of goals of the stakeholders as set by politicians, the above diagram is derived.

Figure 1. A simplified input/output model on PA's function

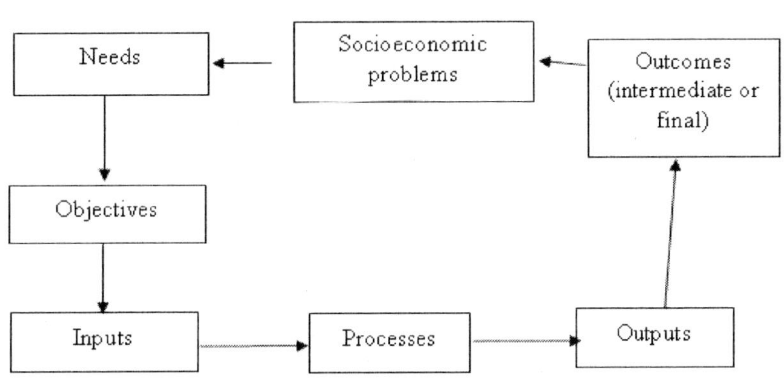

Figure 2. PA's function (different shapes used do not follow any notation, but are used to emphasize differences)

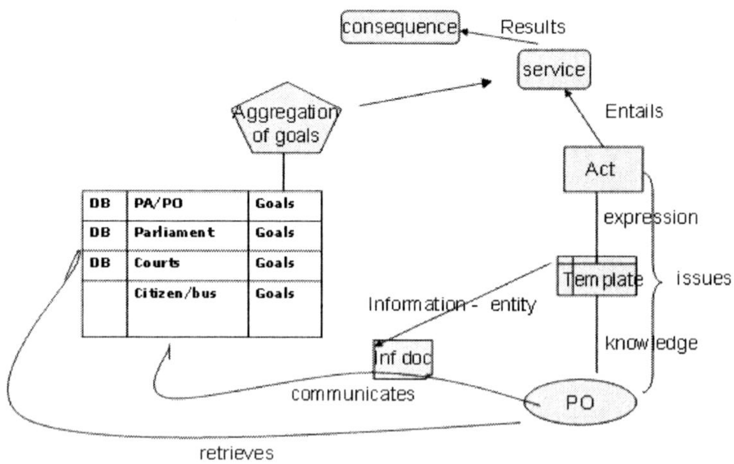

PA comprises Public Organizations (POs). To issue an act a PO has to fill a template, which is a product of a knowledge process and it is incumbent upon the Administrative Law and the rules for the composition of an administrative document (Savvas and Bassiliades, 2008). These rules provide a minimum for the issuing of sound administrative acts that could sustain to objections for typical reasons. In order to fill the template, a PO needs information which may exist either in its own database or in a database of another administrative entity, - i.e. databases of law courts (legal cases), parliament (law) and citizens or businesses (certifications, declarations). To acquire this information a PO performs either retrieval from one of the databases that it can have access to or communicates with other entities (sending an informative document). As a result the information supplied will be in the form of an administrative act or a legal case or a citizen/business document.

To communicate, a PO may use informative documents, documents that ask for information (they are not acts). Triggering of the process described above could be done by: a) Citizen/business applying for a service, b) Law Courts that decide for or against an administrative act, c) Parliament that votes for a new law (that changes or abolishes an older one) and decides probably the provision of a new service or change the preconditions of an older one, or d) Administrative entity seeking for information or orders the provision of a service.

The final act which commands a service is communicated to the entity that triggered the procedure and to the responsible administrative entity which might be different from the one issuing the act, in order to proceed to actions that change the world (e.g. Changes in database records, money transfer).

The Core of PA's Function: The Administrative Act

PA's core function concludes with the issuing of an act. The act has: a) Effects (short term) and b) Consequences (long term).

Service is the effect of an administrative act. A goal of governance is to identify consequences of an administrative act with the aggregation of goals (ideal case).

Public documents are structured by following certain rules and forms. The rationale across

European countries especially between the ones grouping together in the above mentioned models is the same. To illustrate the structure of acts we use the case of the administrative act in Greek PA. The distinction between elements that each public document must possess is as follows (Greek Administrative Process Code, 1999): a) before main text elements, b) main text's elements and c) after main text elements

Especially concerning Decisions of Public Administration (acts), the main text has its own basic rules regarding structure and appearance. The basic characteristics of decision texts are explicitness, accurateness, briefness, rationalism and use of simple language. Decisions are legislative acts and they might be normative or personal/individual. They are divided in two parts: preamble and pronouncement.

In the *preamble*, all the valid legislation that provides the administration with the obligation or ability to produce the act is presented. At the end of the preamble, the reason for which this act is produced is presented, in other words its grounds for decision, either because this is required by the regulation or because it is obligatory by the act's nature. *Reasoning* generally refers to the legislation that regulates the production of an administrative act and its interpretation.

Concluding the above, the composition of an act by a PO follows certain steps (model)

- Administrative process code sets the template
- PA's and Law experts set the law mix
- Law mix sets preconditions (as interpreted by PA experts)
- Preconditions ask for specific information
- Specific and general information is interpreted by civil servant (discretion margin)
- Interpretation forms an aspect

A service is ordered by an act. Act is ruled by a law mix, not a single law and this could differentiate decisions. Law mix is defined in its majority by domain knowledge.

The Triple Functional Requirement/Need of PA

The information that PA is looking for may be distinguished in: General (stable/repeated) or Special (ad hoc/new). Informational need defines the identity of the entities (or databases), which are going to provide the information needed.

Communication can be divided in: Internal (between PA's entities) or External (between PA – Citizen/businesses etc). Communicational need defines the type of communications between the above defined entities.

Finally, expression concerns impressed thoughts and decisions of PA on documents. These documents may be Informative or Acts. They have form (structure) (information apposition) and content (information interpretation) (non-structured). The expressional need relates to the record of the communicated information and the recording of the justification of the decision (reasoning).

Taking a closer look at an act's construction, there is primary need for information. To acquire information there is a need for communication or retrieval and later there is a need for interpretation, for an expression of thoughts.

Act needs three types of information

i. Information for the general provisions or decisions of the preamble
ii. Information about the specific preconditions that the law mix sets
iii. Information for similar cases and legal cases which aids the decision making

First one, (i) is about expert knowledge and law databases. Third one, (iii) is about knowledge and courts databases

For the second type of information:

- A PO is seeking information from other POs or citizens and businesses
- It needs to communicate
- Depending on legal barriers (personal data protection) it might need an intermediary or it can retrieve
- Depending on interoperability barriers (semantic, technical, organizational) it could need an intermediary
- Depending on both (legal and interoperability barriers) it could possibly define type of communication (electronic, fax, personal, telephone etc.)
- Depending on legal barriers it could use data or information
- Depending on all the above centralized or distributed databases will be built

According to what has been discussed above there is a triple functional need/requirement of PA: Informational, Communicational, and Expressional/representational/expressive.

The authors argue that a document (act) is the place where all the knowledge of PA is laid down. Even tacit knowledge of all public servants has to be explicit on an administrative act if provisions of administrative processes code are going to be followed.

The analysis of PA's functional needs apart from a methodological approach underlies a tool for the trace of bottlenecks and warps of its function. Identifying these areas makes it easier to choose or design the strategy of the solution.

STAKEHOLDER GOALS IN CONCEPTUAL MODELLING

To incorporate stakeholders' goals/needs to the whole of a PO's function and every single process, one can start from the effectiveness part. Effectiveness is the measure of achieving goals that are not necessary financial. They could be goals relating to issues such as democracy, equality, etc. and according to contemporary theories they should reflect stakeholders' needs.

Step 1. The effectiveness part. Effectiveness is expressed by two ratios: Effect/Consequence and Output/Effect

i. **Effect/consequence**: The ideal situation is to identify consequences of the administrative action with goals/objectives as set by politicians. These objectives are measurable interpretations of the abstract goals of the stakeholders. Effect in this case is the service in question. The ratio is expressed as actual over prospective, meaning that the service either achieves the goal that the government and the politicians had set, or not. A problem expressed in the values produced by this ratio, reflects for example policy objectives setting and law making problems.

ii. **Output/effect:** This is expressed as act/service. It refers to the number of the acts that actually provide the requested service (note even the denial of a request is considered as a service). It concerns the number of acts that are invalid due to objections or appeals, number of acts that provide service to persons that are not entitled to that and number of acts that provide the service to people who are beneficiaries of a better similar service. Such problems call for changes to the quality of acts (structural and typical matters, matters of interpretation of the legal framework and discretion margins of public servants, matters of dissemination of information.)

Step 2. The efficiency part. Efficiency is described as the ratio of Input/ Output. Acts as outputs need three types of inputs: information, communication and expression as resources. All three are tested versus two variables, time and cost.

Information can be divided in three types as shown in the "triple functional need" section above:

Information in (i) and (iii) concerns administrative knowledge. It deals with knowledge management systems and personnel training, HR culture and development. It addresses the very sensitive matter of discretion margins.

For (ii), information for the preconditions set by the law is required. This information is usually possessed by other POs or by other entities like citizens and businesses. Provisions of laws about personal data protection define direct or indirect access to the information from where it lays (it also help decisions about databases and their data). Technical interoperability issues define direct electronic retrieval or communication between entities through document exchange (informative). For the latter, issues of communication material and type have to be solved (telephone, fax, email, paper etc). There are also matters of organizational and semantic interoperability between either organizations exchanging information or citizens. Such interoperability barriers can further define the type of communication required.

APPLICATIONS

To validate this method a Greek Public Organization was used. To this effect two critical procedures of the Greek PA, as applied in a directorate of one of the country's regions, have been analysed and proposals for the introduction of appropriate systems have been made.

The first procedure refers to the management of the Programme of Public Investments. The method primarily assesses the informational needs of the specific procedure and suggests the development of a database which in a sense is not an integrated e-government initiative. It is used though to illustrate how functional requirements concerning informational needs are determined.

The second project concerns the whole function of a PO's department. Services that the department provides are examined according to stakeholders' needs and are being met by the use of e-government systems. Matters of effectiveness are also addressed here.

A. An application for the management of the Public Investments Programme (PIP) in the Region of Central Macedonia (RCM).

To determine the informational needs of the department responsible for the management of the PIP the acts that the department issues are identified. To fill the template of the acts certain information is needed. For the composition of acts and concerning expressional needs it is essential to form the part of the preamble which lists the laws that rule each procedure. Thus the additional information required relates to (Savvas and Bassiliades, 2008):

- The laws in force and their amendments
- Supporting material of previous acts and case laws (jurisprudence) that will help the public servant to form an opinion based on administrative case based reasoning

Concerning the second (ii) type of information (preconditions set by law) for each act the following procedure takes place.

The documents that the department issues and their relevant informational needs are described below:

1. Proposal to the Ministry of Finance (MoF) for the allocation of the financial resources per project and collective decision. For this act information is required about, (a) the approved payment limits per Collective Decision must be kept (encyclical from MoF), (b) data relating to the implementation level of a project (credits, payments etc) and (c) estimation proposals from the

organization that implements the project for the future course.

2. Notification of all agencies which implement projects for the approved credit per project. No special information except from the addresses of the agencies should be kept.
3. Notification to the MoF for the subsumption of a project to the PIP. Information required for the subsumption in the Regional Operational Programme (acts from the Managing Authority of ROP).
4. Notification of agencies for the subsumption of their projects to PIP. Another organization's (MoF) act, change the world (data of projects) in the database of RCM.
5. Project's credit approval. Information required about, (a) subsumption in ROP, (b) subsumption in PIP, (c) budget, (d) credit, (e) contract details and (f) auctioning details.
6. Order to the Bank of Greece for project financing

Information required in this case is (a) approved credit per project, (b) payment details per project, and (c) requests for financing

From the above six documents only three of them are acts (1,5,6). The other three are notifications required due to communication barriers set by the administration. The above mentioned information required by the acts is described in abstraction. For example for act no. 5 above, (project credit approval), further details for the six points are given below:

- **Subsumption in ROP:** Number of the act of the Managing Authority, date, number of subprojects, time-schedule, budget allocation, agency which will implement, territory (prefecture, municipality), etc
- **Subsumption in PIP:** Number of individual or collective decision of the MoF, date, category, subcategory, special number, number and date of proposal, project's name, etc
- **Budget:** Current, initial, numbers and dates of acts of modifications, number and dates of proposals for modifications
- **Credit:** Current, initial, number and dates of acts of modifications, number and dates of proposal of modifications
- **Contract:** Amount, name of the contractor, number and date of the contract, subcontractors, time-schedule, etc
- **Auctioning details:** Numbers and dates of acts that approved auctioning details, etc

Note that for all of the above categories it is useful to have temporal changes of the information types.

In this way information that should be included in the database to be is defined and the way that it should be retrieved is described.

B. One stop shop for investors in Region of Central Macedonia.

At this moment the department of private investments of RCM provides one service as we can understand from the acts that issues: "*Financing/denial of financing of a proposal for a private investment up to 4 MEURO*".

Starting from the effectiveness part the corresponding ratio effect/consequence, with

effect=service="financing of private investments" and

consequence=rise of the GDP in the whole of the country and regionally.

was checked. This ratio was too high meaning that the service "financing of private investments" was not contributing as much as the government would like to be for the development of the country. Thus, the law that ruled the services provided for investors in Greece has recently changed due to effectiveness deficit.

Afterwards, the ratio output/effect was discovered getting lower identifying another effectiveness problem. This happened due to the fact that the number of rejection acts had been increased. A rejection decision is reached due to either reasons of formality or due to concrete reasons. Reasons of formality refer to incompleteness of the applicants file; while concrete refer to an investor's capability or investment's feasibility.

Both types of reasons originate from the preceding communication.

Potential investors or their consultants were gathering information for investment opportunities by calling RCM, by visiting RCM or simply by reading the law. This was a semiformal type of contact and the provision of information occurred verbally or through brochures. This type of communication is implicated in two kinds of problems. First, it was not so helpful for people conducting RCM and second, it couldn't easily directed to potential investors that had not conducted RCM before (e.g. Foreigners, young and inexperienced investors etc.)

This informal procedure should be formalised and is going to be supported by a portal subsystem, to increase competition and to secure that funding is going to be assigned to the right investors and investments, for the promotion of Development. For the portal subsystem functional needs concerning information provided, expression used and communication recipients, restrictions set by the corresponding stakeholders' goals are groups C,D,E,F,G,J.

Thus following for example group C ("Inclusive") needs, a phonetic portal will also be established to include the whole of the potential investors. Requirement for multi language communication will also be regarded (group D) to attract foreign investors. Also personal data protection, identification – authentication (group E) will be taken under consideration. There would be an attempt to provide all the information needed for a potential investor (group J) not in administrative language but in a way to match investor's (maybe) vague needs to administrative procedures (semantic Interoperability - group F).

Further analysis made is not presented here for brevity reasons.

The establishment of the portal will be a new service, an e-government service, which addresses both efficiency and effectiveness issues. Effectiveness issues are addressed through the publication of the investments' choices that the law provides, in order to increase competition between investors. The right guidance of potential investors through the whole set of funding programmes in region's borders helps so as each potential investor will address the right funding programme. This concerns the quality and abilities of the investors to whom the acts of subsumptions are addressed. It is this effectiveness deficit that brought forth this new service.

Concerning efficiency, if the information given by the portal is accurate and complete, then the number of phone calls or personal visits in RCM will be reduced considerably, allowing servants to work for back office operations like assessments/evaluations.

The e-government project which will be introduced will concern the whole administrative procedure and will address all the three functional requirements of PA. Thus it is going to help the composition of documents, organization, retrieval and accessibility in information and communication with internal and mainly external entities.

Collectively concerning required information, this will be about: a) Completeness of applicant's file (all the necessary documents), b) other PO acts that declare agreement or disagreement to the investment (prefecture agencies, agencies for antiquities etc.), c) database for assessors and inspection committee's d) records of working committee's (keeping and storage), e) government's journal details, f) autopsies' details g) acts for appointment of assessors and inspectors, h) suggestions of assessors and inspectors, i) investors' written reasoning and j) responsible tax offices for every investor.

This information comes from other entities (other POs and potential investors), but mainly from the databases of the RCM itself. Further analysis of the functional requirements detected is made in the same way as in the case of PIP above.

RELATED WORK

Related work presented here relates to a) the identification of stakeholders of their needs and b) modelling of a PA's function.

Architecting PA involves designing PAs to reflect the political and public managers' decisions at a strategic level in operational activities and decisions (Janssen and Hjort-Madsen, 2007).

There are many efforts in modelling PA's processes and function. These attempts are classified in fields like Process Modelling, Business Process Modelling/Reengineering, Enterprise Architecture, and Enterprise Modelling.

Business Process Modelling (BPM) has emerged as an immensely popular theme of conceptual modelling in practice. Research on BPM is based on diverse topics of research methods and covers a wide area including modelling techniques, methodologies, methods and tools, but increasingly also empirical studies related to success factors, complexity drivers, experience reports and success measures (Bandara et.al, 2007).

Glassey (2008) compares three process modelling techniques in order to find common concepts and to identify significant differences. He bases this comparison around three general questions: a) what are the objectives of the organization? b) Who is doing what with which resources? c) How does the organization work?

Bandara et.al., (2006) specifies success factors in BPM. The final validated model employs 15 measures of the three dimensions: Model Quality, Process Impacts and Project Efficiency. PM-success can also be an important independent variable in research that aims to explore causal relations further along the Information Systems Development process value chain.

Simon and Olbrich (2005), argue that the laws themselves specify a process, which can be adapted for the definition of the public processes, which agrees with the authors' perspective.

In their paper Chourabi et.al., 2008 present a new BPM approach. It is based on Business Process Mapping and the UN/CEFACT Modelling Methodology (UMM). The BPMapping provides an overall view of the business processes showing their inputs, outputs and interdependencies.

In King and Johnson (2006) process variety in order to be modeled is described in three dimensions – variety in the range of tasks performed (task variety), variety in the order that these tasks are performed in (sequential variety) and variety in the inputs and outputs of the process (content variety) and suggest that the same approach could be explored using Petri Nets or UML.

Stemberger et.al, (2007), present the business process change methodology suitable for public sector, while the objective of the PICTURE (Becker et al., 2007) modelling approach has been to develop a domain specific modelling method which meets the particular conditions of PAs.

Finally Palkovits and Wimmer (2003) argue that process modelling and process reorganization, have been recognized as being of utmost importance for making e-government implementations success. They present a solution to support PAs in the reorganization and re-engineering of administrative processes towards online service provision.

Modelling of Service Provision mainly refers to administrative and service provision related functions and not other parts of the PA domain like Policy Formulation. In addition this work focuses on conceptual perspectives rather than technical.

Thus, SAP's Public Sector Solution Map Holistic models (2000) and The Government Process Classification Scheme (1996) are not considered here as they don't focus on service provision.

Furthermore, the ONTOGOV service ontology (2005) and the WebDG Ontologies (2003) are not considered as they have far too technical focus.

Other initiatives that could be mentioned are The UK Government Common Information Model (2002), The DIP E-Government Ontology (2004), The Governance Enterprise Architecture (2004), the three spheres in e-Governance (2005), The Gartner Government Performance Framework (2003), A Faceted Classification of Public Administration and Generic Administrative Processes and The Federal Enterprise Architecture (FEA) Ontology.

In this proposal PA has predefined functional areas. Its function is ruled by laws. This stands for all countries, at least those that apply a normative administrative model (e.g. France, Germany, Austria, Spain, Portugal, Italy, Greece etc) (Billiets et. al., 2006). It also has a certain rationale, which is always the same for every procedure/service. PA is trying to fulfil all stakeholders' goals functioning in a certain legitimate pattern. Thus providing services is a different procedure than law making for a service. And furthermore there is no service for PA unless there is a law that orders it. Only then a citizen could claim his rights.

To set functional requirements the common, repeated pattern of each procedure, a generic functional triple need for every procedure of PA has been identified. Through the act that PA issues for this procedure/service, definition of functional requirements is possible. The proposed triple need is generic and applicable to every procedure. Finally, this proposal allows for the consideration of PA as a whole. POs and agencies are instantiations of PA. Entrusting a public service to one of them might be occasional and the rationale that led to it might be revised.

There is though a large body of literature in the area of strategic management which discusses organisations in terms of a stakeholder model (Sharp et al., 1999). The literature suggests examples of stakeholders, but does not provide help in identifying stakeholders for a specific system.

Towards this goal, literature on network approaches to stakeholder analysis in other domains. Social Network Theory and Industrial Networks; work in domain networks and goal reduction may also provide useful techniques. Other work, worthy of investigation, includes algorithms built into space planning software for interior design which incorporate arithmetic for dealing with similar options.

From the works above to more contemporary ones like: Pouloudi and Whitley (1997) and Kaler's (2003) definitions and to Onion models (Alexander and Robertson, 2004) and outcome based approaches and problem decomposition techniques (Wooldridge et.al., 2007), the efforts of defining stakeholders and their needs continued.

In the approach proposed by the authors, stakeholders of PA's function are defined based on the standard PA's environment and the supply and demand side of service provision. The groups that constitute this environment are certain and stable.

Finally in the RE field related work is primarily based on the *i** modelling framework (Yu, 1995) for analyzing requirements. One of the most known and representative work is TROPOS methodology. There is an example of use of the TROPOS methodology in e-government projects, developed for the local government of Trentino (Bresciani et.al., 2004).

In that case a requirements engineer identifies stakeholders for a certain procedure that is examined. This is an ad hoc identification procedure. The identified stakeholders are actually shareholders, not interested in the whole administrative function but only in the specific procedure. As a result there is no stakeholder like government (or public organization), the EU, or public servant. It is much easier to identify requirements/goals of each stakeholder in a domain level (once) and it is sounder to use official documents and views instead of individual perspectives. Stakeholders in any procedure can be seen as instantiations of the stakeholders which are considered for the whole

PA function and their requirements can only be the same or instantiations of the requirements that are presented here.

In this proposal stakeholders are not responsible for achieving goals. PA and its administrative function are. Stakeholders are not part of its function. PA is conceived as a black box. It has no predefined actors (seamless). It only has a rationale. So, which actor (component of an e-government system) is associated with what goals is something that the requirements analysis will show.

SUMMARY/CONCLUSION

A common problem in many organizations in the selection of the appropriate e-government systems for the reorganization of processes and methods, followed in public service provision, is the determination of the whole set of stakeholders and their requirements. E-government systems though have to suit/facilitate a PA's rationale to avoid oversimplifying the process of digitizing services and missing out important features of the service. A goal-oriented method for identifying stakeholder requirements has been adopted. This can be applied through the key functional requirements of a PA. A model to identify these functional requirements at conceptual level has been discussed here. To verify its applicability the method has been put to test through two cases from the Greek PA system. The goals and requirements of the specific services are explored through the administrative act documents which identify scope and stakeholders relevant to the service and the legal framework it abides to. The application of the method culminates in identifying suitable e-government systems that could improve the efficiency and effectiveness of the services under investigation, without overlooking any of the stakeholders' requirements. The results so far have been encouraging and the authors are looking into expanding this work by applying it to cases from other PA in Europe.

REFERENCES

Administration on the net. (2004 June). An ABC guide to e government in Austria Federal staff unit for ICT strategies, Federal Chancellery Austria, Alexander, I., & Robertson, S. (2004). Understanding Project Sociology by Modelling Stakeholders. *IEEE Software, 21*(1), *23–27.*

Axelsson, B., & Easton, G. (Eds.). (1992). *Industrial Networks: A New View of Reality.* London: Routledge.

Bandara, W., Gable, G. G., & Rosemann, M. (2006). Business Process Modelling Success: An Empirically Tested Measurement Model. In D. Straub & S. Klein (Eds.), *Proceedings of International Conference on Information Systems*, Milwaukee, Wisconsin, USA (pp. 1-20).

Bandara, W., Tan, H. M., Recker, J., Indulska, M., & Rosemann, M. (2007). *Bibliography of process modeling: An Emerging research field.* Retrieved from http://eprints.qut.edu.au

Becker, J., Pfeiffer, D., & Räckers, M. (2007). *Domain Specific Process Modelling in Public Administrations – The PICTURE-Approach.* Berlin: Springer.

Billiets, M., et al. (2006). D2.2 – Legislative, Institutional and EU Policy Related Requirements. *IST STREP PROJECT SemanticGov*. Retrieved from http://www.semantic-gov.org/

Bolton, D., Jones, S., Till, D., Furber, D., & Green, S. (1994). Using domain knowledge in requirements capture and formal specification construction. In M. Jirotka & J. Goguen (Eds.), *Requirements Engineering: social and technical issues* (pp. 141-162). Amsterdam: Academic Press.

Boston, J., et al. (1991). *Reshaping the State*. Oxford, UK: Oxford University Press.

Bresciani, P., Perini, A., Giorgini, P., Giunchiglia, F., & Mylopoulos, J. (2004). Tropos: An Agent-Oriented Software Development Methodology. *Autonomous Agents and Multi-Agent Systems*, *8*, 203–236. doi:10.1023/B:AGNT.0000018806.20944.ef

Capgemini. (2006). *Architecture and the Integrated Architecture Framework*. Retrieved from http://www.capgemini.com/resources/thought_leadership/architecture_and_the_integrated_architecture_framework/

European Commission. (2005). *Insights for the European e-Government Community*. e-Government Observatory. Retrieved from http://europa.eu.int/idabc/en/chapter/140

Gartner. (2003). New Performance Framework Measures Public Value of IT. *Research Note, 8*.

Glassey, O. (2008). A case study on process modelling — Three questions and three techniques. *Decision Support Systems*, *44*, 842–853. doi:10.1016/j.dss.2007.10.004

Greek Administrative Process Code. (1999). *Law 2690/1999*.

Janssen, M., & Hjort-Madsen, K. (2007). Analyzing Enterprise Architecture in National Governments: The cases of Denmark and the Netherlands. In *Proceedings of the 40th Hawaii International Conference on System Sciences*.

Kaler, J. (2003). Differentiating Stakeholder Theories. *Journal of Business Ethics Dordrecht*, *46*(1), 71–83. doi:10.1023/A:1024794710899

King, S. F., & Johnson, O. A. (2006). VBP: An approach to modelling process variety and best practice. *Information and Software Technology*, *48*, 1104–1114. doi:10.1016/j.infsof.2006.02.003

Palkovits, S., & Wimmer, M. (2003). *Processes in e-Government – A Holistic Framework for Modelling Electronic Public Services*. Berlin: Springer.

Peters, B. G., & Savoie, D. J. (1998). *Taking stock: Assessing Public Sector Reforms*. Montreal, Canada: Queens University Press.

Pollit, C., & Bouckaert, G. (2004). *Public Management Reform. A Comparative Analysis*. Oxford, UK: Oxford University Press.

Pouloudi, A., & Whitley, E. A. (1997). Stakeholder Identification in Inter-Organisational Systems: Gaining Insights for Drug Use Management Systems. *European Journal of Information Systems*, *6*(1), 1–14. doi:10.1057/palgrave.ejis.3000252

Sabucedo, L. A., & Rifon, L. A. (2006). *Semantic Service Oriented Architectures for eGovernment Platforms*. American Association for Artificial Intelligence. Retrieved from http://www.aaai.org

SAP. (2000). *R/3 System SAP Solution Map*. [cited Dec. 2000]; Available from: www.sap.com.

Savvas, I., & Bassiliades, N. (2008). A Process-Oriented Ontology-Based Knowledge Management System for Facilitating Operational Procedures in Public Administration, accepted for publication in *Expert Systems with Applications*.

Savvas, I., Pimenidis, E., & Sideridis, A. (2007a). Proposing a high-level requirements mapping framework, for testing implementation compatibility in e-government projects. In *Proceedings of ECEG. The 7th European Conference of e-Government.*, Den Haag, The Netherlands (pp. 459-468).

Savvas, I., Pimenidis, E., & Sideridis, A. (2007b). A Review of E-Governance Models in the EU. In *Proceedings of Advances in Computing and Technology (AC&T) Conference*, University of East London, England (pp. 110-119).

Savvas, I., Pimenidis, E., & Sideridis, A. (2008). Using egov systems to remedy public service failings: in search of a transformation algorithm. In *Proceedings of The e-Government Workshop '08(eGOV08)*, Brunel University, UK.

Sharp, H., Finkelstein, A., & Galal, G. (1999). Stakeholder Identification in the Requirements Engineering Process. In *Proceedings of the 10th Int'l Workshop Database and Expert Systems Applications* (pp. 387–391).

Simon, C., & Olbrich, S. (2005). The influence of legal constraints on business process modelling In *Proceedings of The eGovernment Workshop '05 (eGOV05)*, September 13 2005, Brunel University, UK.

Stemberger, M. I., & Jaklic, J. (2007). Towards E-government by business process change—A methodology for public sector. *International Journal of Information Management*, *27*, 221–232. doi:10.1016/j.ijinfomgt.2007.02.006

The Federal Enterprise Architecture (FEA). (n.d.). *Ontology*. Retrieved from http://web-services.gov/fea-rmo.html van Dooren, W. (2006). *Performance Measurement in the Flemish Public Sector: A Supply and Demand Approach*. PhD dissertation, Faculteit Sociale Wetenschappen - Onderzoekseenheid: Instituut voor de Overheid [IO], K.U.Leuven

Weber, M. (1946). Bureaucracy. In H. Gerth & C. W. Mills (Eds.), *Max Weber: Essays in Sociology* (pp. 196-244). New York: Oxford University Press.

Wilson, J. Q. (1993). *Bureaucracy*. New York: Free Press.

Yu, E. (1995). *Modelling strategic relationships for process reengineering*. Ph.D. Thesis, University of Toronto, Department of Computer Science.

Chapter 8
A Holistic Approach to E-Government: Ongoing Research in Oman

Angela Espinosa
University of Hull, UK

Ahmed S. Al-Maimani
University of Hull, UK

ABSTRACT

This chapter presents an example of ongoing action research, which uses systemic approaches to diagnose and design a Knowledge Management strategy for the Public Sector in Oman. Based on previous research concerning contributions of holistic approaches to e-government, it describes a framework for supporting an e-government program, based on generic systemic and cybernetic principles. The primary value of this chapter lies in its account of the ongoing experimentation with innovative holistic approaches to e-government. The results contribute to the practice of e-government by supporting more effective and robust e-government programs. The suggested framework also provides a basis for future empirical studies on the relationship between e-government strategies and organisational and Knowledge Management effectiveness.

INTRODUCTION

In recent years there has been significant interest in electronic government ***(e-government)***, due to the revolution that the Internet has brought into both private and public management practice. Many governments have designed and implemented massive e-government programs, but reported experiences still leave some questions open, for example: is there a clear way of assessing the impact of such programs on the performance of that government? What is the nature of observable improvement in government activities directly related to e-government programs? What are the current failures of e-government implementation? Are they related to cultural and structural constraints? How can common failures in e-government be addressed? What are the main concerns about the efficiency of the various digital initiatives?

We consider that one relevant way to address these questions is to explore more comprehensive holistic approaches that would allow these questions

DOI: 10.4018/978-1-60566-860-4.ch008

to be tacked more consistently and coherently from a humanistic, organisational and technological point of view. Therefore, we focus in this chapter on presenting an example of the use of systemic approaches to the design and implementation of e-government services', with a special emphasis on e-services and e-payment portals. The chapter's reflections are presented in relation to ongoing research on using systemic approaches to support e-government and e-participation in the public sector in Oman.

We first present the background research (a summary of the conceptual and methodological characteristics of the research project), then present the current strategy and method of implementation and finally describe the way the project has progressing, inspired by systemic approaches.

RESEARCH BACKGROUND

Research Context

The idea of e-government loosely relates to the electronic linkage, interaction and communication of a particular government with both its citizens and its trading partners using a variety of gateways, portals and networks. However, there is no clear agreement amongst people coming from the soft and hard approaches to information systems (as in Checkland, 1981) in terms of what is the real coverage and scope on an e-government strategy. Most would recognise that the challenges of designing and implementing an e-strategy constitute a domain of significant complexity where people, processes and technology work together in a highly dynamic and often creative way, usually in complex and interrelated (or "messy" as Ackoff (1981) called them) situations. These are the type of situations where systemic approaches can offer significant benefits to practitioners.

The research introduced in this paper builds on previous research experiences from both authors. Al Maimani (2000) analysed from a critical systems perspective the impact of information systems on decision-making in Oman. He found that during the study period (2000), information systems in decision-making were not properly used, and also found evidence of strong correlations between inefficiencies in information systems and decision-making. This observation has been taken into account and used as the basis for the design and implementation of a One Stop Shop (OSS) (see footnote 4) in (2004): information systems are now properly and effectively used in decision making both for commercial registration and for providing services to the investors in Oman. Espinosa's doctoral research focused on developing an organisational cybernetic framework in the President's Office in Colombia in the 1990s, to align strategic information systems and organisational transformations (Espinosa, 1995).

The ongoing action research project presented in this chapter aims to explain the gap between theory and practice in the design and implementation of e-services still observable in the Public Sector in Oman, and to address it by developing and testing a systemic framework for e-governance, inspired by an Organisational Cybernetics approach to strategic information systems.

Why We Need a Systems Approach to Improve E-Government Practice

McNabb, (2007) defines e-government as consisting of actions to produce and deliver government services to citizens, not in the traditional face-to-face manner, but instead, through the use of communication technology (specially Internet technologies). The Information Technology Authority-Oman (ITA) defined it as; "the transformation of internal and external government processes toward new and better forms of citizen-centred service delivery opportunities which new communications technologies offer" (MNT, 2007a).

Governments aim to develop and enhance their services to maximise quality assurance benefiting from advanced information and communication technology (ICT) and to change their working practices, in order to provide the best and maximum possible service not only through electronic sites and offices, but also through connectivity and delivering services to investors in their premises. Also government entities, using e-government schemes, have opened up communication of conversation and interaction with the customer and the investor.

E-inclusion is the key tool at the disposal of a socially inclusive government. According to the UN report, (2008), e-inclusion goes beyond e-government: it means employing modern ICT technologies to address the issues of access-divide and to promote opportunities for economic and social empowerment of all citizens. Economic and social empowerment today rests on the ability to access, gather, analyse and utilize information and knowledge to widen individual choices for political, economic, social, cultural and behavioural decisions.

Nevertheless the experiences of developing e-government in a particular country have not always been as successful as expected: generally it has not produced the desired improvements in public service effectiveness. Researchers such as Lyytinen & Hirschheim (1997) and Heeks (1998, 1999, 2002, 2003), consider a major reason for such failures in implementation may come from traditional practices being too heavily oriented to designing and implementing technological changes, while not properly linking or embedding them in the respective human and organisational contexts. Heeks (1999), Moon (2002) and Ho (2002) have explained major reasons for e-government failure and they concur in recommending that more holistic, humanistic and organisationally sound approaches to e-governance are urgently required, and it is our aim in this research to explore possible ways of progressing in this direction.

One of the government's e-windows, which the business sector is usually linked with, is the One Stop Shop, which tends to cross departmental and organisational boundaries and therefore impacts significantly on both organisational and process design, human resources and financial issues. We want to explore a research approach that will allow us to deal not only with the technological factors but also with the organisational, social and human factors, when designing and implementing a One Stop Shop for the government in Oman. Such an approach should form the basis for developing a back office platform, with the capacity to firstly, service multiple outcomes in a coherent manner for various stakeholders and secondly, measure the quality of services and their benefits in a clear way. The following sections explain how such an approach is being suggested and developed as part of one of the author's PhD project.

DEVELOPING A HOLISTIC E-GOVERNMENT FRAMEWORK

What Systems and Cybernetics Thinking Can Provide to This Study

The academic literature on Knowledge Management and e-governance usually provides interesting insights and ideas on issues of learning and on improvements in knowledge management; but it fails ultimately to provide clear methodologies to guide successful implementation. One of the problems researchers have observed during the last few decades, when developing organisational and technological transformations, is the lack of alignment between technological and structural changes. Doyle and Galliers -among others- highlighted the lack of systemic approaches to strategic information systems planning and implementation as one of the reasons for failure (Doyle, 1991; Galliers, 1991; Galliers & Baets, 1998, pp. 231-232).

It was not unusual to find that while organisations developed Business Process Re-engineering, Total Quality Management, Knowledge Management, E-Government or similar processes, at the same time, they developed Strategic Information Systems plans–usually technologically rather than organisationally driven - that did not always clearly support ongoing organisational development strategies or did not always encourage organisational learning or improved performance. Galliers and Baets (1998) showed that many Strategic Information Systems Planning and Re-engineering projects begin by implementing innovative technologies that were not always focused on supporting critical organisational processes, and in this way organisations lessen the possibilities for cohesive and synergetic organisational and technological development.

From the late 1970s, the Chaos and Complexity approach began to take shape, and has since suggested that an deep revolution in thinking is needed – as previous systemic approaches have suggested - by understanding and getting inspiration from complex non linear dynamic systems. They have had an impact on several scientific disciplines whether they study natural, human or social phenomena, in particular a big influence in the development of the theory in Knowledge Management. Stacey (2000, 2003) tried to reinvent complexity theory by using its concepts in the service of interpretivism and postmodernism, and has written extensively about organisational complexity and chaos asserting that today's business environment is complex and ever-changing organisations must be Complex Adaptive Systems comprising a large number of interacting units. The idea of Complex Adaptive Systems - nowadays extensively used in management – originated in the understanding of complex natural systems, in particular in the pioneer studies of Mc Cullouch about neural networks- (McCulloch 1965). It suggests that the most effective organisational form to deal with complexity is a self-organised network of semi-autonomous organisations, ruled by very simple rules of interaction and operating in a highly connected, highly efficient networked pattern.

Also inspired by Mc Cullouch's original ideas about neural networks, Beer pioneered this ideas in the 1970, and developed a scientific discipline known today as 'Organisational Cybernetics', which explains the basic theory about complexity, variety management and its implications in management (Beer, 1979; 1981; 1994b). It includes a series of Theorems and Axioms of complexity management, a model of a viable organisation (*the viable systems model or VSM*) and a series of diagnostic and design tools to support organisational transformation and improved performance. This theory and methodology has shown to be very useful for diagnosing and designing organisations, for improving alignment between strategic and structural issues and for supporting Knowledge Management (Yolles, 2006; Leonard, 1999) and for supporting strategic ICT/KM programs (Schumann, 1990, Espejo & Schwaninger, 1993, Espejo, 2002).

In particular, one of the authors suggested detailed methodological guidelines on how to use organisational cybernetics for aligning organisational and technological strategies and offered an application in the President's Office in Colombia (Espinosa, 1995). We shall use this methodological framework as a starting point for the current research project, and adapt it accordingly as in this research project we are interested more specifically in addressing a particular type of Strategic Information Systems portfolio, that is, an e-government program.

Figure 1 presents an overview of the way we see that a more holistic approach to e-government, in particular one building up on these Organisational Cybernetics methodologies and tools, can help to overcome some of current fractioned (*silo*) knowledge management strategies and systems in Oman –as described in next section-. It illustrates the situation before and after building ICT infrastructure and implementing what we will call - following Beer (1981) metaphorically - the 'government nervous system' (GNS). The silos (independent, isolated government institutions) are bridged by

Figure 1. The expected transformation after a Cybernetic intervention to support e-government program's implementation (A. Al-Maimani, 2009)

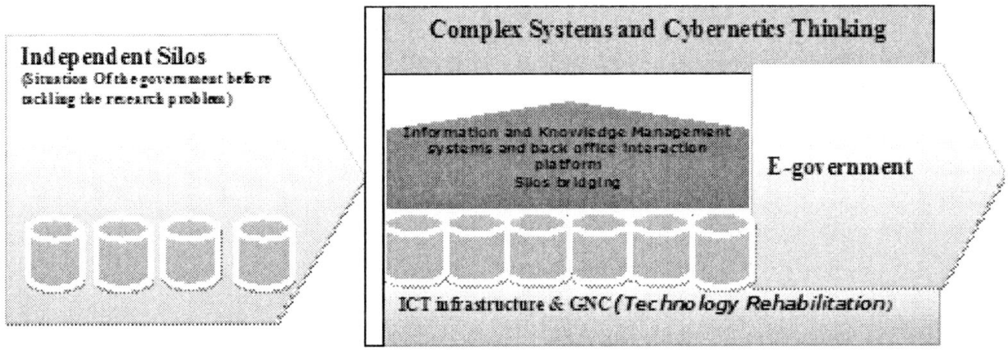

both innovations in their information and knowledge systems, the back office interaction platforms within e-government projects, and the design of more effective structural mechanisms to allow distributed intelligence and self-management at all organizational levels. The following section explains more about Beer's theory and its contributions to the research topic.

Managing Complexity and E-Governance

Managers in public and private sectors face a number of challenges, and most of the time make decisions under the pressure of continuous changes in both their internal and external environments, without having access to sufficient relevant knowledge or information. Beer summarised the situation by saying that the person at the top of an organisation would need a brain weighing at least half a ton to deal with all the information required for good decision making. However, managers have always found ways of dealing with this immense complexity, even if they do not always use the right sort of tools or make the right decisions at the right time (Beer, 1959).

Ashby (1964) described *variety* as the measure of complexity; in social systems this refers to the repertoire of potential behaviours that a system can take in a particular context. He suggested the Law of Requisite Variety - 'only variety can absorb variety'. Beer (1979, 1981) developed further the original insights from Ashby (1964), Wiener (1965), McCulloch (1965) and Von Foerster (1981) among others, concerning self-organisation, learning and complexity management, and explained how these basic ideas can explain behaviour in complex (human) organisations in his Viable System Model (VSM). He explains that viability is closely related to exhibiting 'requisite variety' (the system that controls must have at least the variety of the system it intends to control). He defines a viable system as a system able to adapt and thus to maintain an independent existence as it co-evolves with its environment. It is always embedded in and composed of other viable systems: one of the biological mechanisms for survival is to develop the viability of each one of a cluster of nested viable systems.

According to Ashby's law, the variety that operations have is orders of magnitude less than the environmental variety, while managers' variety is orders of magnitude less than operational variety. There is always some 'residual variety' that is not properly generated/absorbed by operations or managers that needs to be managed in order to keep a stable relationship with the environment/operations. For example, in an organisation, managers make decisions in terms of what information to collect and observe regarding their dynamic interaction with their niche. This also applies to the amount of topics/issues we want to monitor or control from a particular task we are doing: we make choices, intuitively, in terms of how much detail we want to know, which are the most relevant variables to observe, and which communication mechanisms we will use to capture such information; in this way we 'attenuate' this explosive variety.

The VSM offers a language and the required tools to understand the structural invariances of living organisations, co-evolving with their environment, that are a pre-requisite for viability. The focus of VSM analysis is to observe the ability of the organisational system to handle the complexity of the tasks required to fulfil its purpose in the context of a highly complex changing environment: in other words to assess how it manages its knowledge.

A viable system is composed of a set of operations (System 1) which develop the primary tasks, directly related to the organisational identity and a Meta-system (Systems 2, 3, 4 and 5) that provides identity and cohesion to the contained operational elements (see Figure 5 later with an example). Beer (1985) suggested an analytical technique - recursive mapping - that enables the analyst to observe recurrent interactions between viable systems at any level, with the relevant environments. It allows us to perform a rapid but accurate broad-brush diagnosis of very complex systems (e.g. a nation state), as each and every viable system can be studied using the same language and tools (for a more detailed description of the VSM see Walker, 1981; Espinosa and Walker, 2005. Espinosa et al, 2007, 2008).

A balanced structure will foster self-regulation in each of the operational units (Systems 1) while guaranteeing the synergy and cohesion of the whole organisation. In order to do so, there are some basic mechanisms that need to be in place: the autonomous operational parts need to be brought together to build the whole system. The job of providing this cohesion is done by the Meta-system which has four aspects:

- A way of dealing with conflicting interests – System 2
- A way of finding synergistic benefits from working in a co-ordinated, self-regulated way – System 3
- A way of adapting to a changing environment – System 4
- A way of reviewing organisational identity and giving closure to systems operations – System 5

System 2 may use a variety of methods including communication mechanisms to avoid oscillations or conflict between Systems 1, for example by providing shared languages and information tools. System 3 provides synergy between Systems 1, by making sure they all share resources, act according to the same basic rules and fulfil their commitments to the organisation – in terms of the products/services provided. System 3* offers alternative (non formal) mechanisms for collecting information directly from where it is produced, that may 'alert' the meta-system about situations getting out of control. System 4 responds for learning about the 'outside' and 'then' of the organisation: for example, information about anything happening outside that may potentially either benefit or threaten it; about any innovation that may be worth trying, or about any interesting development option. System 5 is responsible for organisational closure and ethos: it reflects on the organisational identity, its core policies and long-term strategies. It makes sure that Systems 3 and 4 are engaged in a balanced interaction that allows both 'down to earth' and 'creative' criteria to be taken into consideration for making crucial strategic decisions.

These basic concepts and the laws for managing complexity are at the core of the Organisational Cybernetic approach and have direct relevance to further our current understanding of Knowledge Management, as we'll explained in the next section. There are an extensive number of applications that have shown the strength of this approach for facilitating organisational transformations. In particular, there have been cybernetic interventions at the level of governments, starting with Beer original intervention in Chile (Beer, 1981, Part II), and following with many applications in different countries (e.g. Espejo & Harnden, 1989; Schumann, 1990; Espejo & Schwaninger, 1993; Espejo et al, 1996; Grabher et al, 2002; Bula, 2004; Espinosa & Walker, 2005, Espinosa, 2007; Stokes, 2008; Perez-Rios, 2008). Eckard-Turke (2009) has recently published his own doctoral work explaining in detail the relevant theory from Beer and second order cybernetics required to explain e-governance, as well as applications in the German public sector.

A Holistic Framework to Support E-Government Design and Implementation

Building on the above theoretical and methodological traditions, we will suggest a framework to support the ongoing e-government program in Oman and to address the previously unseen/unsolved implementation problems. The framework will make use of some input from the Soft System Approach (Checkland, 1991; Checkland & Scholes, 1990), Complex Systems approaches (Stacey, 2000, 2003) with a major input from the Organisational Cybernetics approach (Beer, 1979, 1981; Bowling and Espejo, 1992; Espinosa 1995, 2007, 2008).

To start with we used an ad-hoc framework to support analysis of the e-government strategy in Oman and the required organizational and technological changes for effective implementation. As mentioned before, the suggested stages have been adapted from (Espinosa, 1995) – A cybernetic approach to Strategic Information Systems Planning; this framework also further developed Espejo's VIPLAN method to diagnose organisations including strategic information management. Broadly speaking, the suggested framework includes the following stages:

1. Develop a "rich picture" of the system in focus
2. Agree on "relevant systems" 1(as in Checkland, 1981)
3. Define the TASCOI2 of the agreed relevant system
4. Model the organizational complexity of the system in focus using the Viable System Model
5. Review the e-government strategy in the light of the VSM mapping and identify critical issues for implementation
6. Create the organisational and learning context for the e-government portfolio's implementation

We start by presenting and analysing the current e-government strategy in Oman, already about 80% implemented, and identify major weaknesses and implementation challenges. For the purpose of illustrating the use of the approach in this chapter, we focus on the Oman's e-government strategy related to 'services to investors': this is a critical/strategic e-government project lead by the Ministry of Commerce and Industry, and will be our 'system in focus'3. We then illustrate some examples and findings from an ongoing prototype at the system in focus, aiming to test the holistic framework.

E-GOVERNANCE IN OMAN: CURRENT PROGRAM AND PROJECTS

Current Research Project

The research project described in this chapter starts from the hypothesis that *whilst technology is becoming increasingly flexible and 'fit for purpose', there is evidence that indicates that success or failure is less a technological issue and more an individual and organisational issue.*

In particular, we understand that some of the critical success factors in an e-government project are the ability for effective implementation and it has a close relationship to existing implementation structures and culture. The research team has recognise existing constraints to change public service culture, to motivate public sector workers to new ways of working, to address trade union concerns, to provide adequately skilled and competent management and leadership, and to provide a flexible structural context where learning and knowledge management can better contribute to improved organisational performance.

From this hypothesis, the research work aims to:

- Review the gaps in the current understanding of the e-government strategy formulation and e-government programs' implementation
- Produce a holistic framework for e-government programs, including methodology and tools to aid in the design, and implementation of e-government programs
- Test - through *prototypes* - the suggested e-government framework and learn about required improvements to methods and tools
- Use the framework to improve the e-government services in Oman
- Determine the nature of the impact that this way of approaching an e-government program may have in practice and to critically reflect on holistic ways to approach and understand e-government

The focus is on the following questions:

- Are there specific (complex) systems methodologies or approaches that can provide useful guidance for practice in the design and implementation of e-government strategies?
- How would they help to integrate information, knowledge and technology to serve the desired multiple outcomes in a coherent manner and to meet the satisfaction of multiple stakeholders?

Oman was chosen as a case study for this research due to one of the researcher's closeness to current developments in e-government and knowledge management in the government and his previous research and interests. The next step in the (ongoing) research was to design (and pilot) a holistic framework for the design and implementation of an e-government program in Oman. One of the main issues was to find a way to address the complexity of the required process of transformation in the government, in terms of changing the current structure, technology and culture of citizens and public officers. After preliminary investigation and decisions, it was agreed to focus the research on understanding the e-government phenomenon, for integrated service to offer unity in the provision of public services to the stakeholders, and in particular to develop a pilot study on the provision of e-government services to investors in Oman.

There follows a summary of the Oman *e-government* strategy, and a reflection about the approach used in practice: it shows that previous to e-government, the various parts of the government organisation used to work independently in a very isolated way. A major constraint for successful e-government implementation had been the lack of good *communication* between partners implementing the e-government strategy. It is clear as the starting point of this research that the *organizational context* for e-government implementation was not appropriate, and that there were missing structural mechanisms and tools required to guarantee that each involved organisation would share the same technological standards, follow the same rules for communicating and sharing information and would change from acting as isolated *knowledge silos* into collaborative communities of practice.

In the starting situation, the government organisation's 'silos' were providing their services through independent separate single e-windows. We recognised that a more in depth diagnosis of the existing structures and communication channels, using the suggested (*VSM* based) framework, may be helpful to address problems of implementation and to identify learning constraints.

The E-Government Strategy in Oman

Oman's long-term vision and strategy ('Vision 2020') has focused on the need to diversify its economy (currently based on oil revenues) and consequently, has identified knowledge, technology and research among one of the strategic means of diversifying its revenue. The government organisations are trying to transform Oman into a sustainable knowledge society by leveraging Information and Communication Technologies (ICTs) to enhance government services, to enrich businesses and to empower individuals. In line with this vision, the Seventh five-Year Development Plan (2006-2010) of the Sultanate lays emphasis on upgrading the Information Technology sector by implementing the national strategy for Oman's Digital Society, with more concentration on the basis of e-Government (ITA, 2002, MNT, 2007b).

The first E-government projects appeared in Oman during the 1990s. The Ministry of National Economy established a National IT committee that was meant to design the ICT future strategic plan and to build the foundation for ambitious ICT development plans aiming to convert Oman into a knowledge-based economy and society. Such a plan was needed to face the *complexity* of expected changes, to avoid any major problems during the implementation, and to keep the investments focused upon the most crucial requirements. However, a later research report suggested that 90 per cent of previous investment in ICT was poorly focused upon organisational requirements and failed to contribute to operating or strategic improvements (Mori, 1996).

ICT in Oman consist of two legal bodies; the first authority is the *Telecommunications Regulatory Authority (TRA)*, established in 2002 to liberalize and promote the telecommunications services in the Sultanate of Oman. TRA is committed to develop the telecommunications sector by regulating and maintaining the telecom services and aims to develop the infrastructure and attract foreign investment to promote the Oman economy and to increasing employment prospects for Omani citizens. In 2006, the government set the *Information Technology Authority (ITA)* as an autonomous legal body, responsible for implementing national IT infrastructure projects and supervising projects related to the implementation of the Digital Oman Strategy. ITA, therefore, takes care of planning and implementing a sensitive knowledge-based strategy. It has managed ad hoc services by appointing *Gartner*, an information technology consulting company, who has studied the government requirements for e-Government, and accordingly studied and recommended embracing the development of a *Digital Society (MNT, 2007a)*.

The implementation of e-government and a digital society offer the following opportunities:

- Complementing Vision 2020
- Enabling knowledge-based industries and developing the ICT sector
- Reducing bureaucracy in government services to citizens and business
- Supporting a better competitive environment
- Developing educational and employment opportunities for Omani youth
- Supporting tourism
- Social development and better healthcare
- Making Oman a more attractive destination for foreign investment

Government Organisations are seeking to integrate their services and this study is researching the possibility of building, within the government endeavours for organisation transformation and rehabilitation, a *one-stop shop* that ensures the connectivity and integration of the government back offices. The concerned environment of the study is the socio-economic environment.

E-Government Projects

Figure 2 below summarises the main program and related projects of e-government in Oman. There follows a brief introduction to two of them that have more relevance for this chapter: Table 1 presents more detail on each of the major projects currently under development.

The main Oman government e-government projects are:

- Electronic awareness (eOman)
- Innovation and Support Centre (ISC)
- International Relations (INTIL)
- Disaster Recovery Centre (DRC)
- Electronic law (elaw)
- National electronic payment gateway (NePG)
- Information Security Management Framework (ISMF)
- Convergent Government
- Network (CGN)
- Government electronic services (e-services)
- National IT Training & Awareness (NITTA)
- Standards Framework (STDF)

Government Electronic Services Portal

This Portal is designated to be the main gateway to electronic services offered by the government. In July 2007 an agreement was signed between ITA, Gulf Business Machines (GBM) and NCS of Singapore and a project office was established for this initiative. The e-services portal is in its final stage to provide services online. Various government organisations are able to develop and roll out a wide range of new e-services in a shorter timeframe and to secure the governments' ability to deliver services, Government to Customers, to Businesses, to other Government entities and to Employees. Through the services Portal citizens will also get access to a range of government information and services provided directly online.

Figure 2. Oman e-government projects

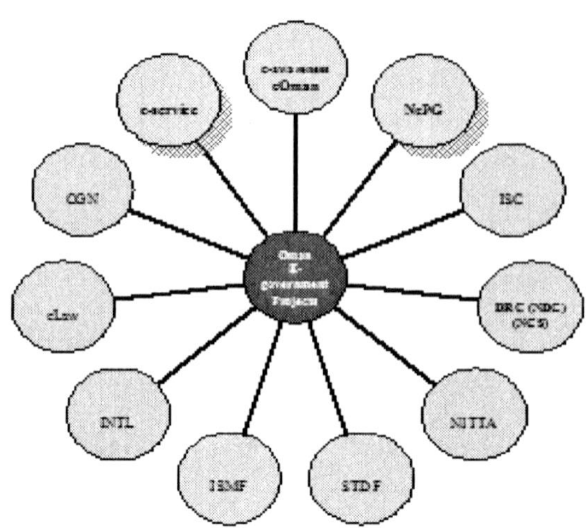

Table 1. A glance at Oman main e-government projects

eOman Awareness (eOman)	This awareness program educates and empowers the public sector employees and the community at large to participate in the knowledge society. It creates an effective link between the public and private sectors for digital initiatives and opportunities for the public and private sector enterprises to interact with the public and to bridge the digital gaps. It also creates ICT awareness and enables people to perceive its benefits.
Innovation and Support Center (ISC)	ISC was the first government initiative of its kind in the Arab world, ITA report, (2006-2007), It is also first ITA practice is to provide ICT incubators of local entrepreneurs and nurture skill and knowledge transfer within the community. ISC offers technical support as well as innovative solutions for electronic projects and the training of Oman citizens in digital innovation and excellence. It also offers standardised training programs to talented Omani youths in the field of digital technology. The centre plays a prominent role in transferring knowledge and skills in modern digital applications from the international markets to the local. The ITA plans to develop the Innovation and Support Centre as a centre of international excellence in IT that will play a prominent role in creating a bright future for young Omanis through its high quality training programs, as well as opening up new horizons for employment while supporting productivity enhancement in both public and private institutions. The center operates at very professional standards and continuously offers training programs. Highly specialised experts from Microsoft conduct these training programs. They focus on improving the skill of the Omani civil employees and increase their technical networking opportunities.
Disaster Recovery Centre (DRC)	To provide business continuity, improved efficiency and access to secure e-services at all times. DRC has several benefits such as providing business continuity, improved efficiency, improving the integration, scalability of applications, providing access to secure e-services at all times and Protecting confidentiality, integrity, availability and privacy of citizen data.
A: National Data Centre	The National Data Centre has been established under DRC umbrella. ITA signed an agreement with Mideast Data Systems (MDS) in the first quarter of 2008. The initiative was put in place: to provide a multi tenant Tier 4 data centre for government IT departments to backup their systems, thus achieving improved business continuity and redundancy for a range of IT systems; to establish a secure common facility for housing government IT systems with a professionally set up infrastructure with Tier 4 availability; to address the current lack of a reliable infrastructure to house critical government IT systems and to sustain the required mission criticality level; to address the need for a disaster recovery strategy that has, to date, been absent in almost all government IT systems; to house ITA-led IT initiatives such as the eServices Portal, Security Infrastructure, the e-Payment gateway and public key infrastructures.
B. The National Security Centre	In October 2007, ITA signed up with an international security provider to establish a national security centre to deliver adequate IT protection to all government entities. This project includes the training of approximately twenty Omani engineers to operate and manage the centre and provide professional security consultancy services to all government agencies. The project supplies security to the unified government network, and provides the equipment necessary to public sector organisations linked to the network.
National IT Training & Awareness (NITTA)	The NITTA initiative is to create sufficient ICT awareness in order to prepare the community for a knowledge society and empower civil services employees under the Ministry of Civil Services with ICT knowledge and skills to enable delivery of public sector e-services. It includes a project to train communities in IT as well as specialised IT certifications for public officers.
Standards Framework (STDF)	STDF documents, reviews and publishes standards framework for ITA and publishes XML messaging Schema guidelines, develops the Standards management Manual and implements the community of Interest (COI) Development Methodology. STDF ensures platform independence and vendor neutrality for ensuring compatibility with different technologies; it also adopts and complies with international standards and related guidelines and specifications and increases flexibility and reduces risk so that applications will become easier to manage, maintain and re-design and makes efficient use of existing resources by ensuring re-use of existing application. It improves communication amongst project teams due to compliance with widely adopted, mature industry and technology standards. Difficulties in implementation led ITA to establish 'The office of Integration & Standards' to outline the standards for application development and define the COI for the Omani E-Government architecture
Information Security Management Framework (ISMF)	The information Security Management Framework is part of the overall ITA standards framework and has been derived based on a structured collection of independent guidelines, processes and practices and primarily from the Information Security Management Systems Standard (ISO 2700). It aims to ensure the protection of information assets from unauthorised access or modification of information, whether in storage, processing, or transit.

Table 1. cont.

International Relations (INTL)	ITA liaises with regional and international organisations in matters related to ICT. It is the focal point of World Summit Information Society (WSIS) under sponsorship of the United Nations (UN). The WSIS body was set up by the UN especially to oversee all ICT related issues and try to bridge the digital divide gap of member countries. ITA is responsible for the technical and logistic arrangements of Oman's representation in WSIS so the ITA plays as a focal point of WSIS. INTL Provides a structured reporting mechanism to international agencies on ICT matters, and compiles with and publishes ICT core indicators, which will help in planning and decision-making. It also provides a comprehensive image of the level of ICT penetration and the use of technology.
e-transactions Law (eLAW)	The law of Electronic Transactions aims to legalise the use of electronic transactions. It is designed to ensure that a transaction is not invalid simply because it took place by means of electronic form of communication. eLaw has several benefits such as providing protection for the various entities in the use of ICT for official and personal communication and transactions and facilitating the use of electronic transaction. It also addresses key issues such as recognition of electronic signatures, admissibility and evidential value of data messages and electronic payments validity besides protecting the privacy of individuals involved in electronic transactions. The Royal Decree No. 69/2008 promulgated the Electronic Transitions Law.
Convergent Government Network (CGN)	This infrastructure involves implementation of a nation-wide telecommunication infrastructure interconnecting government agencies based on the idea of Government Nervous Systems (GNS). CGN links all ministries and government entities and enhances the delivery of a range of e-services provided by them. It lays the foundation of a secure and reliable network, which government agencies can rely upon, and reduces operational expenditure by saving on the cost of training required to manage the network efficiently. The Oman Telecommunications Company (Omantel) is responsible for building and operating the CGN that connects a range of government organisations across the Sultanate. Prior to signing of the agreement, a pilot implementation was undertaken by the Information Technology Authority and Omantel involving approximately thirty different sites across four official bodies. Approximately 121 sites have currently been linked to the Unified Government Network, and IP address migration for twenty-four organisations has been completed.

A number of services and supporting elements will be implemented through the Portal, including electronic forms, which clients can download and complete to access electronic services from government entities. Access to electronic government services will subsequently be available through mobile and interactive phones.

Table 2 presents some examples of e-services.

National E-Payment Gateway (NePG)

To support and handle the activities of *e-payment systems*, the Central Bank of Oman (CBO) and the Oman Telecommunications Company (Omantel) have established a secure communications super-highway among the banking and financial institutions of Oman, called the 'BankNet'. CBO implemented and built Electronic Fund Transfer Systems, which comprise several components such as the Real Time Gross Settlement Systems (RTGS, successfully launched in September 2005) The Automated Clearing House System (ACH, was successfully launched in September 2006); Electronic Cheque Clearing (ECC) through cheque imaging, a new central ATM & Point of Sale (POS) Switch, Delivery versus Payment (DvP) and Payment Versus Payment (PvP).

The effect of the above systems combined with prudent fiscal measures is expected to bring about significant changes in fund management in the Omani economy fuelled by the application of state-of-the-art technologies in the National Payment System (NePG). The National e-payment project enables citizens to make payments online through multiple payment instruments, thereby leading a new era of e-services and e-commerce in Oman. The e-Payment portal also allows commercial establishments within Oman to display and sell their products to customers worldwide through the Internet.

The main characteristics of an e-payment Gateway are as follows:

- A secure transfer of funds via the Internet for online shopping for services or goods.
- ePG is built on MasterCard Internet Gateway Services (MIGS).
- A Payment Server acquires the e-payment and seeks an authorisation of payment from the customer's bank.
- It facilitates the customer's bank to transfer funds to the merchant's acquiring bank. It provides payment services for e-Commerce and e-Government.
- The gateway accepts all international credit cards and debit cards including Visa, MasterCard, Amex and Diners Club cards.
- It supports various electronic channels including the Internet, mobile phones, IVR, Call Centres, etc.

An ePG system is designed to ensure maximum security for payment transaction. Server hosted transactions of the Gateway use a technology called (SSL) Secured Socket Layer to provide secure transmission of sensitive data and encryption, between a customer's web browser and the e-payment server. Additionally, transactions sent from the shopping cart applications to the e-payment Server are channel encrypted to prevent alteration in transit as they are redirected via the customer's browser.

Lessons from the Implementation of E-Government Projects

When the first e-Government projects began in Oman, people in public and private sectors expected to witness results within a year or two: there was clearly a desire from the government to catch up with other countries' developments and to

Table 2. Examples of e-services

One Stop Shop (OSS)	It is a government to businesses initiative to support on-line company registration through MoCI with other ministries and government bodies and to enable a single-window service for Commercial Registration of new companies. It aims to create 'a favourable investment environment that is conducive to the development and prosperity of the Oman economy'. The system has been a huge saving in terms of costs, time and efforts; it is expected to offer e-commercial registration during 2009. There are plans to extend public point-of-access to the OSS beyond the Web to telephone, self-service kiosks and help desks in the Ministry headquarters and at its regional offices. This e-service has simplified the entire process and combined with investor friendly policies have brought efficiency into the operations.
National Registration System (NRS)	(NRS) initiative has expanded to include an electronic purse on its Multi Purpose Smart ID Card. The Royal Oman Police (ROP) introduced this Smart ID card for both Omani citizens and expatriate residents in Jan 2004, being the first country in the region. It is an integrated computer system with archive of accurate information about social events relating to birth, marriage, divorce, Death, Residency & Nationality of all citizens and residents of Oman. This system generates a unique civil number for each individual at the time of registration and supports applications such as driving license, passport, and sponsor details.
Online Visa	Royal Oman Police (ROP) is planning an E-Visa system: a web- based system that integrates different entities involved in visa processing by introducing an electronic communication channel between applicants and visa issuing authorities. The objective of the project is to facilitate visa issuance and processing procedure of visa for resident, visitors and tourists by enabling them to apply for online and get their application processed and approved in a seamless way.
Ministry of Finance (MoF) E-Services	MoF is managing a nationwide network of over 45 ministries and independent units with more than 2500 user accounts, to process the budget, procurement and payment cycle of all civil entities centrally in accordance with the Omani Financial Law. The IFS processes all payment to all suppliers as well as employees across all the civil ministries and also provides centralised audit and budgetary controls through a bilingual (Arabic and English) interface. With the aid of the system it is now possible to formulate the national budget, consolidate public expenditure accounts and manage public debt efficiently.
Ministry of Civil Services (MoCS) E-education service	The citizens now can get the required information by sending the details on the job advertisements to the mobile telephone numbers provided. Through this service queuing is reduced at the ministry. E-education is a unified admission system, processing applications to all higher education institutions of Oman. This system exchanges data electronically and works in close liaison with the Ministry of Education.

achieve and gains the advantages of e-Government by building its knowledge-based industries, as has happened in Singapore. Most people didn't understand that such a program requires a firm foundation, that it should be carried out gradually and steadily and that there would be many challenges for effective and quick implementation.

Reasons for the delays are as follows:

- **Environmental instability and uncertainty:** Characteristic of developing, transitional countries is the relative instability of the social, economic, political environment. Such a situation reduces the value of, and incentives for, learning since often the 'goalposts will have moved' from one e-government project to the next.
- **Difficulties in outlining e-government objectives as stakeholder expectations vary** (Al-Maimani 2000).
- **Difficulties in allocating budgets strategic plans as a substitute plan:** Oman strategic plans commenced before the electronic services era and did not take electronic services and ICT infrastructure into consideration during the planning period.
- **Stakeholder absence:** We have found that when an e-government project ends, key stakeholders have often moved on to other jobs/projects and have no continuing interest in the original project.
- **Resistance to change people, processes and technology:** The key variables involved in the delivery of back-office integration.
- **A tendency amongst the more ambitious projects to fail to fully deliver the anticipated benefits.** This trends increases due to the very high complexity of most e-government projects due to the number of functions and organisations involved.
- **Cultural inappropriateness:** In some organisational or national cultures, it is acceptable to admit and learn from failure. In others, it is not; failures are to be ignored or denied, and are certainly not to be discussed for the purposes of learning.
- **Electronic illiteracy:** One of the most critical issues; for example the eOman education program needed to be a top priority of the e-government strategy.
- **Skewing of incentives:** In some situations there are clear incentives for ongoing failure;
    - With some e-government applications, "success" can mean that the public agency is downsized or loses financial resources because of its efficiency gains.
    - By contrast, failure may mean maintenance of resource levels, or even continuing investment in renewed e-government efforts.
    - Likewise, vendors and consultants may see ongoing income from failures in a way they won't if an application succeeds. In these cases, stakeholders don't want to learn how not to fail; they want to continue failing.
    - Alongside this may be a lack of incentives—such as audit, censure, litigation—that encourage learning from failure.
- **Irrelevance of success/failure:** Some stakeholders might only be interested in association with the high-profile inception of the e-government project, not with the implementation and outcomes. Others might see the project as a means of directing investments to local IT firms, not as a means to achieve e-government goals. In these and similar cases, the outcome - whether the project succeeds or fails - is of no importance to the stakeholders, so they have no interest in trying to learn in order to do better next time.
- **Fear of exposure:** Some stakeholders fear that a learning process will expose their shortcoming; their ignorance about ICTs,

their self-serving behaviours, their corruption, etc. They will thus resist a learning process.

- **Difficulties to defrayal e-government failure costs, such as financial costs:** The money invested in equipment, consultants, new facilities, training programs, and in the time and effort of public servants involved -.
- **Difficulty to deal with intangible costs:** For example, e-government failure increases the barriers for future e-government projects, through loss of morale of stakeholders, particularly e-government champions, who may 'defect' to the private sector or overseas, and through the loss of credibility and trust in e-government approach to change.

PROTOTYPING THE HOLISTIC FRAMEWORK FOR E-GOVERNMENT IN OMAN

In this section, we'll present examples of analyses done during the piloting stage of the action research project with the Oman government. For the piloting project the system in focus was '*e-government services to investors*'. So far the prototype has involved only a small group of experts, guided by one of the authors (Al Maimani), following the stages of the ad-hoc holistic e-government framework, as presented above (section on holistic framework). The piloting stage intended to verify the usefulness of the analytical tools and the ad-hoc holistic e-government framework to support implementation of the e-government strategy. A complete application of the e-government holistic framework is about to start now that the team has satisfactorily finished the testing stage.

The pilot stage's results (summarised below) confirm that the framework helps to develop insightful criteria to diagnose implementation problems at the structural and cultural levels, and to adapt the current e-government strategy and programs accordingly. Also it can be very useful for creating the required mechanisms for improved possibilities of successful implementation.

Rich Picture of the System in Focus

Inspired in Checkand's (1981) 'rich pictures', we invited a group of practitioners to represent their views about the current implementation of the e-government program 'e-government services to investors'. Figure 3 presents an example of a cartoon representation of their views. It is clear from the rich picture –as well as from previous section summary on difficulties on implementation- that issues recognised as problematic in current implementation of the e-government strategy are:

- Over emphasis on technological issues
- Lack of focus on people's issues
- Cultural inappropriateness
- Lack of citizens awareness
- Silo mentality between those in charge of implementing the strategy
- Structural issues resulting from lack of connectivity (e.g. difficulty in financial planning, managing incentives, accountability for projects' results etc)

Defining Relevant Systems

A first attempt to define the relevant systems for the pilot stage resulted in the following:

R 1: A system to improve possibilities of success of the e-government services to investors in Oman.

R 2: A system to improve connectivity, knowledge sharing, decision-making, the e-services to investor area (One Stop Shop).

R 3: A system to improve the 2020 strategy in terms of valuing knowledge management & ICT development in the public sector in Oman, and in particular in e-services to investors.

Figure 3. Rich picture of the system in focus

R 4: A system to link current silo mentality in managing knowledge in the Omani public sector in particular in the Services to the investors' area of the MoCI.

The team considered it useful to continue the process by developing the relevant system R4. There follows the related root definition.

Creating a Root Definition (TASCOI)

There follows a description of the TASCOI developed to describe the system in focus. The TASCOI is a tool suggested by Espejo (1989) adapting Checkland's (1981) CATWOE technique, for describing the core transformation and actors on a human activity system. T means transformation, A means Actors (in charge of implementing the transformation), S are the suppliers (resources needed to do the transformation); C are the customers (who benefit from the transformation's outputs); O are the owners (those with authority to even change the transformation); and I are the 'intervenors' (those with authority to audit externally the transformation).

Table 3. TASCOI of the system in focus

Transformation (T)	A System to overcome the Silo-mentality: From the current situation in the investors sector (services are provided by each organisation separately with each organisation (silo) having it is own regulations; E-government not properly recognised financially or implemented timely or effectively) into the desired situation: E-government services to investors implemented and having impact in fostering the investors sector in Oman. It is assumed that effective implementation of the e-government program will have beneficial impact on investor interests.
Actors (A)	Council of Ministers: Public Sector Government organisations: • Ministry of Commerce and Industry • Ministry of Manpower • Ministry of Environment and Climate affairs • Ministry of Regional Municipalities & Water Resources. • Ministry of Tourism • Royal Oman Police • Muscat Municipality • Public sector employees. Telecommunications regulatory Authority(TRA) & Information Technology Authority (ITA) e-government advisory panels
Suppliers (S): -	ICT (Telecommunications & Information Technology) suppliers. Knowledge Management/e-government consultants Vendors
Customers (C)	Citizens Private Sector: Represented by Oman Chamber of Commerce and Industry. Investors (Business sector) (The aim of this study)
Owners (O)	Council of Ministers Ministry of National Economy Ministry of Finance Information Technology Authority (ITA) IT Department in Government organisations
Interveners (I)	Ministry of legal Affairs Ministry of Justice Ministry of National Economy Ministry of Finance Council of Oman (Majlis Oman) [4] Telecommunication Regulatory Authority (TRA) Communication Organising Authority (ITA) INTIL (UN)

VSM Model of the Organization at the System in Focus

Having decided on a root definition for the system in focus, the prototype team identified the main roles/agencies in the Oman public sector that are directly involved with offering financial services to customers, as part of the e-government strategy. We first agreed on the levels of complexity of the Oman public sector and reflected it in Figure 4, using one of the modelling techniques suggested in the ad-hoc methodology (identifying levels of organisation – see Beer (1985)). The team experimented using this modelling technique and produced several alternative models of complexity for the system in focus. Finally it was agreed to choose the one shown in Figure 4, which uses geographic criteria to unfold complexity, as it was felt that this corresponds more clearly to the way it actually happens currently in the Omani organisations involved. The exercise helped the team to map the complexity of the network of interacting agents involved in the implementation of this e-government program.

On the basis of this exercise, the team then concentrated on doing a preliminary VSM diagnosis of the organisations involved in the system

Figure 4.

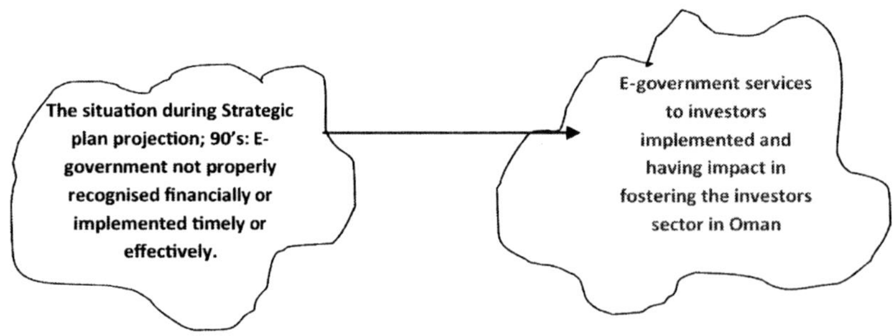

in focus and in particular at the first level of recursion (national to regional). Figure 5 presents the organisations and roles involved, as a VSM diagram, where the Systems 1 are regional agencies in Oman and the S2 to S5 are those decision making mechanisms, and information, communication and knowledge management tools that are an integral part of the way the government currently implements the e-government services at the national to regional levels.

The learning from this stage of the prototype has been so far that by mapping the technological and organisational mechanisms at the right level of organisation it has been easier to reflect on their usefulness and effectiveness. A preliminary finding from this modelling process was the reconfirmation of the importance of implementing a One Stop Shop to coordinate all the technical and informational services that the e-government program for financial services have planned to offer.

Also, the preliminary analysis showed the high variations in terms of the complexity each of the regional organisations have: each one's complexity was measured by proxy indicators (number of employees and amount of services provided) and the results showed important differences among them. The preliminary analysis also suggested that in order to overcome the silo mentality and build an integrated back office, it was necessary to re-engineer some of the core business processes of each organisation and their regional sub systems in issues related to the e-gateway implementation.

Review the E-Government Strategy in the Light of the VSM Mapping to Identify Critical Issues for Implementation

Based on the VSM mapping of the system in focus presented in the last section, and following Beer's criteria on the roles of Systems 1 to 5, we've classified in Table 4, some of the most relevant e-government projects, tools and mechanisms (within the system in focus) for sharing knowledge and information accordingly. This preliminary

Table 4. Examples of e-government projects in VSM language

System 5	Electronic awareness (eOman).
System 4	Innovation and Support Center (ISC) & International Relations (INTIL).
System 3	Disaster Recovery Centre (DRC), electronic law (elaw), National electronic payment gateway (NePG).
System 3*	Information Security Management Framework (ISMF).
System 2	Convergent Government Network (CGN), government electronic services (e-services), National IT Training & Awareness (NITTA), Standards Framework (STDF).

Figure 5. Levels of recursive (public sector) organisation in Oman

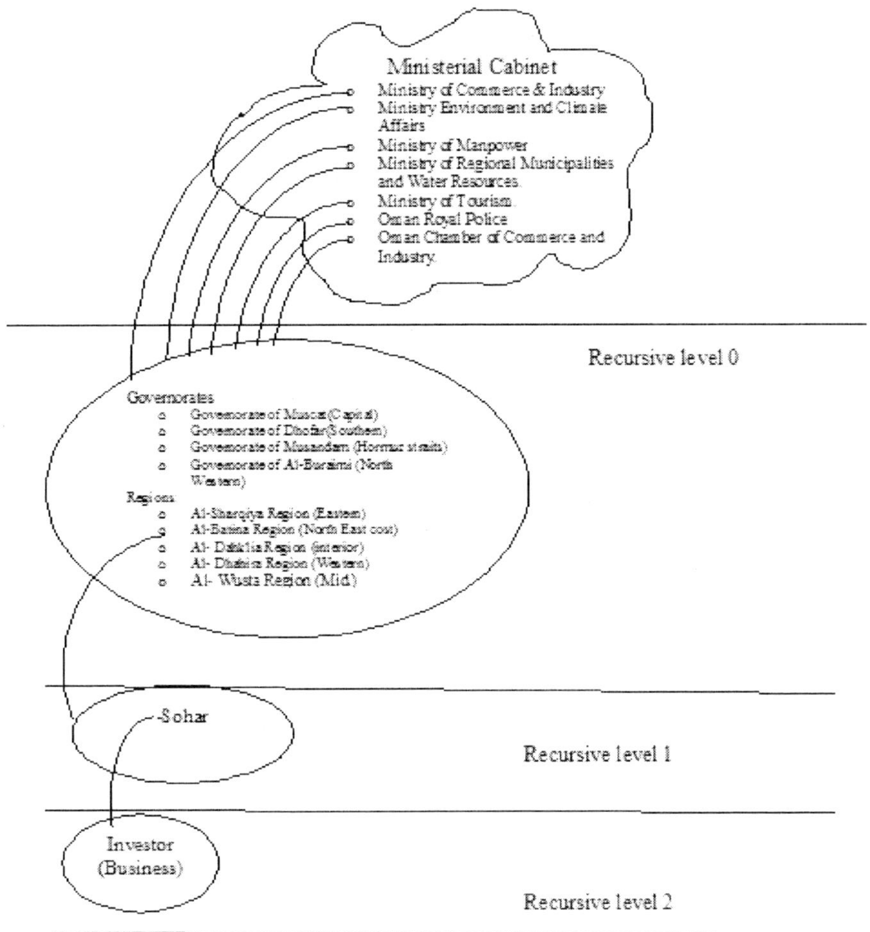

exercise has offered the team criteria to identify gaps in the designed portfolio of e-government projects (e.g. under emphasis on System 3* and over emphasis on System 2 types of projects), as well as to identify structural issues that need to be properly addressed in order to guarantee effective implementation.

Example of gaps are: the need for additional tools for cultural change management (System 2), for monitoring the correct use of the gateway (System 3*) and for penalising illegal uses (System 3 – legal constraints); also for scanning financial trends in the international environment (System 4); and for developing 'ethical and transparency behaviours' in financial exchanges between individuals and businesses (System 5). This type of reflection is helping the team and to identify critical issues for implementing the current strategy, and for more critically assessing its probability of impacting the organisational performance. Next section presents more detailed examples of structural and technological changes or developments required.

Figure 6. VSM of the system in focus

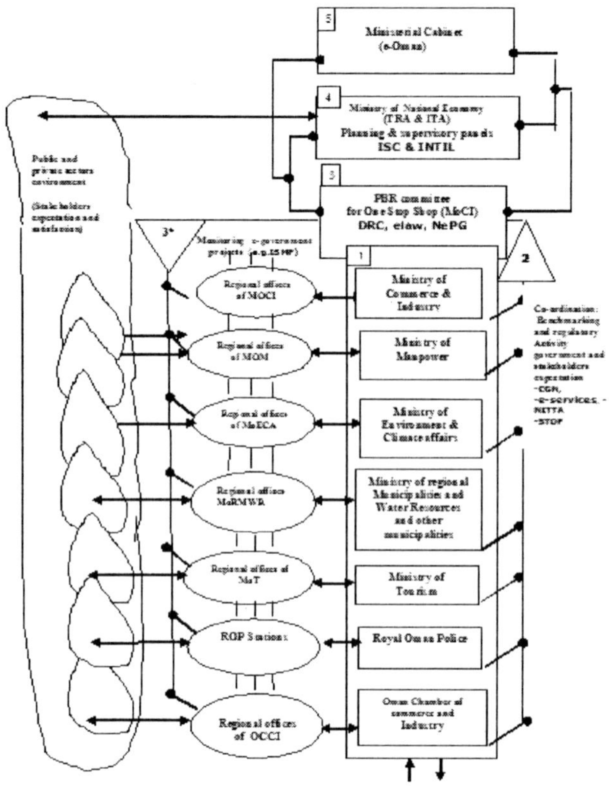

Creating the Organizational and Learning Context for the E-Government Portfolio's Implementation

As a result of this preliminary analysis in the prototype, the team have agreed that a main issue it needs to address, in order to create a proper context for implementation of the e-government strategy, is the lack of operational and technical connectivity between independent agencies (the silo mentality); and it has agreed that the next stage of the research will concentrate on detailing required changes in the strategy and implementing the necessary structural and cultural changes required to more effectively achieve the e-government program's goals.

There follows some examples of actions identified in the VSM diagnosis during the piloting stage:

- **System 1:** In the pilot project we only analysed the national to regional level of implementation for the system in focus (see Table 5 for more details on the System 1). System 1 was identified as the regional offices of the related ministries and government agencies in charge of e-government projects related to the e-gateway or (One Stop Shop) and single window of the ministries that provides services to the investors. A first glance at Figure 5, using VSM criteria, made us realise the relevance of improving current coordination mechanisms between the different regional agencies. This coordination is currently difficult as each agency would firstly receive information from and send information to their 'parent' organisations (e.g. Ministries at the national level). This reinforces the need for firstly implementing the One

Table 5. Details on complexity of each organisation in System 1

Administrative Regions No. of MOCI Administrations No. of Wilayats
Governorates
Governorate of Muscat(Capital) Main 06
Governorate of Dhofar(Southern) 1 10
Governorate of Musandam (Hormuz straits) 1 04
Governorate of Al-Buraimi (North Western) 1 03
Regions:
Al-Sharqiya Region (Eastern) 2 11
Al-Batina Region (North East cost) 2 12
Al- Dahklia Region (interior) 1 08
Al- Dhahira Region (Western) 1 03
Al- Wusta Region (Mid.) 0 04
Total of Administrative Regions (9) and (sub); Admin.: 9 Wilayats: 61
Regional Administrations of:
Ministry of Commerce and Industry (MOCI) Administration
Ministry of Manpower (MOM) Administration
Ministry of Regional Municipalities and Water Resources (MORMW) Administration
Ministry Tourism (MOT) Administration
Ministry or Environment and Climate Affairs (MOECA) Administration
Muscat and Dhofar Municipalities (M&DM) Administration
Royal Oman Police (ROP) Administration
Oman Chamber of Commerce and Industry (OCCI) Administration

Stop Shop as the most natural mechanism for coordination in dealing with financial transactions with the government. Also it clarified the need to review each of the organisations' capacity to implement the e-gateway program, given the big variations between each one's capabilities for dealing with their own tasks' complexity.

- **System 2:** So far most of the projects from the e-government strategy are addressing development of System 2 type of tools, as shown in Figure 5. However there seems to be a need to more clearly address some of the following issues revealed through this study:
  - Lack of electronic connectivity and protocol operation between Ministries
  - Need to improve formal and informal communication, information and knowledge sharing between Ministries and its administrations.
  - Need to complete implementation of the back office integration platform and OSS network (LAN & WAN)
- **System 3:** Current implementation problems show evidence of need to improve existing regulatory mechanisms, as for example:
  - To improve the mechanisms for accountability on the resources used .vs. the benefits of the e-payment services provided
  - To develop clear "intervention rules" to guarantee that misuse of the

  e-gateway system would be clearly punished
  - To improve the capability for resource negotiation and decision making in both the design and implementation of the e-government strategy (e.g. need to align it with the 20 years strategic plan for the country).
  - To develop tools for measuring the impact of the e-gateway project and to monitor its development (e.g. an ITA follow up program -weekly, monthly, yearly).
- **System 3*:** The following needs were identified:
  - Develop mechanisms for auditing/monitoring System 1: monitoring network and infrastructure availability in regional administrations, through the platform of back offices, back office integration.
  - To design a way to collect suggestions for process re-engineering, coming up from analysis and development of e-government projects
- **System 4:** It was agreed to explore the following developments:
  - A system for environmental scanning of the financial markets at the global level
  - Mechanisms to share e-government developmental needs from each organisation in System 1
  - Financial planning tools and improved decision making mechanisms for the design and planning of e-government strategies
- **System 5:** It would be necessary to address the following issues:
  - Development of projects to improve the ethos of transparency, openness and trust required for effective and comprehensive implementation of the e-gateway program
  - Development of mechanisms to reflect on the learning gained through the e-government program's implementation and to re-design any aspects of it found ineffective/inappropriate during implementation (closure).

CONCLUSION

This chapter presented ongoing research on e-government, knowledge management and organisational transformation and development. The literature review introduced core topics for e-government seen from systemic and complexity approaches to management and explained how they may support e-government strategy and implementation. It described a prototype action research project, using an ad-hoc analytical framework to support e-government strategic implementation. By piloting preliminary research using a holistic framework for e-government, we understand the sort of analysis that such an approach would offer – mostly to clarify and align technological and organisational issues in the implementation of current e-government strategy. Several examples from the prototyping stage of the research illustrate the team improved understanding of desirable improvements to the strategy and required structural and cultural changes.

We experienced that taking this approach has provided us with useful guidelines to understand and support the e-government strategy. Although the current e-government strategy covers most of the required technological and informational changes, it doesn't explicitly address the need for cultural/structural changes, which are fundamental for effective implementation. The VSM based framework addressed these issues.

In summary, the preliminary outline of the holistic framework for e-government design and implementation has proved useful, as it has helped to recognise issues of structural complexity, to highlight projects that needed to be reinforced in order to create a proper learning context (e.g. the One Stop Shop to improve connectivity) and to address issues of the user's involvement and participation in the design and implementation processes. The main research project will take place in the next few years and will undoubtedly provide a significant contribution to current research on systemic approaches to e-governance.

Forthcoming stages of the action research project include conceptual and methodological developments of the ad-hoc methodology, new prototype projects, and implementation of some of the recommended adjustments to strategy, structure and cultural changes. They are expected to conclude in the next three years.

This project contributes in strengthening the loop between theory and practice in the application of systemic methodologies and tools to support to e-government programs. The learning from mapping the current e-Government program in the Sultanate of Oman with the current institutional networks responsible for implementation makes us feel this an important research path to further developments in all theory, methodology and applications. Undoubtedly other systemic approaches would contribute to develop critical thinking and strategic thinking, and finding ways of benefiting from them at different stages of the research process would be another desirable path for future research.

ACKNOWLEDGMENT

We would like to acknowledge the support of Dr. Salim Al Ruzaiqi, CEO of ITA, and Zumuroda Al-Belushi, Mohammed Othman and Tamadher Al Wahaibi, from Ministry of Commerce and Industry whose help enabled us to gathering some of the information. Also to Dr Jon Walker for his help and valuable comments when proof- reading the chapter.

REFERENCES

Ackoff, R. (1981). The Art and Science of Mess Management. *Interfaces*, *11*(1), 20–26. doi:10.1287/inte.11.1.20

Al-Maimani, A. S. (2000). *Information Impact on socio-economic decision-making (national economy of Sultanate of Oman).* MBA dissertation, the University of Hull.

Ashby, W. R. (1964). *An Introduction to cybernetics*. London: Methuen.

Beer, S. (1959). *Cybernetics and Management*. London: English Universities Press.

Beer, S. (1979). *Heart of Enterprise*. Chichester, UK: John Wiley & Sons.

Beer, S. (1981). *Brain of the Firm (2nd Ed)*. Chichester, UK: Wiley.

Beer, S. (1985). *Diagnosing the System for the Organisation*. Chichester, UK: Wiley.

Bowling, D., & Espejo, R. (1992). An intervention with the Cybernetic Methodology in Regent Engineering. *Transactions of the Institute of Measurement and Control*, *14*(1), 17–28. doi:10.1177/014233129201400104

Bula, G. (2004). Observations on the development of cybernetic ideas in Colombia: A tribute to Stafford Beer. *Kybernetes: The International Journal of Systems and Cybernetics*, *33*(3-4), 647–658. doi:10.1108/03684920410523625

Checkland, P. (1981). *Systems Thinking, Systems Practice*, Chichester, UK: John Wiley & Sons.

Checkland, P., & Scholes, J. (1990). *Soft Systems Methodology in Action*. Chichester, UK: John Wiley & Sons.

Doyle, J. R. (1991). Problems with Strategic Information Systems frameworks. *European Journal of Information Systems*, *14*(1), 273–280. doi:10.1057/ejis.1991.48

Espejo, R. (2002). Systems and the Information Society: Requisite Organisation and Problem Solving. *Systems and the Information Society Network*. Retrieved November 2004 from http:\\ www.itsy.co.uk

Espejo, R., & Harnden, R. (Eds.). (1989). Outside and then: An interpretative approach to Viable System Model. In *The Viable System Model: Interpretations and Applications of Stafford Beer's Viable System Model*. Chichester, UK: John Wiley & Sons.

Espejo, R., Schumann, W., Schwaninger, M., & Billelo, U. (1996). *Organisational Transformation and Learning*. Chichester, UK: John Wiley & Sons.

Espejo, R., & Schwaninger, M. (Eds.). (1993). *Organisational Fitness: Corporate Effectiveness through Management Cybernetics*, Frankfurt and New York: Campus Verlag.

Espinosa, A. (1995). *Strategic Information Systems at the Colombian President's Office; A Managerial Cybernetics' Perspective*. Doctoral Thesis, Aston Business School, Aston University, Birmingham, England, U.K.

Espinosa, A. (2007). A cybernetic re-evaluation of socio-economic development programs. *Kybernetes*, *35*(1/2), 30–44. doi:10.1108/03684920610640218

Espinosa, A., Harnden, R., & Walker, J. (2007). Beyond Hierarchy: Stafford Beer revisited. Invited paper. [Special issue]. *Kybernetes: The International Journal of Systems and Cybernetics*, *36*(3/4), 333–347. doi:10.1108/03684920710746995

Espinosa, A., Harnden, R., & Walker, J. (2008). A Complexity Approach to Sustainability: Stafford Beer revisited. *European Journal of Operational Research*, *187*, 636–651. doi:10.1016/j.ejor.2007.03.023

Espinosa, A., & Walker, J. (2005). Environmental Management revisited. Lessons from a cybernetic intervention in Colombia. *Cybernetics and System. International Journal (Toronto, Ont.)*, *36*(7), 573–589.

Galliers, R. D. (1991). Strategic Information Systems Planning: myths, reality and guidelines for successful implementation. *European Journal of Information Systems*, *1*(1), 55–64. doi:10.1057/ejis.1991.7

Galliers, R. D., & Baets, W. R. J. (1998). *Reflections on BPR, IT and Organizational Change, in Information technology and organisational transformation*. Chichester, UK: John Wiley & Sons.

Grabher, D., Schere, R., & Walser, M. (2000). One region, three networks: Processes of self organization as a strategy for sustainable development. In *Proceedings of the 40th European Congress of the Regional Science*, Barcelona, Spain, Aug-Sept.

Heeks, R. (1998), Information Age Reform of the Public Sector: The Potential and Problems of IT for India', In *Information Systems for Public Sector Management*, working Paper Series, Paper no, 6.

Heeks, R. (1999). *Reinventing Government in the Information Age: international Practice in IT-enabled Public Sector Reform*. London: Routledge.

Heeks, R. (2002). Information Systems and Developing Countries; Failure, Success, and Local Improvisations. *The Information Society*, *18*, 101–112. doi:10.1080/01972240290075039

Heeks, R. (2003). Most e-government–for-Development projects Fail: How Can Risks be Reduced? *iGovernment Working paper Series*, 14.

Ho, T.-K. (2002). Reinventing local governments and the e-government initiative. *Public Administration Review*, *62*(4), 434–444. doi:10.1111/0033-3352.00197

ITA. (2002). *Summary report of Information Technology Authority (ITA): Towards Digital Oman*. Oman.

Leonard, A. (1999 August). A Viable System Model: Consideration of Knowledge Management. *Journal of Knowledge Management Practice*.

Lyytinen, K., & Hirschheim, R. (1997). Information systems failure; A survey and classification of the empirical literature. *Oxford Surveys in Information Technology*, *4*, 257–309.

McCulloch, W. S. (1965). *Embodiments of Mind.* Cambridge, MA: MIT Press.

McNabb, D. E. (2007). *Knowledge management in the public sector: A Blueprint for innovation in Government.* London: M.E. Sharpe.

Ministry of National Economy (MNE). (2007a). *e-Oman Bulletin*, Oman.

Ministry of National Economy (MNE). (2007b). *Oman Digital Society Report*. Oman.

Moon, M. J. (2002). The evolution of e-government among municipalities: rhetoric or reality. *Public Administration Review*, *62*(4), 424–433. doi:10.1111/0033-3352.00196

Pérez Ríos, J. (2008). *Diseño y Diagnóstico de Organizaciones Viables. Un enfoque sistémico*. Valladolid, Spain: Iberfora.

Schumann, W. (1990). Strategy for information systems in the film division of Hoechst A.G. *Systems Practice*, *3*(3).

Scott Morton, M. S. (1994). The state of the art of research. In F. W. McFarlane (Eds.), *The Information Systems Research Challenge Proceedings* (pp. 13-45). Cambridge, MA: Harvard Business School Press.

Stacey, R. D. (2000). *Complexity and management: Fad or Radical Challenge to Systems Thinking?* London: Routledge.

Stacey, R. D. (2003). *Strategic Management and organisational Dynamics (4th ed.)*. Harlow, UK: Prentice-Hall.

Stokes, P. (2008). *The Viability of Societies: Governance and Complexity Today*. Saarbruken, Germany: VDM Verlag.

Sundbo, J. (2001). *The strategic Management of Innovation*. Cheltenham, UK: Edward Elgar.

Sydänmaanlakka, P. (2002). *An Intelligent Organization*. Oxford, UK: Capstone.

Turke, R. E. (2008). *Governance: Systemic Foundation and Framework*. Berne, Switzerland: Physica-Verlag.

United Nations. (2008). *Public Administration and e. government network*. UNPAN Reports.

Von Foerster, H. (1981). Notes on an epistemology for living things. In H. Von Foerster (Eds.), *Observing Systems* (pp. 257-271). Salinas, CA: Intersystems Publications.

Wiener, N. (1965). *Cybernetics or Control and Communication in the Animal and the Machine (2nd ed.)*. Boston, MA: MIT Press.

Yolles, M. (2006). Organizations as Complex Systems: An Introduction to Knowledge Cybernetics. Yolles, M. (Ed). *ISCE Book Series: Managing the Complex Series*. Editor(s): M. Lissack, K. Richardson. ISCE Research.

ENDNOTES

1 Stages 1 and 2 use analytical tools suggested by Checkland (1981)

2 Modeling technique to define organizational identity – suggested in Espejo & Harnen (1989)

[3] The e-government program in Oman includes a single window services (One Stop Shop) which is leaded by one of the authors, who is One Stop Shop BPR & steering committee chairman, at the Oman's Ministry of Commerce and Industry.

[4] Council of Oman consist of; The State Council (Majlis A'Dawla) and The Consultation Council, citizens representative council (Majlis A'Shura)

Chapter 9
Electronic Government Implementation Projects with Multiple Agencies:
Analysis of the Electronic Invoice Project Under PMBOK Framework

Alvaro Junqueira
Secretariat of Finance of State of São Paulo, Brazil

Eduardo Henrique Diniz
Fundação Getúlio Vargas, Brazil

Marcelo Fernandez
Secretariat of Finance of State of São Paulo, Brazil

ABSTRACT

Contemporary public administrations have been suffering several ways of pressure to promote modernization in their structures and services. One of governments' options to meet that demand lies in the potential use of information and communication technology for the benefit of better service to citizens and greater state apparatus efficiency. This contemporary movement particularly connected with personal computing and the Internet arrival has been called Electronic Government (e-Gov). One of its actions relates to building government electronic services that integrate several government agencies in a collaborative format. However, e-Gov projects demanding integration to this degree have an implementation complexity that is greater than traditional projects, which reflects on low success rates worldwide. This paper investigates the good practices identification in multi-agency e-Gov project management. Using the PMBOK framework, this paper presents an in-depth study of Nota Fiscal Eletrônica - NF-e (Electronic Invoice) project implementation. This project was based on a nationwide integration of multiple agencies (26 State Tax Administrations, the Brazilian Federal Revenue Service and nineteen large companies) and faced major technical and management complexity. The results found suggest that the project was implemented in an informal way, however with impressive results. There were identified two complementary management models. The first model was nationwide, focused in Scope and Communication Management. The second model was local to each institution, focused in Cost, HR and Procurement Management.

DOI: 10.4018/978-1-60566-860-4.ch009

INTRODUCTION

Information and Communication Technologies (ICT) have been used as an instrument to make structure changes viable in public management (Osborne and Gaebler, 1992; Bresser Pereira, 2005). In the last few years, as the internet use evolved in all sectors of economy, including the public field, the concept of Electronic Government (e-Gov) has consolidated itself (Reinhard e Dias, 2005). E-Gov is currently considered an alternative to improve services delivery (e-Services), increase Public Administration's internal efficiency (e-Administration), as well as support citizens' greater participation in the political process (e-Democracy) (Medeiros, 2004).

While it's one of e-Gov's most visible aspects, the e-service concept faces some implementation difficulties, among which the following: a lack of digital access in the majority of the population, specially in developing countries; a resistance by government itself in overcoming cultural barriers to paradigm change; finally, a lack of integration among government agencies, reflected on information systems, yet originated in the very structure of government. This last aspect of e-service, i.e. government agencies' internal integration, is very important to prevent external citizen integration from frustrating expectations (Ho, 2002).

In the public sector, collaboration among organizations via ICT is associated with Joined-Up Government and Government in Network concepts, which commend a government that is integrated among its departments and agencies, from the conception of public policies to their implementation and maintenance. In order for that integrated government to become reality, especially in providing electronic service, there must be an e-Gov project implementation that promotes integration among those multiple government agencies.

However, ICT project execution in the public sector still shows spectacular failures, even in developed countries (Hazlett and Hill, 2003).

Considering developing countries, Heeks (2004), from the analysis of 40 reports on real cases in a research with eGoverment for Development Information Exchange1 2members, there is an alarming figure of only 15% of e-Gov projects being considered as being successful. The concept used by Heeks for e-Gov project success refers to situations where the majority of stakeholders have their main goals achieved and don't experience undesired results (Heeks, 2004, p.2).

Project Management Body of Knowledge—PMBOK, from the Project Management Institute—PMI is a guide that consolidates the best practices on the subject of project management, and it's one of the most comprehensive guides in the market (Neto e Bocoli, 2003, Charbonneau, 2004). While it's not ICT projects-specific, a complementation to PMBOK called Government Extension to PMBOK (PMI, 2006) focuses on public sector project peculiarities.

Understanding the aspects that favor e-Gov project management proves relevant particularly when providing citizens with integrated services, where there's no need to identify which government agency they are dealing with. Good management practices, such as those proposed by PMBOK, are even more critically important in particular for e-Gov projects that deal with complexities referring to the integration among multiple government agencies,

Therefore, the purpose of this paper is to analyze Nota Fiscal Eletrônica – NF-e (Electronic Invoice) implementation project, performed by 26 states and the Federal District, in cooperation with the Brazilian Federal Revenue Service, from the conceptual, theoretical references in A Guide to the Project Management Body of Knowledge, Third Edition (PMI, 2004) and Government Extension to the PMBOK Guide Third Edition (PMI, 2006).

A brief review of the Literature on Project Management will be presented next, as well as a NF-e case description. Subsequently, methodology aspects of the research on NF-e project implementation will be presented. After presenting the research results, a few conclusions and recommendations will be discussed.

LITERATURE REVIEW

E-Gov Project Implementation

Since the 1980's the public sector has suffered a "revolution" in providing services (Skelcher, 1992 apud Hazlett and Hill, 2003), but since the 1990's, as Internet became disseminated and personal computer resources available, those changes gained a new momentum. One of its main points is putting the citizen as the central focus of its works, promoting better public service rendering (Hazlett and Hill, 2003).

In the case of developing countries, research performed by Heeks (2004) shows that 85% of IT projects in the public sector are considered as being partial or total failures. About this great failure to implement projects, the poorer the country, the more severely it occurs (Heeks, 2003 apud Heeks, 2004). Hence the relevance of studying and understanding further those aspects that favor e-Gov project management, particularly in developing countries such as Brazil.

Nidumolu et al. (1996) identified three major perspectives influencing the implementation of an e-Gov project in developing countries: one related to functional aspects, the second to policy perspectives and third, to society. The functional aspects refer to objectives provided by the information system that motivate individuals to participate in the process: reduction of cost, to make information access easier etc. Policy perspectives focus on personal interests and power relations, in contrast to functional issues which are more technically oriented. Finally, the social perspective, when individual perceptions of the projects are influenced by their social group's perceptions. In their research on a large scale project of the Egyptian government, Nidumolu et al. (1996) showed that these three groups of perspectives influence the project in different ways. Particularly in the implementation phase, political and social issues were prominent.

Gil-Garcia and Pardo (2005), have related critical success factors - FCS of e-Gov projects, with the mapping of a set of relevant variables that influence their successful implementation. Particularly in this study, five major issues that must be observed in e-Gov projects were identified:

1. **Information and data:** Aspects related to data management, from capture or generation, to its use, spread and share
2. **Information technology:** Issues related to technology itself, such as usability, compatibility, complexity, among others
3. **Organization and management:** Organizational aspects, as organizational structure, management model, strategic alignment between business and IT, besides individual behavior
4. **Legal and regulatory:** Regulatory aspects that affect the mechanisms of action in the public sector
5. **Institutional and environment issues:** Related to the environment where organization operates, as well as the influences and interests of various political actors

To support e-Gov projects that demand integration among multiple agencies, the White Paper of the Electronic Government Interoperability for Latin America and the Caribbean (ECLAC, 2007) have introduced the concept of interoperability platform, developed from the experiences of projects in Brazil, Chile, Colombia, Mexico and Trinidad and Tobago. The interoperability issues have been divided in organizational aspects (establishment of collaboration procedures), semantic aspects (guarantee for keeping communication meanings) technical aspects (technological solutions and platform) and governance aspects (multilateral mechanisms for communication and deliberation).

Project Management and PMBOK

Project Management refers to the application of knowledge, skills, tools and techniques to project activities in order to meet the requirements of the project (PMI, 2004). Carayannis, Kwak and Anbar (2003) define the latest developments in project management in four main stages:

1. **Before 1958:** It was characterized mainly by speeding time spent by project implementation, given the emergence of new technologies as automobiles and communications. According Jugdev (2004), project management at that time had a strong focus in operations.
2. **From 1959 to 1979:** The commercial use of computers has strengthened the tools of project management, including the dissemination of project offices, used to disseminate information. There was then a criticism of the excessive emphasis on the rational approach and, consequently, spreading studies on leadership and organizational structures on project management (Jugdev, 2004).
3. **From 1980 to 1994:** Project management tools have become more accessible, given the gradual migration of platforms, from mainframes to PCs. Easier to use, the spread of PCs and local area networks has enabled project management systems to be used for other managers and not only by computing engineers.
4. **From 1995 up to now:** With the Internet, project management tools began to be used to disseminate information and implement processes. The concept of project office now works over the Internet ignoring physical and geographical boundaries.

Among several project management models, PMBOK has been developed through a consensual, voluntary process, with specialists from the project management field. Practically, PMBOK is a publication by the Project Management Institute (PMI), a non-profit organization who has over 230,000 associate professionals in over 125 countries. Therefore, PMBOK is an accumulation of the knowledge those professionals have, which includes both traditional and innovative initiatives to provide better project management effectiveness (PMI, 2004). Besides spreading its knowledge, PMBOK also aims to standardize terms used in the project management area (Neto and Bocoli, 2003). PMBOK is accepted as a project management standard by ANSI – American National Standards Institute, and by IEEE – Institute of Electrical and Electronics Engineers (PMI, 2007).

PMBOK (PMI, 2004, p.5) defines a project as a temporary enterprise that is used to create a specific product, service or result. So projects are different than operational works in that, in the later, work is repetitive and without a clearly defined purpose.

PMBOK (PMI, 2004) also defines three main stages in a project's life cycle. The "initial" stage, normally with smaller costs and human resources, is responsible for identifying the projects feasibility, establishing the project's management plan and its final scope. The "intermediate" stage details the project plan and effectively executes its several activities. In this stage, human and financial resources are normally higher. Finally, the "final" stage is characterized by products' delivery and approval by the project sponsor, with consequent dismissal of resources.

In government projects particularly, according to the Government Extension to PMBOK (PMI, 2006), there are three stages that are quite specific in a project's life cycle when it incorporates contracting third party companies. The first one, called "origin," is the stage in which documents to be used in obtaining budget financing are made. The "planning and project" stage is normally related with selecting choices of execution and obtaining execution proposals. Finally, the "contracting" stage, which is normally established in public legislation, defines the project's acquisition and execution activities by a contracted company.

PMBOK (PMI, 2004) also recommends project management through process management. Each process is defined as activities that receive information and generate specific results. Processes are normally classified into five specific groups: Initiator Processes, Planning Processes, Execution Processes, Monitoring and Control Processes and Closing Processes. In that context, project management is subdivided, according to PMBOK (PMI, 2004) proposition, in nine knowledge areas and 44 work processes. These processes are not recommended to just any project, but as a general guideline that should become particular within the characteristics of the specific project in question.

Moreover, those processes are defined within the above mentioned five big groups according to their feature: It's important to stress that PMBOK itself (PMI, 2004) makes it explicit that the mere use of their guide in this knowledge areas aren't enough for an effective project management. In that sense, a description is made of four expertise areas to be treated by the project management team together with those processes proposed in PMBOK:

1. **Knowledge of segment specific features, patterns and regulation:** The team must know the particular features for the segment the project is in, e.g., regarding its technical, functional structure aspects. Such knowledge comprehends existing patterns (consensus established by a recognized technical body) and regulation (government impositions specifying process, service or product characteristics).
2. **Knowledge of the project's environment:** It considers the necessity for the team to adequately put the project in context within its environment; with that purpose, PMBOK subdivides it in three groups:
   2.1. **Cultural and social:** Project impacts on people are foreseen, while people's influence over the project is observed. This may require understanding of economic, demographic, educational, ethnic, religious among other aspects of people affected or with an interest in the project;
   2.2. **Political and international:** It refers to costumes and legislation on international, national, regional or local level, as well as considerations on time zone differences, holidays, logistic restrictions to personal attendance meetings
   2.3. **Physical:** Knowledge on ecologic and geographic impacts, especially in projects that incorporate physical changes to the environment
3. **Knowledge of management in general:** Understanding of those main areas of administration that are necessary to organizational operations directly or indirectly related with the project, such as finance, sales, marketing, contracts, logistics, ICT, organization structure etc
4. **Interpersonal skills:** Capacities to interact with other individuals, such as: effective communication, leadership, motivation, influence capacity, negotiation, conflict management, problem solving etc

It is important to emphasize that PMBOK is not the only framework with a focus on management by processes, even when talking about software development. Sherer and Thrasher (2005) made a comparison between PMBOK and Capability Maturity Model Integration - CMMI, indicating that both models are applicable to project management and have complementary characteristics. PMBOK is more comprehensive that CMMI because is not related to project type or organization size. Moreover, the CMMI is more structured for software engineering processes, although limited to use in small and medium organizations. Related to training, PMBOK focuses on project manager professional development through its Project Management Professional Certification - PMP, while CMMI addresses processes improvement and development through levels of maturity of the organization. To improve results, the authors suggest using a composite method. Although there are other project management methodology options, the use of PMBOK allows for an interesting approach to study the implementation of Nota Fiscal Eletrônica (NF-e) project. Beyond a system construction, the project related, on a national level, with other issues, such as: modification of legal rules, cost, acquisitions, human resource management etc. Therefore, the comprehensiveness provided by PMBOK has proved an adequate theoretical base, when compared to other specific frameworks in the systems area.

NF-e Case in Brazil

NF-e is a project to transform paper invoice into an electronic one, with no physical counterpart. The project was conducted on a nationwide level through the Encontro Nacional de Coordenadores e Administradores Tributários Estaduais (National Congress of State Tax Coordinators and Administrators) – ENCAT, a forum with participation of the State Tax Administrations for tax administration's best practice integration and exchange. The NF-e project started in February 2005 at SEFAZ3/SP and its production commencement took place from September 2006. In this first stage, the project was implemented in six State Tax Administrations (the states of Bahia, Goiás, Maranhão, Rio Grande do Sul, Santa Catarina and São Paulo), the Federal Revenue Service, as well as in 19 large companies.

In addition to building the system, during that period a national electronic document issuing model was defined and changes were performed in the national tax legislation. Since the projects main goals were achieved, the project was considered as successful. (SEFAZ/SP, 2006)

One of the project's features is an intense collaboration among the federation units that participated and the Federal Revenue Service, in order to make a national standard viable. System information are exchanged online by tax administrations, through a program developed by State Tax Administration in Rio Grande do Sul called TED-DIST. Some State Tax Administrations, like SEFAZ/SP, are currently changing that scheme so as to use webservice, an inter-system communication standard based on internet protocol.

In spite of that, it's important to stress that some specific objectives were not achieved, such as the voluntary mass use strategy. The 200,000 electronic invoices issued in the state of São Paulo until June 2007 were still well below initial estimations, considering the project's infrastructure dimensioning to provide, still in the first semester of 2007, up to 10 million NF-e/month (SEFAZ/SP, 2006). However, in April 2007, a new mass use strategy was introduced, as the use of NF-e became mandatory from April 2008 to fuels and tobacco sectors, according to Protocol ICMS #10/07, changing Protocol ICMS #03/07, both from CONFAZ, the forum responsible for establishing tributary rules. Later changes to this Protocol ICMS established three new ways for the obligation to issue NF-e; from December 1, 2008, nine new economic sectors were obliged to issue NF-e; from April 1, 2009, it will be over twenty five sectors, and from September 1, 2009, fifty four new sectors in the economy. At the end of this period, it's estimated that issuing businesses will be more than 15,000 in the state of São Paulo alone.

For its innovative aspect and for benefits it proposed, NF-e project won e-Business Award's "Prêmio Padrão de Qualidade em B2B 2006" (B2B Quality Standard Award 2006) in the "National Electronic Commerce Incentive" category, as well as "Prêmio Mário Covas 2006" award in the "Public Resource Use Efficiency and Debureaucratizing" category, awarded by the state of São Paulo (SEFAZ/SP, 2007)

It's also important to stress that the name NF-e is inappropriately used sometimes by initiatives that don't eliminate the paper counterpart, but simply have the taxpayer forward the tax authority some kind of electronic document (SEFAZ/SP, 2006). One such project is Bahia's state government's Electronic Invoice for public procurement, also called Compra Legal, which, rather than eliminate paper, has in practice created a new obligation to the taxpayers who sell to the government, having them type the paper invoice information into the State Administration website.

This NF-e project is presented as a case study that is particularly relevant for two main reasons: its integration among multiple agents and its high management complexity. 2

Regarding its multi-agent integration, the project comprehends both federal and state tax administration in designing a national standard. So its management practices incorporate non-trivial issues in communication, priority and scope; for example, the fact that they're physically distant, yet demanding intensive communication in order to define new standards required a non-conventional communication management process. Moreover, these managers would gather at ENCAT, a relatively recent forum that was only created in 2001. So, systematic understanding of these groups' interaction on a national level through this forum, as well as its impact on the project development in São Paulo, are relevant aspects to the goals of this research.

The second aspect deals with the project complexity, from both technical and management perspectives. In the IT aspect, the project was innovative as it used technologies such as: digital certification, XML, webservices and online communication on a 24X7 basis. Identifying how this technical complexity was managed by São Paulo's group and even nationally standardized are interests of this research. As to management complexity, it's important to stress that NF-e has just composed the core of Federal Government's Programa de Aceleração do Crescimento (Growth Acceleration Program) – PAC (Governo Federal, 2007).

By the features presented above, the NF-e project can be considered, according Yin's definition (1994) for case studies, as a rare case, in which the difficulty to find it in other sources and the contribution from studying it justify it as unique case.

METHODOLOGY

Theoretical Model Applied to the Research

Initially, a bibliography gathering was made about the theory of ICT Project Management, focusing, whenever possible, the public sector.

A conceptual model was then produced based on PMBOK's nine knowledge areas (PMI, 2004 and PMI, 2006), as in Figure 1. The Government Extension to PMBOK (PMI, 2006) didn't change the processes themselves, but adequate them to the public sector reality.

From each process definition, this model enabled the Nota Fiscal Eletrônica (Electronic Invoice) project management analysis. It became possible then to identify how each of these processes was treated in the project's daily course. Analyses were also made on each of these activities' repercussion to the project and its influence over PMBOK management processes.

Data Collection Procedure

From among the features of study case, there's the possibility to collect data through multiple sources. Yin (1994) argues this is one of the strengths in this methodology, which has the ability to deal with a large variety of evidence, including documents, artifacts, interviews, observations and also participation in researched events. From the six possible evidence sources described by Yin (1994), this research has used two: documents and interviews.

Documents

Particularly in this study, the researcher's initial plan comprehended the project public document analysis. However, since the researcher is authorized by SEFAZ/SP to execute this research, the State Tax Administration internal documents were also made available, except that it could not incur in taxpayer tax secrecy break. So, as to the documental aspect, 41 project meeting minutes were collected, as well as 20 miscellaneous documents.

Interviews

Interaction with Nota Fiscal Eletrônica project participants who might provide support to the researcher's analysis and hypothesis was foreseen. To perform the interviews, a semi-structured form was designed, with a research basic protocol for each profile. Questions study a few of the interviewee's basic details (education, experience, project role), project history and management-related factors based on research conceptual model. So interviews were performed with 11 profession-

Figure 1. Conceptual model applied to the research. Source: PMI (2004) and PMI (2006).

PMBOK Knowledge Areas		
1. Integration Management	**2. Scope Management**	**3. Time Management**
1.1. Develop Project Charter 1.2. Develop Preliminary Project Scope Definition 1.3. Develop Project Management Plan 1.4. Direct and Manage Project Execution 1.5. Monitor and Control Project Work 1.6. Integrated Change Control 1.7. Close Project	2.1. Scope Planning 2.2. Scope Definition 2.3. Create WBS 2.4. Scope Verification 2.5. Scope Control	3.1. Activity Definition 3.2. Activity Sequencing 3.3. Activity Resource Estimating 3.4. Activity Duration Estimating 3.5. Schedule Development 3.6. Schedule Control
4.Cost Management	**5. Quality Management**	**6. HR Management**
4.1. Costs Estimating 4.2. Costs Budgeting 4.3. Costs Control	5.1. Quality Planning 5.2. Perform Quality Assurance 5.3. Perform Quality Control	6.1. Human Resources Planning 6.2. Acquire Project Team 6.3. Develop Project Team 6.4. Manage Project Team
7. Communication Management	**8. Risk Management**	**9. Procurement Management**
7.1. Communications Planning 7.2. Information Distribution 7.3. Performance Reporting 7.4. Manage Stakeholders	8.1. Risk Management Planning 8.2. Risk Identification 8.3. Qualitative Risk Analysis 8.4. Quantitative Risk Analysis 8.5. Risk Response Planning 8.6. Risk Monitoring and Control	9.1. Plan Purchase and Acquisitions 9.2. Plan Contracting 9.3. Request Vendor Responses 9.4. Select Sellers 9.5. Contract Administration 9.6. Contract Closure

als. The interviews sought to embrace the main participants in the project management model on a national level, with ENCAT's coordination, as well as the project management on São Paulo's level. In the first case, ENCAT coordinators, main participant states' representatives, a corporate representative and a bookkeeping representative were interviewed. The interviewees from SEFAZ/SP were the former State Tax Administration Coordinator in São Paulo, the Modernization Program Coordinator, the NF-e project leader and the system managers.

MANAGEMENT MODEL ANALYSIS THROUGH PMBOK PERSPECTIVE

This section presents the NF-e project's management model analysis based on each of PMBOK nine knowledge areas.

Integration Management

Neither in collected documents nor in the interviews was a formal management mechanism identified as having been used by ENCAT in NF-e project. Such absence reflects on defining both project management specific methodology and the necessary support to that use. Nowhere in the project were PMBOK-suggested practices found, such as: the use of a methodology, information systems for project management and experts to support project management. This lack of management structure is recognized Technical Coordinator of ENCAT:

We're really in want of a more scientific project management methodology. We don't project control with a specific tool, like an MS-Project and other tools. What we sought to use was the project management concept where we sought to integrate and document all decisions. [...] We have knowledge of PMBOK methodology, but we follow it in a much more informal way. Alvaro Bahia, ENCAT Coordinator and NF-e Project Leader in SEFAZ/BA.

In fact, ENCAT's conduction of NF-e project is recognized to be a relatively informal environment. That characteristic is considered by the interviewees as a favorable aspect to NF-e project's quick execution, since while project controls existed, they didn't require a great effort.

That very lack of formalism is indicated by some members as a necessity to be overcome by ENCAT, because of the harm they may cause to the project. In one interview, it was commented that no history of the project documented in details currently exists, and even ENCAT's management team don't have all the project's meeting minutes. Moreover, the decision process isn't explicit, nor is the teams' priority definition.

In summary, the lack of a formal methodology to conduct the NF-e project appears to have been overcome due to participants' high integration ability. PMBOK-suggested process group is eventually executed in a relatively informal way, yet with commitments being assumed and fulfilled by participants at each meeting. This may be observed through the meeting minutes, which contained, in their great majority: actions to be performed, a responsible person clearly defined and delivery date.

If the lack of formality might bring project conduction risks, on the other side it strengthened the need for an environment where participants' mutual trust would become established and would so remain, given the project teams' low turn-over. Another highly relevant aspect was the participation of companies incorporated into the project, and their contribution to enhancing management mechanisms. Tax administration's very model of organization in order to have a relationship with those companies appears to have improved based on the experiences of its own organization.

From the moment companies organized themselves, they forced the tax administration into organizing to a same extent, raising the project overall level. An example of this evolution was the definition that meetings with the companies should have a previously defined agenda. Another management mechanism was establishing a meeting with the states' technical teams the day before the meeting with the companies, in order for the tax administration to discuss and take one stand on themes to be discussed.

Scope Management

The definition of a single national scope was considered by interviewees a fundamental factor to implement the NF-e project. Its management, despite not having followed PMBOK's techniques, presented two quite noticeable stages. The first one refers to defining the initial scope of the project through multiple tax administrations with different interests. That stage was close, to some extent, to PMBOK's Scope Definition process (PMI, 2004). The second stage refers to the maintenance of that scope, including some specific mechanisms related with the Scope Control process.

The NF-e project initial scope definition occurred as a result of I ENAT, in July 2004, after a work group was created to deal with the subject. However, because SEFAZ-SP developed a model from the Chilean project, two possible scopes were created for the national project. Definition was given through the presentation of models at the XV ENCAT meeting, where tax administrators chose to adopt the model originated from the Chilean project.

As a result, even with no deliberate use of PMBOK guidelines, two techniques suggested by that methodology were used: Alternatives Identification and Stakeholder Analysis. When the groups presented different choices to tax coordinators, the project's sponsors on state tax administration level, the administrators' choice was based on a fairly comprehensive choice range. The fact that technical teams performed their preliminary studies about similar projects also brought in a wider range of alternatives presented. From the moment sponsors themselves chose one of the alternatives, understanding on desired scope was unified.

As for its change control, while there wasn't a formal process, the NF-e project availed itself of a few important instruments.

The first instrument was ENCAT's technical meetings. With the technical team assembled, the main deliberations on change needs were made from opinions that were heard about whether the subject was convenient to state representatives or not. Because states had already been investing in its construction, and they were pressed as to implementation deadlines, resistance to non-priority scope changes was strong.

Therefore, a couple of factors appear to have created an adequate change control balance. If state entities had a certain autonomy regarding their projects, on the other side, the project's scope was only one, and for this reason, their positions couldn't be unilateral. That promoted a discussion that regulated the volume of its alterations.

Another relevant instrument was the approval of a legislation before expected. The team's forecast for December 2005 was brought forward to September, due to a great arrangement among participant State Tax Administration Coordinators. Moreover, another relevant instrument was having its changes made formal through technical team meeting minutes. Records were being made of discussions that were normally not re-discussed in later meetings.

Thus the junction of these factors promoted a greater control over these changes, which made the project implementation viable, even with some delay, still in 2006.

Time Management

Although the national NF-e project conduction didn't have a formal methodology, its schedule was one of the documents treated with great attention by ENCAT's project team. Even though it wasn't formally structured with all of PMBOK-proposed stages (PMI, 2004), such as formal activity definition (Activity Definition), Activity Sequencing and resource quantity and profile estimating (Activity Resource Estimating), one may perceive that Activity Duration Estimating, Schedule Development and Schedule Control were ENCAT routine activities.

It's perceived, since the minute of the first meeting, that the project target regarding deadlines was already set by the high administration. The deadline senior managers initially set, January 2006, was considered by the very leader of project in São Paulo a non-realistic target, yet relevant to set the pace of the project.

One perceives from the minutes that schedule control was a practice in every technical meeting held by ENCAT. In spite of the great established targets in place (deadlines for legislation passing, project pilot implementation etc.), there's the schedule control at meetings, mainly about small activities agreed upon. Activities to be developed were being recorded into minutes, along with responsible person and deadline normally defined. Results were quarterly presented by the team at the ENCAT meeting with tax administration coordinators.

The national project schedule was normally a joint construction, as there were the interests of several states in implementing a common project. However, since each state had its own singularities as to priorities, structure and availability, each worked with own schedules, seeking to incorporate them into the project's schedule.

At times, some states took the initiative to execute stages having no consensus with the other entities; one example is the definition of the project's implementation in production, in September 2006, when the schedule had been having postponements, and systems of a few states were still presenting problems.

Cost Management

Because ENCAT is not a constituted entity with its own resource source, its structure is subject to the "good will" of participant entities to execute its works. Since no financial resources are directly bound to the project, expenses are born by participant tax administrations.

For example: the Tax Administration of the state of Rio Grande do Sul developed a system of file exchange among states and the Brazilian Federal Revenue Service; that very state tax administration has been providing the system's maintenance. As a result, despite being used by all federation units participating in the project, costs inherent to that product are exclusively incorporated into SEFAZ/RS' project costs. Consequently, there is no cost sharing structure in the project, which creates dependencies that increase its risk. In the example above, in case SEFAZ/RS decides to discontinue its tool for whatever reason, that decision will have a nationwide impact. Given public sector's greater political instability when compared with the private sector, that susceptiveness to change becomes a significant risk factor.

Nevertheless, it's necessary to note how the participants' commitment to project result seems to overcome that limitation, which is shown through the responsibilities assumed by tax administration representatives regarding project activities.

It's important to stress that, from the moment those costs become operational and involve higher sums, the states express the need to share costs. However, this subject still hasn't gone further in a more accelerated way, as it bumps into the cost sharing discussion.

Quality Management

No clearly defined products or processes were identified regarding NF-e national project quality management. With the use PMBOK's three processes as a reference, no Quality Planning, Quality Assurance or Quality Control systematized mechanisms were found, which resulted in processes with non-standardized quality, in spite of ENCAT's good environment to minimize such problems.

[Quality of products created] isn't uniform. Even because there aren't any ENCAT standards. Each organ has its own methodologies, standards and rules.

Yet, integration in an environment like ENCAT's, as well as experience exchange, also contributes to improve that. NF-e Project National level Team integrant.

Although there isn't a quality control standard in place for products created by NF-e project, whether for information systems built, for legislation passed or for lectures delivered, the project used a few documents in order to level participants' understanding. One such document was the Manual de Integração do Contribuinte (Taxpayer's Integration Guide), which detailed aspects of the information systems to be built. As a result, both companies and states managed to establish a minimum standard to build their systems. The guide featured, for example, all possible mistakes and respective messages, as well as communication protocol and digital signature technical specifications.

A contribution to production start security came also from the adoption of a "parallel" stage, so that companies would use tax administrations' system as though in production, but with no legal validity and no systematic changes made to paper invoice issuing. In this stage, many issues were minimized regarding companies' insecurity feeling as to tax administration system reliability, response time and support structure.

This mechanism made problem spotting relatively reliable, as more than one company was "testing" the tax administrations' system; in case one of them pointed a problem and others didn't, that would probably be attributed to the system of the company in question and not to the tax administration. On the contrary, if one company with the same system failed a transmission to the tax administration and others performed it successfully, the problem would probably be located in a certain state tax administration.

In addition, the companies themselves presented improvements to the quality management process. One company, for example, presented their performed tests report through a dashboard. This report was then included in the project quality management, and other companies also were suggested to produce a similar report.

In spite of all problems indicated, the final result was considered by interviewees to be generally fairly reasonable, mainly because of this rather close tax administration-company participation.

Human Resource Management

Since ENCAT has no budget of its own, no human resource contracting or training processes were identified on a national level. Therefore, this type of action was limited to each tax administration's local environment.

However, one aspect interviewees unanimously pointed as of high relevance to project success referred to leadership and management ability of state leaders in the project's first stage, especially project leaders in the states of Bahia, Rio Grande do Sul and São Paulo. That characteristic was stressed by interviewees through several different words: "technical capacity," "the will to do it," "interested," "negotiation ability," "dialogue ability," "competence of colleagues," among others.

As to PMBOK (PMI, 2004) aspect on segment specific feature, standards and regulation knowledge, it's important to stress that the state managers who were chosen to lead the project in their respective states were most usually public service career employees from the tax inspection sector, with a large professional experience in tax issues.

Still, in order to adequately know project specificities, these professionals first availed themselves of other tax administration experiences for comparative analysis. The São Paulo and Rio

Grande do Sul teams traveled to Chile in order to learn about "Factura Electrónica," a project that's conceptually similar to NF-e.

As to knowledge of project environment, tax inspection experience also appears to be an important aspect, as there is a dimensioned knowledge of cultural, social, political and legal impacts based on these managers' previous experiences. Therefore, they are senior professionals who had already experienced government transitions and technology project implementations.

As to the knowledge on the matter in general, all members had a college degree, although in curiously diverse areas, such as computer science (SEFAZ/Rio Grande Do Sul representative), business administration (SEFAZ/Bahia representative), law (SEFAZ –Federal Revenue Service), administration and engineering (SEFAZ/São Paulo). In addition, the project's main managers held a college degree in some area related with administration, i.e., they had management formal knowledge.

Finally, the interpersonal skills aspect interviewees referred was also found to be features of the leaders involved in the project.

In summary, the state leaders who participated in the project had, in the four expertise areas, capacity necessary to execute activities. In the majority of cases, that knowledge was previous, whereas in segment knowledge it was built in the course of the project through technical visits to existing projects.

Such leadership ability may be considered a project risk, as a certain dependency on people exists until today. However, this risk was minimized to a great extent, since the project's already become institutionalized in many aspects, for example: with production processes, instituted legal rules and defined responsibilities.

Communication Management

Because it's one of the critic processes when it incorporates multiple agents in one project, communication management was one of the most formal, PMBOK model-adhering (PMI, 2004) aspects .

At ENCAT, prior project communication channels used until then were adopted first: electronic messages (e-mail) were used and decisions were made formal through meeting minutes. Although there was no formal document in place to define the management model, nor the level of details as proposed in PMBOK's Communication Planning process, this subject was agreed upon among those who participated in the projects initial meetings.

E-mails were used for quick communication, and the purpose was to pass information and clarify sporadic doubts. In cases where more discussion was necessary, mainly to decide on complex subjects, channels chosen were personal attendance meetings or video-conferences. From 04-07-2005 to 03-28-2007, 24 personal attendance meetings were and 10 video-conferences were held, the later more intensely from February 2006. Interviewees consider this channel to be more object than personal attendance meeting, and so it was more intensely used in the project implementation stage.

One perceives that the states most involved in the project participated more often in personal attendance meetings, sending their representatives to practically all meetings, which was the case, for example, of Bahia (no absence) and São Paulo (four absences). One may also perceive that alignment meetings were held in order to provide other states the opportunity to participate in the project and to validate the conceptual model.

Due to technical project complexity, another very relevant channel was the Manual de Integração do Contribuinte (Taxpayer's Integration Guide) (ENCAT, 2006). That document, designed by tax administrations, its first version becoming available in January 2006, concentrated the project's technical details.

For the PMBOK-described Information Distribution process, the above described channels were generally used; however, there is no consensus among the interviewee as to the communication mechanism. Some interviewees believe the communication process was adequate and met the project need, while the bigger part believes there are faults, especially in distributing the information, with many cases being reported of people not getting information. In addition, the project legislation's very publication was an important instrument to formally divulge NF-e project business rules. Through legislation, in spite of its technical language, the electronic tax document's concepts are presented, as well as its legal validity and use conditions.

However, the lack of a project document storing repository was identified. Apparently, that problem was minimized for public documents due to a project's national website that was available, in which there is miscellaneous subject document section comprehending: Legislation, Taxpayer's Integration Guide, XML schemes etc.

Performance Reporting and Stakeholder Management processes were normally made by the project's Technical Coordinator directly with ENCAT's Technical Coordinator, who in turn kept tax administrators informed at ENCAT quarterly meetings. State leaders also updated their own coordinator and State Tax Administrators about work developments, as well as the other technical teams.

Risk Management

No documents were identified that could demonstrate a formal risk analysis in the national environment. However, it can be perceived that risk identification was a concern for participant managers. In one meeting minute, for example, ENCAT's very General Coordinator mentions the necessity to define project risks, and signals that consultants might be occasionally contracted for that purpose.

Apparently, main risks were identified not in a structured way, but from leaders' project perception at some of the meetings. The Chilean model itself had already signaled that the team should pay attention to the graphic sector, one that was harmed in process of replacing paper invoice for electronic documents. Another concern was that NF-e was seen by companies in a negative way, as an additional tax control, both to issuing parties and buyers.

Another risk identified by the project team, one that was also exposed by the SEFAZ/SP's leader, refers to federal and state government transitions that took place in 2006 and 2007 with the possible change of heads of State Tax Administration, tax administrators and the ENCAT coordination itself. Companies helped the administration about that risk, through letters sent to new heads of state tax administration suggesting NF-e technical teams to be maintained as well as Eduardo Almeida as ENCAT's General Coordinator.

It can be perceived that project risks were dealt with in the course of the project generally with attitudes that sought to anticipate problems. Although companies joined in at a time when project initial risks had already been identified, they also collaborated, pointing risk spots which then had not been identified. Again, companies' participation also helped in a project management aspect.

It is possible to perceive that the project team also identified the failure of the project as a risk for their own institutions, as described below.

If states weren't mature enough to create national range solutions [...] perhaps they would be doomed, or maybe we are already, with a Congress process coming 'down from above' to change ICMS (a tax on gods and services that's charged by each Brazilian State – T.N). Either states will be mature enough as to seat, talk and create solutions to make the companies' life easier, or society as a whole and companies will press the Congress and the politicians to change this tax administration process, which is what's already happening today. Ricardo Neves, Former Project NF-e Leader at SEFAZ/RS.

Procurement Management

Unlike a company or a public entity, ENCAT is not a legal person. Consequently, it doesn't have a budget nor does it perform procurement or contracting processes in its own name.

Therefore, in ENCAT's case, there is no need to think of structured contracting processes, or its comparison with PMBOK models. However, the interviewees recognize that such impossibility leads to limitations to ENCAT's own management structure.

The solution doesn't appear to be simple, though, since ENCAT's constitution as a person under public or private law tends to bring in complications to the proposed management model, based on state sharing. That lack of structure is compensated by the solution sharing culture promoted by ENCAT itself.

Consolidation of Main Results

It's possible to summarize the main aspects identified in the project related with PMBOK through table 1, shown below.

CONCLUSION

Findings

It's interesting to observe that NF-e project routinely availed itself of two complementary management models. The first one, coordinated

Table 1. Summary of NF-e project main aspects through a PMBOK perspective. Source: the authors.

PMBOK Areas	NF-e Project main aspects
Integration Management	• No specific methodology used; • Informal environment; • Great participant integration, resulting from high level of trust and commitment, as well as low *turn-over*; • Companies favored management mechanism;
Scope Management	• Project scope definition through alternative choosing; • Made by team who developed knowledge about the subject; • Focus on "achievable" scope development; • Control through meetings and narrow deadlines; • Control strengthened through legislation publication
Time Management	• Deadlines as a main project aspect; • Deadline control through technical coordinator each meeting; • Actions of entities not backed by the rest created deadline competitiveness.
Cost Management	• Absence of ENCAT own resources; • Non-existing cost sharing in ENCAT's structure; • Reliance on each entity's "good will" to build common solutions;
Quality Management	• In want of formal quality control mechanisms; • Usage of a few documents to level knowledge and quality standards; • Companies collaborated to testing diverse entities' systems; • "Parallel" stage usage made system-adequate testing viable.
Human Resources Management	• Non-existing ENCAT Human Resource contracting process; • Strong local leadership, often directly chosen by higher administration.
Communication Management	• Wide range of channels used (meetings, video-conferences, e-mails, documents, portal, etc); • Large number of personal attendance meetings; • Legislation used also as a communication mechanism.
Risk Management	• Non-structured way identified; • Dealt with during project course.
Procurement Management	• Because it hasn't its own budget, ENCAT doesn't procure directly; • Resource sharing culture minimizes that need.

by ENCAT, used more informal instruments to manage their activities. The management model was mainly directed to communication management (very relevant to integrate so many people involved), scope management (capital to establish a single national scope) and time management (to conciliate project activities). Based on the research performed, it's possible to perceive that other knowledge areas were not despised, yet they were less intensively treated.

As for the management model managed by each State Tax Administration, the focus was building information systems and acquiring goods and services necessary to implement the project. So, focus was placed on the following knowledge areas: procurement management (materials acquisition according to formal procedures), cost management (in order to measure project costs) and human resource management (in order to contract resources to develop systems or to build internal teams). Time management, although it was focused in the national management model, it was also locally focused in an intensive way, given time milestones were agreed upon at ENCAT, yet their details were set by each State Tax Administration.

Therefore, it is possible to perceive that models are complementary, with an emphasis in the positive aspect of each of them. Weaknesses occasionally identified were sorted out through companies' participation, who, in a short time, promoted significant changes to the management model, as described in the previous chapter. It was possible to observe that such improvements took place mainly in those aspects related with NF-e management model on a national level, since Scope, Communications, Time, Quality, Integration and Risk were subjects of their direct interest. The other subjects, such as cost, HR and procurement, were particular to each participant entity, including companies.

This management model complementariness is perceived in a consolidated way through figure 2 below:

It is also interesting to notice that the design of NF-e has adopted many of interoperability best practices advocated by ECLAC (2007), in its White Paper on Electronic Government Interoperability for Latin America and Caribbean, among them:

- **Principle of Legality:** NF-e established national and regional legal frameworks
- **Transversal Services and Applications:** NF-e established communication among different departments, being transparent for the user citizen
- **Homogeneous quality for services:** Service were established for operating 24x7 nationwide
- **Evolutionary Adoption Levels:** SEFAZ Virtual was the solution adopted by those states that did not have interest or infrastructure in the beginning
- **Definition and adoption of technical standards and open source for a variety of platforms:** The adoption of XML standard adheres to market standards
- **Lack of centralization:** Institutional autonomy was maintained for each participant member of the federation
- **Cooperation and reuse:** Best practices for reuse of solutions were adopted
- **Digital certification use:** Digital signature is the legal basis for the transactions made

Thus, the adherence of the NF-e with interoperability standards can contribute to a greater expansion of the model to other Latin American countries, in order to exchange international trade transactions.

RECOMMENDATIONS

The NF-e project, in spite of all positive aspects that were identified in the initiative, still wants improvements in some areas, which may be considered the contribution of this research to the

Figure 2. Management model of NF-e project. Source: the authors

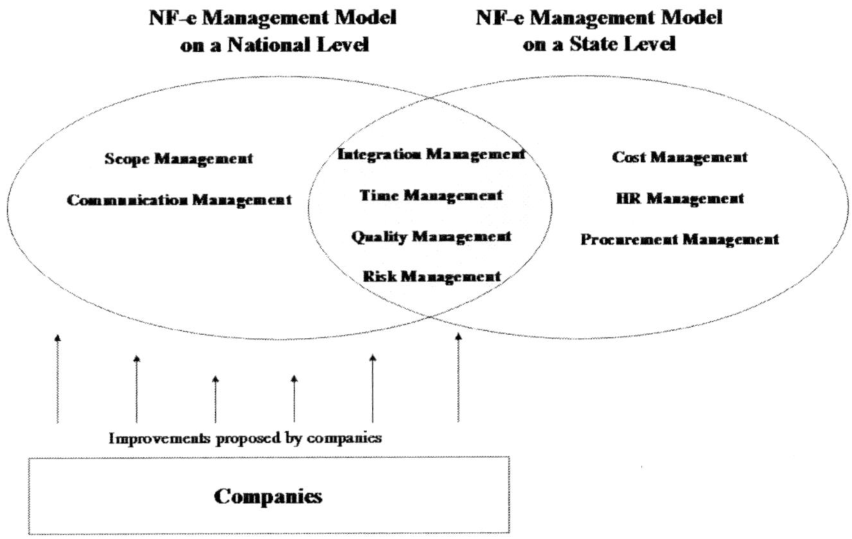

project in question, as well as to other projects that have similar features.

In the previous chapters, NF-e project management structuring was observed to have been made based on a model with great dependency on project leaderships and team. As indicated by interviewees themselves, the difficulty to form teams with same capability and integration as verified in NF-e's current team may result in a greater slowness or in the unfeasibility of other projects with this size and importance. ENCAT's very management model base, leadership created through consensus and a trust relationship, is still dependent on its General Coordinator figure. Hence, in order to minimize that deficiency, it may be useful to professionalize tax administration's teams who participate in government strategic projects. Through professionalizing, not only the idea of PMBOK-proposed expertise areas, but also availability and support for such activities are infused.

Another aspect that's also relevant about NF-e project management model is the lack of previous structuring to its management processes. Many actions executed, documents built and processes established were defined along the course of the project, when situations emerged to teams. Because it included experienced people, the team's adaptation capability enabled course corrections and process improvements that resulted in a performance regarded as successful. Therefore, it may be useful to design a minimum process set that allows for smaller dependency on people, so as to allow more projects to be simultaneously executed with uniform results. Processes shouldn't be expected to replace people, quite the contrary. As stated by Pollit (2003, p.41) "structures by themselves don't promote cultural changes, since processes and attitudes also require attention," which indicates the need for a joint action between people and processes. What is expected is that people with less experience in the management model could also achieve satisfactory results.

At last, NF-e project structure limitation is its dependency on a management model that's exclusively focused on sharing solutions among participants. When faced with situations where the investment to be performed is more substantial, as well as operational cost situations, this factor brings obstacles to more project-adequate solutions. Difficulty to define the way costs should be divided among entities has hampered the adoption of solutions on a national scale. ENCAT's very management structure, since it doesn't have its own budget to contract human resources fully dedicated to the organization, is another limiting factor to management model growth and replication. Therefore, it may be useful to create a national structure to support inter-agency project execution. Facility and quickness to that cost division would favor the appearance of initiatives with greater sharing, both in solutions and infra-structure, such as systems, physical facilities, expertise, operational activities, etc.

FUTURE RESEARCH

Given the nature of this research, it is possible to consider a number of future studies to broaden and deepen the analysis on the project management model employed by the NF-e implementation. Some questions of interest are:

1. What are the similarities and differences between the project management model adopted at the NF-e implementation and other models not related to tax services inside the public sector?
2. Which management models for e-government are more effective among the many used in projects the public sector?
3. What are the critical success factors in e-government initiatives in Latin America when compared with projects in other regions? What are the main reasons for failure?

REFERENCES

Bresser Pereira, L. C. (2005). *Construção do Estado e Administração Pública: uma abordagem histórica.* Relatório do GV Pesquisa, Fundação Getulio Vargas de São Paulo – EAESP/FGV, 27.

Carayannis, E., Kwak, Y., & Anbari, F. (2003). *The Story of Managing Projects: an interdisciplinary approach.* Quorum Books.

CEPAL. (2007) Libro Blanco de interoperabilidad de gobierno electrónico para América Latina y el Caribe. Versión 3.0, Septiembre.

Charbonneau, S. (2004). *Software Project Management: A Mapping between RUP and the PMBOK.* Rational Software. Retrieved fromhttp://www-128.ibm.com/developerworks/ rational/library/4721.html

ENCAT. (2006). *Manual de Integração do Contribuinte: padrões técnicos de comunicação. Encontro Nacional dos Coordenadores e Administradores Tributários Estaduais* – ENCAT. Available at www.nfe.fazenda.gov.br.

Gil-Garcia, J. R., & PARDO. (2005). T. A. *E-Government success factors: Mapping practical tools to theoretical foundations. Government Information Quarterly, 22*(2), 187–216. doi:10.1016/j.giq.2005.02.001

GOVERNO FEDERAL. (2007). *PAC: Programa de Aceleração do Crescimento.* Sítio Institucional. Retrieved fromhttp://www.brasil.gov.br/pac

Hazlett, S., & Hill, F. (2003). E-Government: the realities of using IT to transform the public sector. *Managing Service Quality, 13*(6), 445–452. doi:10.1108/09604520310506504

Heeks, R. (2003). *Most eGovernment-for-Development Projects Fail: How Can Risks be Reduced?* IDPM i-Government Working Paper no.14, University of Manchester, UK.

Heeks, R. (2004). *eGovernment as a Carrier of Context*. Institute for Development Policy and Management, iGovernment Series, Working paper, no. 15. Retrieved fromhttp://idpm.man.ac.uk/publications/wp/igov/index.shtml

HO. (2002). A. *Reinventing Local Governments and the E-Government Initiative. Public Administration Review*, *62*(4), 434–444. doi:10.1111/0033-3352.00197

Medeiros, P. H. (2004). *E-gov no Brasil: aspectos institucionais e reflexos na governança.* Dissertação de mestrado em Administração da Universidade de Brasília (UNB), Brasília.

Neto, J. J., & Bocoli, F. S. (2003) *Sucesso*^{SW}*=CMM*$_2$ *+ PMBOK*. PMI Journal, Seção Rio Grande do Sul, 5 (May).

Nidumolu, S., Goodman, S. E., Vogel, D., & Danowitz, A. K. (1996). Information technology for local administration support: The governorates project in Egypt. *MIS Quarterly*, *20*(2), 197. doi:10.2307/249478

Osborne, D., & Gaebler,T. (1992). *Reinventing Government – How the Entrepreneurial Spirit is Transforming the Public Sector*. Addison-Wesley Publ. Co.

PMI. (2004). *A Guide to the Project Management Body of Knowledge: PMBOK Guide Third Edition*. Project Management Institute Inc., Pennsylvania, USA.

PMI. (2006). *Government Extension to the PMBOK Guide Third Edition*. Project Management Institute Inc., Pennsylvania, USA.

POLLIT. (2003). C. *Joined-up Government: a survey. Political Studies Review*, *1*, 34–49. doi:10.1111/1478-9299.00004

Reinhard, N., & Dias, I. (2005) Categorization of e-Gov initiatives: a comparison of three perspectives. *X Congreso Internacional del CLAD sobre la Reforma del Estado y de la Administración Pública*, Santiago, Chile, 18 - 21 Oct.

SEFAZ/SP. (2006). *Prêmio Mário Covas: Nota Fiscal Eletrônica, a modernização do fisco à serviço da sociedade*. Secretaria da Fazenda do Estado de São Paulo. Internal Document.

SEFAZ/SP. (2007). *Projeto Nacional da Nota Fiscal Eletrônica, NF-e: resultados obtidos em 2006*. Secretaria da Fazenda do Estado de São Paulo, 2007. Internal Document.

Sherer, W., & Thrasher, S. (2005). *Contrasting CMMI and the PMBOK.* CMMI Technology Conference & User Group, November.

Skelcher, C. (1992), *Managing for Service Quality*, Longman, Harlow.

Yin, R. K. (1994), *Case Study Research, Design and Methods (2nd ed.)*, Sage Publications, Beverly Hills, CA.

ENDNOTES

1 EGovernment for Development Information Exchange is a project coordinated by the Institute for Development Policy and Management at University of Manchester and it consists of an on-line discussion network on e-Gov in developing countries

2 SEFAZ is the acronym for *Secretaria da Fazenda*, the State Tax Authority in each Brazilian State.

Chapter 10
Beyond E-Procurement:
A Framework to Develop E-Government Services for Small and Medium Enterprises (SMES)

José-Rodrigo Córdoba-Pachón
University of London, UK

Juan-Gabriel Cegarra-Navarro
Universidad Politécnica de Cartagena, Spain

ABSTRACT

In this chapter the authors develop a framework to allow them to facilitate further development of e-government services to small and medium enterprises (SMEs). It is inspired in the idea that benefits for SMEs can be derived if we consider more fully e-government services as sources of SME knowledge. Benefits obtained by the use of these services can be enhanced in practice with the help of systems thinking and systems methodology use. Although the authors do not claim that knowledge is the only vehicle that SMEs could use to improve their situation, they suggest that awareness on how SMEs renew their knowledge using e-government as a source is an important element. In other words, by promoting the idea that SMES can become more aware of their external environment and respond to them, they will benefit themselves and others.

INTRODUCTION

There is little doubt nowadays about the worldwide importance of knowledge as an intangible but valuable asset to support business development. This assertion spans across different types of organisations including the Small and Medium Enterprise (SME) sector (Sparrow, 2005; Wong & Aspinwall, 2005). Although in this sector knowledge management has been seen as as a 'foreign' type of practice that pertains more to bigger companies (Riquelme, 2002), many SMEs have invested in technologies (computers, internet) to enhance their capabilities to absorb, capture and use knowledge to enhance their competitiveness (Costa, 2001; Jeffcoate, Chappell, & Feindt, 2000). SMEs have also the possibility of widening the scope of of their markets, their products and their networks if they are able to codify and integrate knowledge that could be anywhere. With existing business models changing into loose and

DOI: 10.4018/978-1-60566-860-4.ch010

dynamic alliances (Tapscott & Williams, 2007), SMEs could provide services and products anywhere in the world if they knew who needs what they are able to provide.

Whilst SMEs strive to assess and incorporate the value of technologies, partnerships and alliances, other external organisations like local governments have similar concerns about delivering online services to SME that they could use effectively. The benefits of using and integrating such external sources by SMEs *as sources of knowledge* have not been fully assessed. Models of e-government portray the development of one-stop shop portals to enable administrative and commercially oriented services (Ballestero-Diaz & Font-Jaume, 2004). In portals companies can find information and more recently can participate in providing services to others in what has been called e-procurement (Ballestero-Diaz & Font-Jaume, 2004; Locke, 2006). However it is not clear how such portals are to benefit more directly *the inside of* small and medium organisations, their activities and ultimately their survival in a competitive environment.

The availability of a number of portals to facilitate procurement of services is growing and companies find that they can use these portals to perform a number of administrative transactions. Whilst these portals seem to streamline process of tendering[1], we believe that their portfolio of services could be extended if we knew how else they could benefit SMEs apart from automating the loading, consultation of information and follow up of tendering contracts.

In this chapter we develop a framework to allow us to facilitate further development of e-government services to SMEs. It is inspired in the idea that benefits for SMEs can be derived if we consider more fully e-government services as sources of SME knowledge (Cegarra, Dewhurst, & Briones-Peñalver, 2007; Locke, 2006). Benefits obtained by the use of these services can be enhanced in practice with the help of systems thinking and systems methodology use. Although we do not claim that knowledge is the only vehicle that SMEs could use to improve their situation, we suggest that awareness on how SMEs renew their knowledge using e-government as a source is an important element. In other words, by promoting the idea that SMES can become more aware of their external environment and respond to them, they will benefit themselves and others. This is particularly important in the current economic climate which requires SMEs to be more flexible, inventive and 'daring' than ever so as not to depend on protection from bigger companies or governments worldwide.

To build our framework we initially make use of concepts of 1) Organisational Memory (OM) and 2) Exploitation of Knowledge (EK). We then suggest that the development of these two concepts in practice can help SMEs to convert information related to e-government services into knowledge. Moreover, develolping such concepts can be helped by using a variety of systems methodologies. In this way we expect to help SMEs and also governments by enabling joined up thinking, cooperation and collaboration, in other words to improve existing relationships to give a variety of benefits to parties. Results from an initial empirical investigation being oriented by our framework suggest that there is a need for fostering joint design opportunities for e-government services.

We start our chapter by providing a context for the use of e-government services. We then review in detail the concepts of organisational memory and exploitation of knowledge. We relate them to to the available types of e-government services so as to foster cooperative thinking between SMEs, government actors and other stakeholders. We conclude by highlighting some managerial implications of the use of this framework for future developments in e-government services for SMEs.

SMES IN THE INFORMATION SOCIETY

With the emergence of a network paradigm in society (Castells, 2001), organisations like SMEs are part of chains oriented to make the flow of materials and information more efficient across the globe (Christopher, 2005). Regardless of their physical location, SMEs can provide bigger businesses with expertise and knowledge to increase productivity whilst enabling businesses to develop into new areas and markets. They also provide flexibility and innovation capabilities, and produce solutions for problems which companies do not have the required expertise in house (Tapscott & Williams, 2007). The internet and other technologies enable them to exchange information, to search for opportunities and provide solutions to emerging problems elsewhere.

Currently SMEs account for a significant percentage of countries' GDP in many countries. They also constitute a sector that supports employment either formally or informally. In the current global situation, SMEs have a new set of challenges, given that the businesses they serve could decide to reduce the scale of their operations or adopt standards that SMEs need to comply to continue being in business. These and other challenges pose opportunities for SMEs but also threats if they cannot get adequate resources and training to be able to develop change (Storey, 2004).

To address these challenges, online technologies have been employed as they can provide efficiencies in costs and quality. Electronic commerce (E-commerce) has enabled SMEs to address the issues of flexibility and scope of their operations in an efficient and cost-effective way (Costa, 2001; Jeffcoate et al., 2000). E-commerce integrates processes (e.g. purchase, payment, delivery) through the internet and combines other types of systems (e.g. electronic identification or verification; procurement) to support core business operations. With electronic commerce, companies could improve their value proposal to customers by streamlining and connecting activities through networks of suppliers, manufacturers, distributors and end customers (Christopher, 2005).

The success of e-commerce in industrial sectors has also meant that it is now being imported into government (Ballestero-Diaz & Font-Jaume, 2004). Technologies are being customised and applied in government affairs so that they support the delivery of government information and services (Dunleavy, Margetts, Bastow, & Tinkler, 2006; Locke, 2006). There are differences though in the ways in which e-commerce technologies are to be adopted in organisations. Their use becomes a form of 'public trade' that needs to be socially inclusive, more structurally dependent (on government organisations) and accountable by the public as a main shareholder (Carter & Bélanger, 2005; Grimsley & Meehan, 2007). The purpose is twofold: To achieve efficiencies in service delivery, and and to contribute to transformations in government institutions (Ballestero-Diaz & Font-Jaume, 2004). By fulfilling these two purposes, it is expected that e-government systems will empower individuals through facilitating their access to information and decision-making (Irani, Love, Elliman, Jones, & Themistocleus, 2005; Tan & Pan, 2003), and facilitate more proactive interactions between them and their governments (Banister, 2004; Criado & Ramilo, 2003; Grimsley & Meehan, 2007; Norris & Moon, 2005; West, 2004).

MAIN FOCUS OF THE CHAPTER

E-Government and SMEs

In principle E-government systems can be conceived of as inter-organisational systems, which could help different organisations to achieve their strategic aims and serve well their customers (Galliers, 1999; Ward & Griffiths, 2002). SMEs are part of the general public and so should be able to have more control and influence over the shaping and nature of e-government services, so that their perception of the public 'good' could also be enhanced (Grimsley & Meehan, 2007; Phang, Sutanto, Kankanhalli, Tan, & Hock-Hai, 2006). Moreover, they should be able to enhance their own activities and processes of decision making as a way of empowering them to become better players in the information society.

What we find typically on an e-government portal is a one stop shops with electronic commerce facilities which in principle could help governments and users transform public service delivery via electronic sharing of knowledge among public and private partners (Jeffcoate et al., 2000). In portals, businesses (including SMEs) can access and share information, get training and employment resources, seek business opportunities, collaborate with other organisations and register in electronic business directories. An example is the city of Newcastle's city portal (www.newcastle.com, accessed February 2009); it has business section which offers guidance information for council services like advertising, business rates, support and advice, commercial property, commercial waste, grants, planning, trading standards and the like. The structure of the portal is similar to many other city council ones, in which service information predominates at the expense of more interactive online-services. In these main portals though, SMEs can pay social security contributions from employees and taxes. They can register legally as firms, provide custom declarations and access public procurements (Rambøll Management, 2004).

According to Locke (2006), of properly conceived of, these e-government portals can often help SMEs to quickly acquire information (business-related or environmental) and therefore save them valuable time, contribute to their intelligence generation, generation of new business activities (i.e. profit). This could to some extent to improve SME's understanding of societal issues (e.g. regulatory, environmental) which could affect their operation. If these benefits are accrued, organisations and people in general could in turn contribute to the enhancement of their credibility and trust in their governments in the long-term (Carter & Bélanger, 2005; West, 2004). Online facilities for business transactions (i.e. tax payments, licences issuing, and procurement as explained in the above figure) can contribute to have more transparent, efficient, reliable and ultimately better services (Mincomunicaciones, 2007).

More dedicated sites that SMEs are using are those called e-procurement portals (Ballestero-Diaz & Font-Jaume, 2004). These enable SMEs to register themselves as potential service providers; bid for funding (provision of products / services) contracts; follow up the tendering process and if successful, prepare, send and receive contracts. A generic model of e-procurement is shown in the following figure (Figure 1).

In these portals, what we can see is that 'trading' aspect is the focus of attention[2], and it is assumed (so far) that SMEs are able to produce and deliver if they are awarded a contract. What is far from clear is how SMEs go about challenges of use and incorporation in their own activities. At a general level, these challenges include overcoming barriers related to perceived systems usefulness and ease of use (Grimsley & Meehan, 2007; Phang et al., 2006), overcoming unidirectionality in communication (Criado & Ramilo, 2003) and ensurig that electronic services are secure to access and use (Ballestero-Diaz & Font-Jaume, 2004). At a more deeper level of analysis of e-government portals, their services and their management, there is a more complex set of challenges that can be addressed to transform relationships between individuals and the state (Dunleavy et al., 2006). These include lack of trust in government-related institutions (Carter & Bélanger, 2005; West, 2004), or enabling a harmonious and co-operative degree of information sharing between local and national institutional actors (Heeks, 2006). The complexity of challenges (functionally and beyond) lead us to conceive of the importance of planning and designing adequate transition plans to move

Figure 1. A generic model of e-procurement

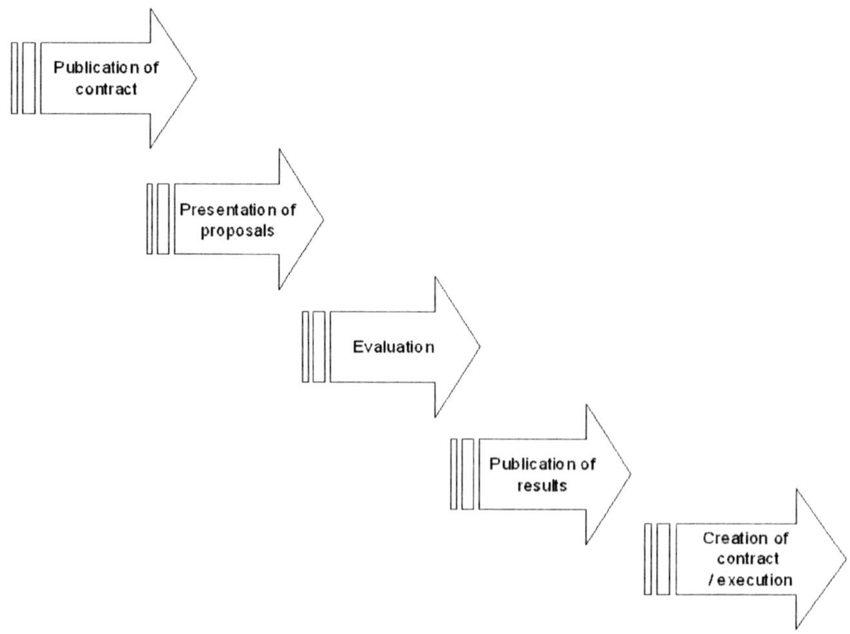

e-government services from just being 'highway billboards" to become transactional facilities and beyond that, interactive dialogue spaces with the public (Banister, 2004; Locke, 2006; Norris & Moon, 2005; West, 2004).

Transition would also require overcoming perceptions that although e-government has been benefitial for administrative government staff, benefits on cost reduction and efficiencies still remain to be achieved (Jae-Moon, 2002). For SMEs as actors in the e-government service provision process, they are users[3] *and* active shapers of such services. Adequate transition for them would require not only ensuring that e-government services respond to various types of SMEs and their needs (Locke, 2006), or considering the politics that need to be dealt with to facilitate exchange of knowledge (Heeks, 2006; Jeffcoate et al., 2000). It would also require adequate definition of services and management so that SMEs get support and participate in the transfer of skills and knowledge necessary to enabling their uptake of services, with a view of continuously providing them with service benefits and opportunities. In this way, we think we could better assess how SMEs could make use of e-government services by incorporating them as a kind of internet-based technology to improve their own business activities in different contexts (Costa, 2001; Jeffcoate et al., 2000), and then inform the definition and implementation of such services.

Figure 2 provides a synopsis of the arguments above. In the next sections of the chapter we present the concepts of organisational memory (OM) and the exploitation of knowledge (EK). These will allow us to link SME learning with e-government services as external sources of learning.

Organisational Memory (OM) and the Exploitation of Knowledge (EK)

We now use two concepts that will allow us to assess the value of e-services being provided to SMEs. These concepts are *organisational memory* and *exploitation of knowledge*. We choose them for two reasons. First, they allow us to assess and

Figure 2. Enhancing e-government services for SMEs

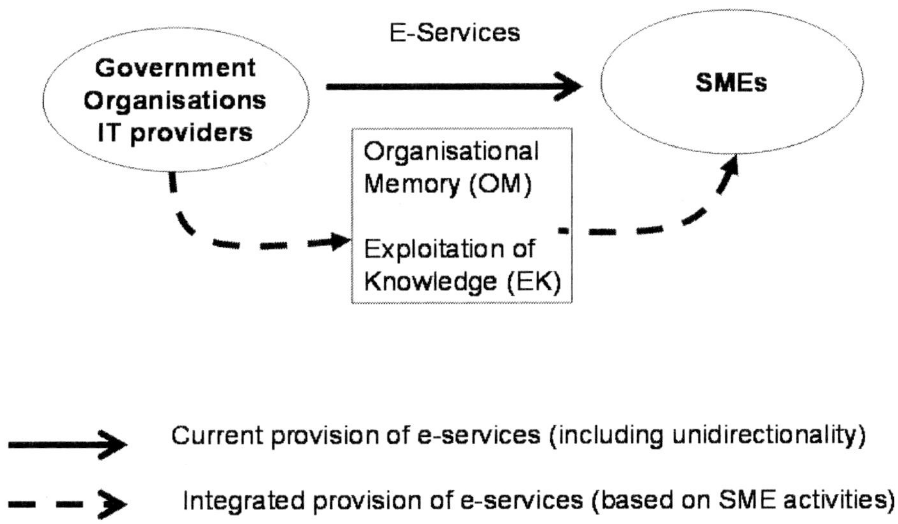

improve the use of knowledge by SMEs and their processes of learning. Second, they allow us to relate current e-services provision to SME activities by considering the former as external sources of knowledge to be used and exploited.

'Organisational memory' (hereafter OM) is a shared and tacit cognitive map, a kind of collective mental model (Senge, 1999) that is used to guide decision making (Megill, 1997). There are many (and occasionally competing) definitions of OM, so we will start by briefly describing its usefulness and move on to detail OM general features. Megill (1997) for instance, defines OM as all active and historical information about an organisation that is worth sharing, managing and preserving for later reuse. OM is also called 'corporate knowledge' by Hamel and Prahalad (1994), as it refers to the repository where knowledge that has been learnt is stored for future use and advantage. Davenport and Grover (2001) confirm the usefulness of OM by asserting that an organisation's long-term success and growth become dependent on the successful use of its OM across its business processes.

OM includes not only 'hard-information' e.g. numbers, facts, words and figures but also 'soft-information', that is, individual or social information with meaning e.g. expertise, experiences, anecdotes, critical incidents, stories and details about strategic and operative decisions. Selnes and Sallis (2003), for instance, suggest that relationship memory includes idiosyncratic routines in the form of encoded formal and informal procedures and scripts including how the parties have learned to do things. The goal of soft-OM is that all members of the organisation are aware of where the useful complementary abilities (e.g., who knows what? Who can help with that? Who can exploit new information?) reside. In this regard, Walsh and Ungson (1991) assert that soft-OM is 'represented' by many diverse aspects of an organization including for example: the organization's culture, transformations (production processes and work procedures), structure (formal organizational roles), ecology (physical work setting) and information archives (both internal and external to the organization).

In SMEs, OM can take a variety of forms including operational systems, organisational structures and daily routines. The most obvious structures for encoding these memories include information systems such as corporate manuals, databases, filing systems, etc. These systems are continually being updated and analysed and are thus capable of generating new streams of information, thereby expanding knowledge (Zuboff, 1988). Existing organisational memory (OM) on e-government can be seen as usually based on what has worked in the past, in other words what has been proved to be (un) successful (e.g. face-to-face government and related affairs with government institutions). This type of knowledge is useful when it allows for its practical exploitation, what Nonaka and Takeuchi (1995) call transforming tacit knowledge into explicit knowledge.

SMEs have a variety of purposes when engaging in knowledge management initiatives (Sparrow, 2005), and their knowledge management strategies go through different stages of formality and informality (Edvardsson, 2006; Sparrow, 2005). Those SMEs with a view of using knowledge as a collective resource to inform their business decisions engage in communities of practice and best practice networks, in other words they *externally* acquire knowledge via "learning from partners/ alliances, networks and trade bodies" (Sparrow, 2005:142). From the possibilities and benefits given by knowledge management practices, it can be said that developing SMEs organisational memory (OM) can be enhanced by acquiring and incorporating 'know-how' that resides elsewhere (Nonaka & Takeuchi, 1995), and use existing informal knowledge sharing strategies to complement more formal strategies (Edvardsson, 2006). In the case of e-government services, their use can become a valuable source of OM acquisition and learning. SMEs could incorporate e-government knowledge into their learning processes by engaging with external groups that have it or are involved in its creation.

However, as Sparrow (2005) argues, tacit knowledge might be difficult to elicit, renew or update in SMEs because it operates unconsciously and is already embedded in routines and ways of thinking. Sparrow (2005) also argues that SMEs are these routines are often influenced by lack of resources or the owner management styles. Moreover, development of knowledge in SMEs via formal training methods has not yet proven to produce positive results in performance (Storey, 2004). This consideration can help us explaining why certain e-government websites have difficulties in getting SMEs formally engaged in using e-services. It might the case that they offer little possibilities for the integration of sources of knowledge into SME business processes, or that they do not accommodate flexibility to incorporate 'secondary' or 'informal' sources of knowledge.

Therefore in order to develop OM capabilities in SMEs via e-services, more informal mechanisms, secondary sources of information (i.e. trade journals, sector research, conferences and professional magazines) or personal contacts (with suppliers, customers, or bank employees) (Smeltzer, Fann, & Nikolaïssen, 1988) rather than solely 'formal', ICT-based or more expensive mechanisms need to be considered. Implementation of e-services should then pay less attention to 'formal' knowledge (i.e. laws, regulations, plans, computer systems) and more to alternative sources of knowledge (i.e. experiences with the laws and regulations; word of mouth, networks of contacts) to enhance SME OM capability. Ideally, these resources could also be integrated into learning processes in SMEs that take into account individual and group understandings of the organisation in 'knowledge' terms (Sparrow, 2005). Such integration of knowledge will also help those responsible for e-government implementations at the government end to get customer feedback and act upon it (Irani et al., 2005).

As much as developing organisational memory (OM) can help organisations integrate its various sources of knowledge, by exploiting it organisations can gain long-term advantages (Davenport, 2001; Nonaka & Takeuchi, 1995). March (1991), Nooteboom (1999) and Volberda and Lewin (2003), among others, introduce the dichotomy of *exploration – exploitation* in organizational learning. They call learning from direct experience 'exploration' (hereafter EK). This includes things captured by terms such as variation, risk taking, experimentation, play, flexibility, discovery, and innovation. The 'exploitation' process includes refinements, choice, production, efficiency, selection, implementation or execution.

According to Carroll (1998), Exploitation of knowledge (EK) supports learning because it reduces uncertainty. It tells employees about their learning – what is working (do more of this) and what is not (do less of this). Carroll (1998) combines four processes of organizational learning that form a EK cycle: (1) 'observing' (i.e. noticing, attending, heeding, tracking); (2) 'reflecting' (i.e. analysing, interpreting, diagnosing); (3) 'creating' (i.e. imagining, designing, planning, deciding); and (4) 'acting' (i.e. implementing, doing, testing), where acting affects observing, and so forth. Carroll (1998) also claims that EK takes place through various kinds of work activities, e.g. meetings, peer visits and exchanges of best practices.

As noted above, EK can be fostered by *re-arranging* existing OM sources. It can also be fostered by making existing knowledge more explicit (including knowledge external to an organisation) (Nonaka, 1994; Nonaka & Takeuchi, 1995), and by building and reviewing situated knowledge-theories (March, 1991; Wenger, 1999). With processes of knowledge exploitation, SMEs could improve the use of their knowledge-in-action, reduce transaction costs and increase cooperative efficiency. These benefits are said to also be obtained by using e-government services (Locke, 2006; Thompson et al., 2005).

The above rearrangement of knowledge sources needs as Spender (1998) and Malhotra (2000) suggest, focusing on individual processes of 'unlearning', or challenging our current use of existing sources of 'raw' data or perceptions and our subsequent (e.g. intersubjective) interpretations of it (Ackoff, 1981; Checkland & Holwell, 1998; Clarke, 2007; Jackson, 2005). To enable EK to occur via individual unlearning, Cegarra and Dewhurst (2006) propose the following three phases:

- Identification of problems
- Changing cognitive patterns
- Incorporation of new measures of control

These phases are similar to those used by Pan et al (2006) to explain how an e-government initiative was taken to successful completion, where people had to be engaged in a process of 'unfreezing–changing–refreezing' their commitment in order to take alternative courses of action different to those leading to failure. Similar process-view orientations on e-government are presented by Mcloughing et al (2006) and Heeks (2006). They account for the different 'moments' or junctures in e-government projects, so that in those moments projects can be re-defined in terms of their scope, participants involved and the roles of ICT. The idea is to exploit knowledge sources that are available within projects in order to address current problems. Unfortunately, these frameworks seem still to be focused on the government end, and little is said about e-services users (i.e. SMEs).

We then use Cegarra and Dewhurst (2006)'s classification to develop a framework to assess and develop exploitation of knowledge (EK) via e-government services by SMEs. With the framework SMEs can exploit knowledge by 1) Identifying issues that affect identification of problems and opportunities with e-government services 2) Facilitating the changing of cognitive patterns towards e-government and 3) Integrating new measures for SME monitoring their exploitation of knowledge on e-government. Figure 2 provides a synopsis of a process that could be used to enable better design and use of e-government services and provision to SMEs via EK.

As shown in figure 3 below, there are three main areas of inquiry, and the arrows indicate levels of e-government knowledge exploitation. Departing from an identification of problems (diagnosis), we establish ways of developing OM capabilities to address these problems. Such problems might be related —in principle, with the use of e-government as an external source of knowledge. Development of OM capabilities requires changing individuals' cognitive patterns, and should be based on facilitating interaction between individuals as well as using formal and informal sources of knowledge on e-government. This can result in organisations establishing mechanisms for the continuous renewal of 'e-government-based' knowledge, as well as monitoring its acquisition, use and maintenance.

Solutions and the Role of Systems Methodologies

Considering the above figure, we argue that it could be possible to develop action in all these three fronts simultaneously with the help of systems methodologies and concepts. As Jackson (2005) suggests, there is a variety of systems methodologies 'off the shelf' which can help organisations in developing their processes of knowledge acquisition, sharing and renewal. In the next section we detail each of Cegarra and Dewhurst (2006)'s stages of unlearning with contributions of systems methodologies.

Identification of Problems

It can be said that any form of unlearning begins with failures that generate events which do not conform to expectations in individuals or organisations (e.g. not being able to find the right e-service and information) (Schein, 1993). Failures result in processes in which old ways of thinking and behaving are discarded and new ways can be accommodated. Consequently, organisational culture should foster the belief that employees have different points of view, some of which might be more valid or accepted than others. Despite this possibility, group members can become narrowly focused and fail to adopt alternative perspectives. Janis (1982) calls it 'groupthink' and relates it to the tendency for members of the highly cohesive groups to lose their critical abilities and adopt narrow definitions of problems. To address this issue, Janis (1982) suggests the development of strategies that involve personal rotation among different functional areas. This aims at permitting an organisation to have different members at each

Figure 3. Areas of enquiry to improve e-government services for SMEs

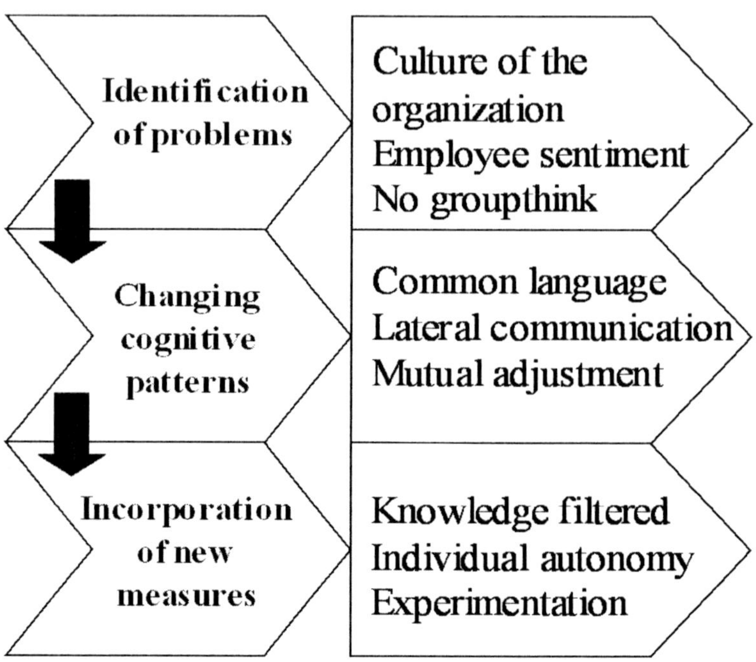

meeting, allowing a multidisciplinary perspective, a greater fluency of knowledge and facilitating the creation of conflicts.

Problem identification about e-government could make SMEs aware about the existence of cultural features in organisations that enable or inhibit alternative perspectives on government services to existing ones. At both the SMEs and government ends, a possible way to identify problems is to enable people to participate in different organisational activities related to the provision and use of e-services, and involve other stakeholders (i.e. technology vendors). Jackson (2005) proposes that knowledge dialectics, where different individuals from inside and outside the organisation confront their knowledge to go beyond what is currently know (Nonaka & Takeuchi, 1995) can be supported by the Strategic Assumption Surface Testing (SAST) Methodology (Mason & Mitroff, 1981), Critical Systems Heuristics (Ulrich, 1983) and Soft Systems Methodology (Checkland, 1981). These systems methodologies enable the identification and exploration of different viewpoints between stakeholders (i.e knowledge elicitation). They also allow people to share their appreciations and reflect on the value of the assumptions that support their decisions. The use of these methodologies could support the development of creative (and dialectic) discussions about existing knowledge on e-government, as well as in identifying issues of concern that inhibit a more full incorporation of e-services into SME daily activities. Their use can involve a variety of stakeholders, and encourage collective discussions among them about the nature, purpose and scope of e-government services.

Changing Cognitive Patterns

In order for SMEs to develop their OM and exploit it as a source of knowledge, they would need to review and reflect on their cognitive patterns, or in other words, they would need to incorporate new 'know-how' to allow e-government be part of their business. Pfeffer (1998) suggests that some key obstacles for developing new knowledge in organisations are related to the existence of authoritarian management styles and hierarchies, something that Sparrow (2005) also identifies as a factor influencing the adoption of knowledge initiatives by SMEs. To overcome this, Nonaka and Konno (1998) suggest intensifying dialogue among individuals, groups and organisations. Hedlund (1994) argues that three fundamental catalysts are required to facilitate the flow of individual knowledge throughout an organisation: common language, lateral communication and mutual adjustment. This can result for instance in the design and sharing of 'boundary objects' (Wenger, 1999) among individuals who see themselves as part of a wider learning community. The design of such objects requires according to Brown and Duguid (1999), creating a climate of mutual trust, which is in turn based on mutual adjustments between different communities or teams. In the case of e-government, these objects could be designed around the idea of having learning communities with SMEs (and why not the government institutions).

Systems methodologies have been also used to design alternative scenarios for organisations and their environments (Ackoff, 1981; Checkland, 1981; Checkland & Scholes, 1990). This requires establishing dialogue processes with stakeholders on what constitutes improvement(s) for them (Churchman, 1970; Jackson, 2003; Midgley, 2000; Ulrich, 1983). According to Jackson (2005), methodologies like Soft Systems Methodology (SSM) (Checkland, 1981) could enable the articulation of tacit knowledge to create shared models of action. Such models can be used to compare and contrast different scenarios and support the individuals' internalisation of explicit knowledge via agreements on feasible and desirable changes, thereby facilitating systemic learning processes. The definition of knowledge-based visions can be supported by another systems methodology called Interactive Planning (IP) (Ackoff, 1981).

Using this methodology, stakeholders focus their attention on the ends they would all like to pursue, and develop creatively a set of idealised designs of their organisations to meet the ends stated so that the design becomes continuous.

For SMEs using e-government services, the use of systems methodologies would help stakeholders to elicit and 'visualise' alternative scenarios for action in which they incorporate acquisition of external knowledge into their activities. SMEs could take advantage of e-services as means to make these scenarios technologically feasible as well as culturally desirable (Ackoff, 1981; Checkland, 1981). To support the design of cognitive activities that use information technologies, systems methodologies can also be used to enquire about the boundaries of the systems to be engineered (not only those aligned to organisational 'government' boundaries) (Córdoba & Midgley, 2006), and the knowledge objects that will enable co-ordination and communication between SMEs and other stakeholders (Córdoba, 2005). In short, the use of systems methodologies can also enable SMEs and other stakeholders to develop creative processes of design of new scenarios where e-government services are incorporated into business activities, as wel as facilitating continuous acquisition and renewal of knowledge.

Incorporation of New Measures

From an individual's point of view, adopting e-government requires investing time and effort in learning. Nonaka and Konno (1998) assert that the implementation of new measures requires individual autonomy, but it will only occur if this is fostered by management. Schein (1993) suggests that employees should be motivated to take risks, innovate and come up with creative solutions to problems to facilitate unlearning. It is then necessary to extend the range of indicators and measures that SMEs normally have (i.e. without e-government) and incorporate new measures (individual and collective) that enable people to assess their learning and performance. Again, this also presupposes that SMEs can incorporate e-services as part of their business strategies.

Looking again at the contributions of systems thinking to knowledge management, Jackson (2005) suggests that other systems methodologies like Beer (1985)'s Viable System Model (VSM) or Team Syntegrity (Beer, 1994) would enable individuals to diagnose existing organisational contexts and structures, with a view to reflect on the autonomy needed to share and acquire knowledge effectively. Using VSM, people can determine structural changes that need to be made in organisations to facilitate flexible adaptation with their environments by enabling more democratic dialogue and effective debate. With these and other structural adjustments put in place, the definition of measures of achievement and development can be better integrated into activities at different levels that facilitate the use of knowledge for decision-making.

For SMEs, the diagnostic and design of their organisational structures would enable them to place appropriate roles to their owner's management style, as well as fostering creative and autonomous thinking to develop their collective knowledge capabilities (Sparrow, 2005). It would also facilitate the integration of measures of knowledge acquisition and sharing into the SMEs' businesses processes.

A Framework to Assess and Develop E-Government Services (E-Services) by SMEs

Having considered how 'unlearning' about e-government can be fostered, and how systems methodologies can help, we now link unlearnin with assessment of provision of e-government services. To do so we first classify 'levels' of e-services by adopting the model proposed by Amit and Zott (2001) on e-business. They characterise such a model as having three dimensions to develop: 1) Transaction content; 2) Transaction structure and

3) Transaction governance). Following these, we classify e-services into these levels:

- Level one (1) refers to transaction content i.e., the goods or information that are being exchanged and the resources and capacities that are required to enable the exchange. For SMEs, these e-services include core and non core information on e-government (Locke, 2006). SME can visit e-government websites to acquire up to date information about regulations, opportunities and requirements related to e-services.
- Level two (2) refers to the transaction structure i.e., parties that participate in the exchange and the ways in which these parties are linked. For SMEs, this level includes two-way interactions (Criado & Ramilo, 2003), like dealing with government officials or units to clarify information, paying taxes, registering companies or working with other organisations (i.e. other SMEs) to negotiate contracts, grants or benefits.
- Level three (3) refers to the transaction governance i.e., the ways in which information sources and flows are managed or controlled by relevant parties. For example, SMEs could be supported by search engines or simulation tools to help them self-assess their tax contribution, or define products and services with other SMEs. This level of governance can enable SMEs to use their existing knowledge to use *and modify* an e-service. It can also allow them to gain more knowledge that could help their decision-making processes in the future, for instance by exerting accountability on the provision of services by their governments (West, 2004).

With the above classification we now provide a framework to assess and develop e-services by SMEs as shown in Figure 4. Systems thinking can help developing processes of 'unlearning' as discussed before. The two-way arrows between upper (SME) and lower (government) elements indicate the relations that could exist between 'un'learning processes at the SMEs end and, the support they should have at the governments end via e-services. The figure shows that e-services

Figure 4. A framework to develop and assess e-services for SMEs

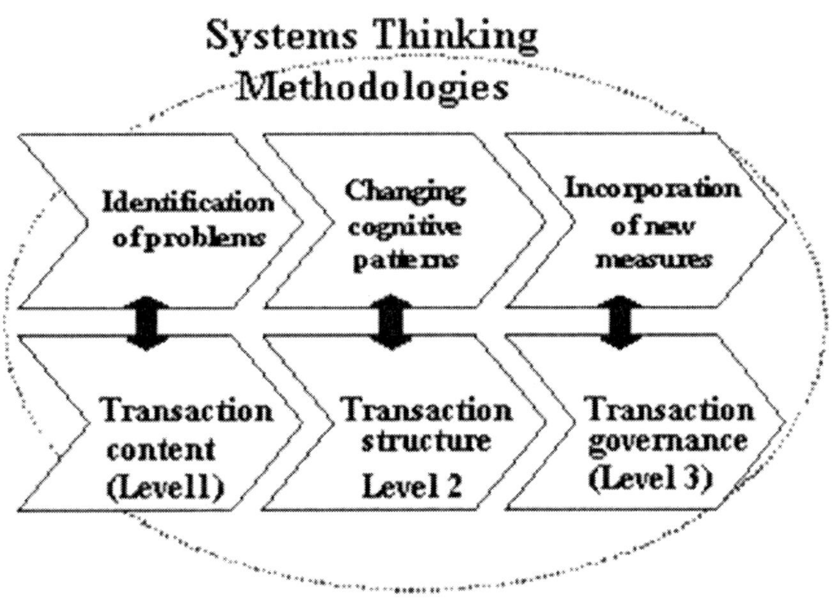

can provide content to aid in the identification of problems at the SME (level 1). Problems could be related to existing ways of 'doing things' which are stored in the SME OM. E-services at this level can help SMEs to challenge these ways by providing up-to-date information on new ways of operating. At the SME end, the identification of problems can also trigger reflection on how e-services content could facilitate or inhibit *unlearning* about e-government. A good example of this is the existence of e-government websites that provide the same information as in physical documents. Although this can be seen as an efficient (i.e. unified) way of keeping information up to date (i.e. by government institutions or agencies), this type of representation of their own OM could become a *fixed* source of routinization. Exploitation of knowledge at both government and SME ends without practices of renewal could become an inhibitor of 'unlearning' if it is not continuously renewed (Levinthal & March J, 1993). These findings support the views of organizational learning strategy literature which show that firms more often converge (exploit) rather than reorientate (explore) due to a variety of reasons such as organizational inertia (Hannan & Freeman, 1984; Milliken & Lant, 1991).

In this level 1 of our framework, we see that the use of systems methodologies mentioned above to facilitate knowledge dialectics should be encouraged, with a view of reviewing and re-designing the content of e-government services. The strategies and purpose of e-government services as seen by government institutions, SMEs and other stakeholders could be elicited and debated.

For SMEs, the existence of level 2 e-services that offer transaction facilities (for instance online payment or approval of information) (West, 2004) can become a source of *change*. Change could happen at the SME end if among other factors, the availability of e-services enables people to abandon existing routines and patterns of interaction, and engage in developing new relationships which require the acquisition of new knowledge about technology or technology itself (Criado & Ramilo, 2003; Dunleavy et al., 2006). For instance, e-services can become available wherever and whenever they are needed, so that SMEs can access and use them to transform their activities. E-services that enable people to ask questions and receive answers, to communicate efficiently with government institutions and to facilitate exchange of information could also contribute to enhance SMEs' own business strategies.

At this level 2, the use of systems methodologies that facilitate collective action among stakeholders by facilitating identification of issues of concern, design and definition of appropriate agendas for change. Among others, these methodologies include Soft Systems Methodology (Checkland, 1981) and Idealised Design (Ackoff, 1981). They could help SMEs and government among other stakeholders to design systemically feasible and culturally desirable changes.

We now discuss some managerial implications of our framework which we deem as relevant to inform practice of e-government project managers and other people involved or affected by the implementation of services to SMEs.

MANAGERIAL IMPLICATIONS

Organisations in general have become concerned with the value that they deliver to their customers. From our framework, we suggest that any public organization wishing to implement an e-government service should make efforts to first understand what the sources of value in their users are. SMEs are no the exception. They continuously seek for new opportunities and ways of complying with existing service or norm regulations if not going around them. How to identify ways of providing value to SMEs with services is a complex question, since the differential values that users have are intangible and heterogeneous elements and, furthermore, the management of these elements will be different, depending on the type of e-service being provided. This leads us to reinforce an idea previously discussed (Cegarra et al., 2007), that in the case of many governments (e.g. the Spanish and possibly other EU member states ones) there might be a high degree of investment of web sites to support e-government but under-investing in knowing users' expectations and perceptions. Therefore, the first managerial implication of this chapter is to question the existing models which relate e-government and SMEs. To the extent to which we assert, following Cegarra et al (2007), that although e-government represents an important source of competitive advantage in competitive markets, only if aligned with user demands can this potential source becomes a powerful tool for success.

A second managerial implication from our framework is that there could be key enablers leading to effective e-government services utilization by SMEs. Although any new model that claims to be a new paradigm, idea or framework must be tested, the model shown in Figure 4 finds support on previous research (Cegarra, 2006; Cegarra et al., 2007). The model denotes a perspective where valuable knowledge is conceived as being captured and codified from individuals, packaged and transmitted and processed though the use of three frameworks (i.e. identification of problems; changing cognitive patterns; and incorporation of new measures of control) and, hence, disseminated and used by e-government in other contexts. For managers, this means that in the case of a SME the company needs both to explore new possibilities and exploit old certainties (i.e. organizational memory) to help people to consider and access valuable sources of knowledge necessary to implement e-government. Otherwise, we may end up with managers who often confuse experience with believing that they always know better than anybody what e-service does work and does not work. If this is the case, E-services might then lose their appeal even before they are to be used by SMEs. Managers can focus on enabling individuals to venture, innovate and challenge existing practices. They could well see e-services as a resource to enable these activities to develop further.

A third implication of our framework derived from the previous two is that governments should enable e-government users (SMEs) to participate more actively in the design and implementation of e-services. This can help generating win-win situations between parties involved (governments, project teams, SMEs) whilst facilitating the exchange of knowledge between them. Governments should review how dominant their normative perspective on e-government is, and how they balance it with a more innovative and knowledge-based one that could help them improve communication with their users, technology suppliers and ultimately the public in general.

The above implications give us some insights as to how to proceed in the future development of e-government services (e-services) for SMEs. We now present some preliminary findings that confirm the possibilities of enabling better knowledge-based e-government practices.

CONCLUSION

This chapter has provided a framework to assess and develop e-government services to SMEs. E-government can be seen as a set of activities that can enable organisations to acquire and exploit knowledge by transcending existing organisational boundaries. In doing so, it can be a way of enabling the development of SME 'un'learning, and therefore can be seen as a 'system' in which different stakeholders are involved in facilitating such learning. The perspective developed in this chapter emphasises the use of SME organisational memory (OM) to achieve better and improved use of knowledge in organisations through knowledge exploitation (EK). Achieving this via better design and provision of e-government services requires rethinking existing approaches to e-services design and use, as well as their levels of governance by SMEs.

In our framework, assessing e-government services (e-services) requires SMEs to continuously and appropriately 'unlearn', to think 'outwards', to acquire and exploit knowledge, and to use e-services as means to support knowledge exploitation. For those organisations currently providing e-government services (e.g. government institutions, ICT providers), our framework should encourage them to have a better and more flexible degree of involvement with SMEs in the design of such services to facilitate continuous knowledge creation, use and renewal.

We aim to challenge constructively existing assumptions and values about government services by SMEs, so that those responsible for these services look into more detail into their users and their potential benefits accrued from service use. We join Jackson (2005)'s claims about the possibilities to facilitate knowledge practices with systems methodologies. A variety of methodologies is available to facilitate inquiry into the identification of problems (by facilitating knowledge dialectics), the changing of cognitive patterns (by enabling design of shared scenarios of action) and the incorporation of new practices and measures in the use of e-government services by SMEs and other stakeholders (by developing appropriate contexts for knowledge acquisition and renewal).

Although in our research we have started to get positive indications on the connections between SME capabilities and the use of e-services (Cegarra et al., 2007), there are still unanswered questions. For instance, how to assess the influence of e-services provision to the development of SME organisational memory (OM) and exploitation of knowledge (EK), or how to integrate e-services more effectively with SME business strategies still remain areas to explore. Furthermore, how to develop SME governance with e-services (level 3 of our framework) so that ultimately governments learn to trust their users (not only users trusting their government!) is a key challenge to address in the near future. With the ideas contained in our framework, we have taken a first step towards providing ways and approaches to make appropriate transitions in the development of more useful, interactive, and accountable e-government services to SMEs.

REFERENCES

Ackoff, R. (1981). *Creating the Corporate Future: Plan or to be Planned for*. New York: John Wiley and Sons.

Amit, R., & Zott, C. (2001). Value creation in e-business. *Strategic Management Journal*, *22*(6-7), 493–520. doi:10.1002/smj.187

Ballestero-Diaz, F., & Font-Jaume, A. (2004). El despertar del e-procurement en las administraciones públicas. *Información Comercial Española (ICE) . Revista de Economía*, *813*, 73–88.

Banister, F. (2004). *Deep E-government*. Paper presented at the European Group of Public Administration (EGPA) Annual Conference, Ljubljana.

Beer, S. (1985). *Diagnosing the System for Organisations*. Chichester, UK: John Wiley and Sons Ltd.

Beer, S. (1994). *Beyond Dispute: The Invention of Team Syntegrity*. Chichester, UK: John Wiley and Sons.

Brown, J., & Duguid, P. (1999). *The Social Life of Information*. Cambridge, MA: Harvard University Press.

Carroll, J. S. (1998). Organizational learning activities in high-hazard industries: The logics underlying self-analysis. *Journal of Management Studies*, *35*(6), 699–717. doi:10.1111/1467-6486.00116

Carter, L., & Bélanger, F. (2005). The utilization of e-government services: citizen trust, innovation and acceptance factors. *Information Systems Journal*, *15*(1), 5–25. doi:10.1111/j.1365-2575.2005.00183.x

Castells, M. (2001). *The Internet Galaxy: Reflections on Internet, Business and Society*. Oxford, UK: Oxford University Press.

Cegarra, J. G. (2006). *Linking e-government with organisational memory through exploitation of knowledge*. In M. A. Sicilia (Ed.), *Competencies in Organizational E-Learning: Concepts and Tools*. Hershey, PA: Idea Group Publishing.

Cegarra, J. G., & Dewhurst, F. W. (2006). Linking shared organisational context and relational capital through unlearning: An initial empirical investigation in SMEs. *The Learning Organization*, *13*(1), 49–62. doi:10.1108/09696470610639121

Cegarra, J. G., Dewhurst, F. W., & Briones-Peñalver, A. J. (2007). Factors affecting the use of e-government in the telecommunications industry in Spain. *Technovation*, *27*(10), 595–604. doi:10.1016/j.technovation.2007.03.003

Checkland, P. (1981). *Systems Thinking, Systems Practice*. London: John Wiley and Sons.

Checkland, P., & Holwell, S. (1998). *Information, Systems and Information Systems: Making Sense of the Field*. Chichester, UK: John Wiley and Sons.

Checkland, P., & Scholes, P. (1990). *Soft Systems Methodology in Action*. Chichester, UK: John Wiley and Sons.

Christopher, M. (2005). *Logistics and Supply Chain Management: Creating Value-Added Networks* (3rd ed.). Harlow, UK: Prentice Hall.

Churchman, C. W. (1970). Operations research as a profession. *Management Science*, *17*, b37–b53.

Clarke, S. (2007). *Information Systems Strategic Management: An Integrated Approach* (2nd ed.). London: Routledge.

Córdoba, J. R. (2005). Communities and evaluation of e-government services. In S. Clarke & E. Coakes (Eds.), *Encyclopaedia of Communities of Practice in Information and Knowledge Management* (pp. 32-34). Hershey, PA: Idea Group Publishing.

Córdoba, J. R., & Midgley, G. (2006). Broadening the boundaries: An application of critical systems thinking to IS planning in Colombia. *The Journal of the Operational Research Society*, *57*(9), 1064–1080. doi:10.1057/palgrave.jors.2602081

Costa, E. (2001). *Global E-commerce Strategies for Small Businesses*. Cambridge, MA: MIT Press.

Criado, J., & Ramilo, M. C. (2003). E-government in practice: An analysis of web site orientation to the citizens in Spanish municipalities. *International Journal of Public Sector Management*, *16*(3), 191–218. doi:10.1108/09513550310472320

Davenport, T. H., & Grover, V. (2001). General perspectives on knowledge management: Fostering a research agenda. *Journal of Management Information Systems*, *18*(1), 5–21.

Dunleavy, P., Margetts, H., Bastow, S., & Tinkler, J. (2006). *Digital Era Governance: IT Corporations, the State, and E-Government*. Oxford, UK: Oxford University Press.

Edvardsson, I. (2006). Knowledge management in SMEs: the case of Icelandic firms. *Knowledge Management Research and Practice*, *4*(4), 275–282. doi:10.1057/palgrave.kmrp.8500111

Galliers, R. (1999). Towards the integration of e-business, knowledge management and policy considerations within an information systems strategy framework. *The Journal of Strategic Information Systems*, *8*(3), 229–234. doi:10.1016/S0963-8687(00)00023-8

Grimsley, M., & Meehan, A. (2007). e-Government information systems: evaluation led design for public value and client trust. *European Journal of Information Systems*, *16*(2), 134–148. doi:10.1057/palgrave.ejis.3000674

Hamel, G. P., & Prahalad, C. K. (1994). *Competing for the Future*. Cambridge, MA: Harvard Business School Press.

Hannan, M., & Freeman, J. (1984). Structural inertia and organizational change. *American Sociological Review*, *49*, 149–164. doi:10.2307/2095567

Heeks, R. (2006). *Implementing and Managing Egovernment: An International Text*. London: Sage.

Irani, Z., Love, P. E., Elliman, T., Jones, S., & Themistocleus, M. (2005). Evaluating E-government: Learning from the experiences of two UK local authorities. *Information Systems Journal*, *15*(1), 61–82. doi:10.1111/j.1365-2575.2005.00186.x

Jackson, M. C. (2003). *Creative Holism: Systems Thinking for Managers*. Chichester, UK: John Wiley and Sons.

Jackson, M. C. (2005). Reflections on knowledge management from a critical systems perspective. *Knowledge Management Research and Practice*, *3*(4), 187–196. doi:10.1057/palgrave.kmrp.8500067

Jae-Moon, M. (2002). The evolution of e-government among municipalities: rethoric or reality? *Public Administration Review*, *62*(4), 424–433. doi:10.1111/0033-3352.00196

Janis, I. (1982). *Victims of Groupthink* (2nd ed.). Boston, MA: Houghton Mifflin.

Jeffcoate, J., Chappell, C., & Feindt, S. (2000). Attitudes towards process improvement among SMEs involved in e-commerce. *Knowledge and Process Management*, *7*(3), 187–195. doi:10.1002/1099-1441(200007/09)7:3<187::AID-KPM86>3.0.CO;2-6

Levinthal, D. A., & March, J. G. (1993). The myopia of learning. *Strategic Management Journal*, *14*, 95–112. doi:10.1002/smj.4250141009

Locke, S. (2006). E-local government strategies and small businesses. *The Journal of American Academy of Business*, *10*(1), 21–30.

Malhotra, Y. (2000). Knowledge management for e-business performance. *Information Strategy: The Executives Journal*, *16*(4), 5–16.

March, J. G. (1991). Exploration and exploitation in organisational learning. *Organization Science*, *2*(1), 71–87. doi:10.1287/orsc.2.1.71

Mason, R., & Mitroff, I. (1981). *Challenging Strategic Planning Assumptions: Theory, Cases and Techniques*. New York: John Wiley and Sons.

McLoughlin, I., & Cornford, J. (2006). Transformational change in the local state? Enacting e-government in English local authorities. *Journal of Management & Organization*, *12*(3), 195–208.

Megill, K. A. (1997). *The Corporate Memory: Information Management in the Electronic Age*. London: Bowker & Saur.

Midgley, G. (2000). *Systemic Intervention: Philosophy, Methodology and Practice*. New York: Kluwer Academic/Plenum.

Milliken, F., & Lant, T. (1991). The effect of an organization's recent performance history on strategic persistence and change: The role of managerial interpretations. In J. Dutton, A. Huff, & P. Shrivastava (Eds.), *Advances in Strategic Management* (Vol. 7, pp. 125-152). Greenwich, CT: JAI Press.

Mincomunicaciones. (2007). *Agenda de Conectividad - Gobierno en Línea y Plan de Acción 2007*. Bogotá: Ministerio de Comunicaciones.

Nonaka, I. (1994). A dynamic theory of organizational knowledge creation. *Organization Science*, *5*(1), 14–37. doi:10.1287/orsc.5.1.14

Nonaka, I., & Konno, N. (1998). The concept of "ba": Building a foundation for knowledge creation. *California Management Review*, *40*(3), 1–15.

Nonaka, I., & Takeuchi, H. (1995). *The Knowledge Creating Company*. Oxford, UK: Oxford University Press.

Nooteboom, B. (1999). The dynamic efficiency of networks' in Interfirm networks. In A. Grandori (Ed.), *Organization and Industrial Competitiveness* (pp. 91-119). London: Routledge.

Norris, D., & Moon, M. J. (2005). Advancing e-government at the grassroots: Tortoise or hare? *Public Administration Review*, *65*(1), 64–75. doi:10.1111/j.1540-6210.2005.00431.x

Pan, G., Pan, S., Newman, M., & Flynn, D. (2006). Escalation and de-escalation of commitment: A commitment transformation analysis of an e-government project. *Information Systems Journal*, *16*(1), 3–21. doi:10.1111/j.1365-2575.2006.00209.x

Pfeffer, J. (1998). *The Human Equation: Building Profits by Putting People First*. Boston, MA: Harvard Business School Press.

Phang, C. W., Sutanto, J., Kankanhalli, A., Tan, C., & Hock-Hai, T. (2006). Senior citizen's acceptance of information systems: A study in the context of e-government services. *IEEE Transactions on Engineering Management*, *53*(4), 555–569. doi:10.1109/TEM.2006.883710

Rambøll Management. (2004). *User Satisfaction and Usage Survey of eGovernment Services*. EU: Top of the Web Survey.

Riquelme, H. (2002). Commercial internet adopting in China: Comparing the experience of small medium and large businesses. *Internet Research: Electronic Networking Applications and Policy*, *12*(3), 276–286. doi:10.1108/10662240210430946

Schein, E. (1993). How can organisations learn faster? The challenge of entering the green room. *Sloan Management Review*, *34*(2), 85–92.

Selnes, F., & Sallis, J. (2003). Promoting relationship learning. *Journal of Marketing*, *67*(July), 80–95. doi:10.1509/jmkg.67.3.80.18656

Senge, P. (1999). *The Fifth Discipline: The Art and Practice of the Learning Organisation*. London: Random House.

Smeltzer, L. R., Fann, G. L., & Nikolaïssen, V. N. (1988). Environmental scanning practices in small business. *Journal of Small Business Management*, *226*(3), 52–62.

Sparrow, J. (2005). Classification of different knowledge management development approaches of SMEs. *Knowledge Management Research and Practice*, *3*, 136–145. doi:10.1057/palgrave.kmrp.8500053

Spender, J. C. (1998). The dynamics of individual and organisational knowledge. In C. Eden & J. C. Spender (Eds.), *Managerial and Organizational Cognition: Theory, Methods and Research* (pp. 11-39). London: Sage.

Storey, D. (2004). Exploring the link, among small firms, between management training and firm performance: A comparison between the UK and other OECD countries. *International Journal of Human Resource Management, 15*(1), 112–130. doi:10.1080/0958519032000157375

Tan, C., & Pan, S. (2003). Managing e-transformation in the public sector: An e-government study of the Inland Revenue Authority of Singapore (IRAS). *European Journal of Information Systems, 12*, 269–281. doi:10.1057/palgrave.ejis.3000479

Tapscott, D., & Williams, A. (2007). *Wikinomics: How Mass Collaboration Changes Everything*. London: Atlantic.

Thompson, D., Rust, R., & Rhoda, J. (2005). The business value of e-government for small firms. *International Journal of Service Industry Management, 16*(4), 385–407. doi:10.1108/09564230510614022

Ulrich, W. (1983). *Critical Heuristics of Social Planning: A New Approach to Practical Philosophy.* Berne, Switzerland: Haupt.

Volberda, H. W., & Lewin, A. Y. (2003). Co-evolutionary dynamics within and between firms: from evolution to co-evolution. *Journal of Management Studies, 40*(8), 2105–2130. doi:10.1046/j.1467-6486.2003.00414.x

Walsh, J. P., & Ungson, G. R. (1991). Organizational memory. *Academy of Management Review, 16*, 57–91. doi:10.2307/258607

Ward, J., & Griffiths, P. (2002). *Strategic Planning for Information Systems*. Chichester, UK: John Wiley and Sons.

Wenger, E. (1999). *Communities of Practice: Learning, Meaning, and Identity*. Cambridge, UK: Cambridge University Press.

West, D. (2004). E-government and the transformation of service delivery and attitudes. *Public Administration Review, 64*(1), 15–27. doi:10.1111/j.1540-6210.2004.00343.x

Wong, K. Y., & Aspinwall, E. (2005). An empirical study of the important factors for knowledge-management adoption in the SME sector. *Journal of Knowledge Management, 9*(3), 64–82. doi:10.1108/13673270510602773

ENDNOTES

1. One example is Alito (http://scms.alito.co.uk/, accessed in February 2009), a procurement portal that offers SME opportunities for contracting services with the councils of the UK region of Yorkshire. Another example is Adquira (http://www.adquira.es/adquira/servletAdquira, accessed February 2009), a procurement portal set up by Spanish companies (e.g. Telefonica, Iberia) to enable businesses in LatinoAmerica and Spain to trade their services to each other.
2. A reason that is cited for a higher use of e-procurement portals is the involvement or private companies (Moon, 2002).
3. More recently we have seen that some government portals include statistics of service use and surveys about user satisfaction as a way of involving citizens and users of services. An example of this is the Colombian online government initiative (www.gobiernoenlinea.gov.co, accessed in March 2009). This could be a starting point to ensure that citizen's feedback is taken into account.

Section 3

Emerging Forms of Community Interaction:

Groups and Tools

Chapter 11
Online Communities:
People and Processes

Cecilia Loureiro-Koechlin
Oxford University, UK

ABSTRACT

The chapter describes the basic tenets of online communities, their typology and their fundamental elements with a view of informing ways in which people can get to participate online, something that those individuals or organizations dealing with e-governance initiatives also aim at. A key consideration is the importance of socializing as an ingredient that motivates individuals and which can offer them a degree of freedom to discuss government related matters.

INTRODUCTION

The Internet is a computer network that removes barriers to access information and communication by connecting millions of people worldwide. Since its creation the number of Internet users has grown exponentially, and so have the forms of participation and intentions of use. Modern forms of Internet tools include Weblogs, Online Forums, and Social Networking sites, such as Facebook and YouTube. Institutions and people from all backgrounds use these tools including government departments and community areas, public institutions, local governments as well as individual citizens. They create websites and weblogs, and participate in forums to express their opinions to the online world. Examples of these are "The Gordon Brown Blog", Madrid City Council Portal and "CRNUK"- Community Recycling Network UK web site.

Within internet environments, groups of people come together to share their ideas. Online communities emerge when these groups grow in numbers and engage into active participation. These communities are kept alive thanks to their members' desire to participate and share experiences, their common interests, and the time devoted into developing their relationships. In addition, interesting, good quality content, which is updated regularly, is needed to keep the members interest as well as to attract new ones. Content can be provided by the "owner"

DOI: 10.4018/978-1-60566-860-4.ch011

of the community (e.g. the institution who owns the forum, the blogger, etc.), the moderators or by the members themselves. Norms of behaviour are designed within each group to ensure that the interactions within the communities run smoothly, without misunderstandings. Some common behaviour norms include, avoiding inappropriate language and focusing the content of posts on a topic (or topics) for which the forum was initially created and intended. There are numerous ways to encourage members to follow community norms including moderation and rating.

The use of new improved web technologies and the various forms that the mentioned community aspects can take allow users to create and apply different styles of participation. These forms of interactions can have an impact on social processes, such as leadership, decision making and community administration which can affect (positively or negatively) the capability of the communities to regulate their activities. The aim of this chapter is to explore the aspects of online communities' dynamics which are likely to have an influence on their successful self administration and therefore on the achievement of their goals. I expect that this discussion will help readers to draw some connections between the development of online communities and e-governance efforts, as both of them aim at engaging people in online participation. Successful online communities could be used as models for e-governance initiatives as they have managed to motivate and get the interest of their online participants, and are obviously achieving their goals. This chapter will explore some definitions of online communities, types, members' roles and technologies. A final discussion will highlight some overlaps between online communities' dynamics and e-governance efforts.

WHAT ARE ONLINE COMMUNITIES?

Online communities are formed by groups of people who interact through online or internet tools. The name "online community" itself refers to two important aspects of these kinds of groups. The "online" aspect represents the use of networked computers and software tools used for communication. The following definition of online community by Lee et al. (2003) emphasises the use of technology:

a cyberspace supported by computer-based information technology, centered upon communication and interaction of participants to generate member-driven contents, resulting in a relationship being built up (Lee et al., 2003)

The "community" aspect represents people with similar interests, who gather for entertainment, to discuss or ask for help. The second definition of online communities, by Rheingold (1994), focuses on the human and social processes emerging from the participations:

social aggregations that emerge from the Net when enough people carry on those public discussions long enough with sufficient human feeling, to form webs of personal relationships in cyberspace. (Rheingold, 1994)

In relation to this second aspect of online communities, it is important to stress that the mere act of gathering people online does not imply that a community will be created. Online communities cannot be engineered or manufactured. The best technological tools could be used but if users do not engage then communities do not emerge. Indeed most of their success depends on human and social rather than technological factors. In this section I highlight some of these factors.

Online communities emerge from the interaction of people who have the "desire and need to share problems, experiences, insights, templates, tools and best practices" (APQC (2004) quoted in Coakes and Clarke (2006)). This also involves a shared interest or common purpose. For instance, an online community can be created for citizens to discuss the "effects of the credit crunch in the local community". Some online communities are grounded on their "occupational communities" as they share "similar goals, work practices, beliefs interests, and value systems" (Elliott and Scacchi, 2003). Members of actual physical organisations or institutions can start online communities to allow extra forms of communications between them. These organisations have distinctive goals and so their online communities.

Online community goals are usually aligned with the ones of their owners or creators and new members join because they share those goals. However, goals cannot be imposed on online members. New goals or purposes need to be decided in democratic consultations. Members who do not agree with the direction the online community is moving to are always free to leave.

Another community success factor is the sense of membership that the users achieve. This is shown by members' commitment to participation and solidarity to other members. For example, a high number of replies to an enquiry would indicate a high the level of commitment and solidarity of the group with its members. This will also suggest that members identify with the group. Usually the core members are the ones who define and reflect the identity of the community, as they are the ones who participate more often. (This will be discussed a bit more in the next section.) Also, the way outsiders or newcomers are treated and the differences between them and insiders reveal the identity of the group. An outsider will have to learn how to integrate into the community and gain his/her membership by establishing relationships with other members and by earning their respect.

As in real-world societies norms are created to show the users of online communities how activities are carried out there (Elliott and Scacchi, 2003). These norms, in a way, reflect the nature of the communities, the nature of their members and their objectives. Norms are established to ensure that the community goals and members needs are being met. This also involves members feeling comfortable with what is said there, with the way their fellow members treat them, and that no one gets offended or abused by anyone. Figure 1 shows a screenshot of a message posted by a site's moderator specifying the rules of a forum. This is an example of how a moderator (or any other member) can let others know about the community's norms. Online members can challenge these rules if they wish by replying to the original post. A thread of conversation can emerge here until users reach an agreement. However, with time, new requests may be added as new members join in or as new events affect the activities of the community.

In addition, as the online world is mainly a written medium there are some "conventions of communication" (Jacobson, 1996) or "chat codes" (Greenfield and Subrahmanyam, 2003) that participants need to adopt. These are not formal norms but unwritten rules that people have to follow to make online communication possible. Some conventions are well spread through the cyberspace whereas others are kept within the online communities that created them. Some conventions mimic real-world conventions from oral or written speech, but others are created especially for the online world. For example, to stress an utterance it can be highlighted by using a different colour. To call the attention of a specific person (in a group conversation for example) his name can be indicated within brackets, as in "Olivia: [Bruce] what is your opinion?" Feelings can be articulated by using especial characters like <<Daniel is concerned>> or by using emoticons which can be graphical or which can be made up of computer characters. Some emoticons are characteristic of western societies:) others are more popular in the east (-_-;). Common expressions can be abbreviated. For example YMMV, which means Your Mileage May Vary (or your experience could be

Figure 1. Rules of Use posted by an online forum moderator. Source: StudyOptions www.studyoptions.com

different). People can raise their voices by using CAPS. However that would be considered rude in most online environments. Online members themselves would let the poster know that he/she cannot use caps in their forum. When new members join online communities they need to learn these codes and learn when and how to use them. In other words they need to learn the community's communication conventions to become members and to be accepted by the others. This learning process can be done through guidance from older members and by reading the logs of community communications.

Changes occur within the communities as new members join in and as they learn from their interactions with others. Topics of interest change according to events inside or outside the communities. New members become core members and core members may go into a passive mode or leave. People who have been part of other communities bring with them all their knowledge and experience when they join a new one. Additionally new versions of the communities' software replace the old ones and new communication tools become available. New tools allow users to create new ways of interaction thus changing the social processes of their communities. This is how online communities evolve: through changes in their social processes and changes in their software. Evolution implies as well the challenge of the community norms and their further adaptation to fit the new needs of their members (Rheingold, 1994).

Over the years, the evolution of online communities has influenced the way users perceive the online environment. While before people used to go online to search for information, now they go online to work and socialise. For them the online world is not an alien setting anymore but another manifestation of their real life (Grefter, 2006), a life in which online relationships are built and thought as genuine (Parks, 1996).

The next section will present a classification of online communities and their members. The classification is not exhaustive but tries to put a bit of order into this vast ever changing spectrum of online communities.

A CLASSIFICATION OF ONLINE COMMUNITIES AND ONLINE MEMBERS

Online communities can be open or close communities. Open communities are created upon personal or professional interests, religious or political views, or any kind of entertainment. Members are geographically dispersed and rarely meet each other but nevertheless develop ties between them (Millen and Dray, 2000, 167). The access to these environments is not difficult as they welcome new members who share their interests. In practical terms this means that members do not need to sign up for an account, or if needed, there are no restrictions as to who can open one. To communicate, these communities use public online settings such as bulletin boards, online forums or mailing lists which are sponsored by public or private organisations and which usually run in a browser.

Closed online communities are grounded on physical organisations (Millen and Dray, 2000) like for example an NGO, and their membership is restricted to the physical organisation members. Their online environments are usually held within the intranets or internal systems provided by the organisation. Being all part of the same organisation, members will interact with people they either know in person or by other means such as telephone or email. Connections could also be established with people who are known by people they already know (friends of friends). Most of the activities carried out within these online environments are work related. The purposes of these communities can be: 1) to perform or coordinate task related activities, 2) to broadcast information and 3) to socialise (Kettinger and Grover, 1997).

In either kind of community (open or close) members can be classified according to their role and level of participation. The following classification is based on Millen and Dray (2000) and McDaniel et al (1996) typologies of participants:

- **Core members or frequent contributors:** They usually dominate (McDaniel et al., 1996) the conversations by initiating more conversations and by sending more replies than the others
- **Sporadic contributors:** Participate when "a topic of personal interest is discussed" (Millen and Dray, 2000)
- **Marginal contributors:** They have low levels of participation. They could be new community members or members who wish to stay in the peripherals of the community
- **Lurkers:** They read without responding (Parks, 1996)

The above list defines a spectrum of the online members' levels of participation, having the senders on one side and the readers on the other. This agrees with Marcoccia's (2004) taxonomy of participation as well. A role that was not mentioned in the classification is the host or monitor. Depending on their personal approach to monitoring, hosts could think of themselves as community members or as outsiders. Hosts act as the moderators of conversations and have power over the other members (e.g., they can ban members when they do not follow the norms). Some hosts participate in discussions (sender end of the participation spectrum), some just stay in the background checking on the members behaviour (reader end of the participation spectrum.) Core members do not have the privileges of the moderators, but they are influential opinion leaders and as such they are respected by the other members.

They are closer to the sender end of the participation spectrum. Core members tend to dominate topics of conversations. Their ideas and opinions are the ones that prevail and the ones which are portrayed by the community. So in a way it could be said that the opinion of an online community is the opinion of its core members, or that what we perceive is the community's ideology is only a reflection of its core members' ideas. However these ideas are only a partial view of the totality of the community. Sporadic and marginal contributors, may have opposed or different opinions but because of their lack of participation, their views are not that visible. Unfortunately, due to their lack of participation, lurkers' perspective and behaviour cannot be observed. However, their role could be thought of as supportive of the community. Comparing their participation with real world situations, lurkers are like the people who attend meetings but do not talk. In spite of this their presence is seen by the rest. In online settings their presence can be noted by the times the member has logged into the community, and if the community does not have authentication system then by the visit counters.

Despite the scepticism of some scholars who thought that online gatherings were a "socially-impoverished domain" (Baym, 1995) people are using the internet more to engage in online interactions. There is evidence to prove this. For years now research has been carried out to study the nature and dynamics of different kinds of online communities. This research shows that online communities are a phenomenon that is growing in importance, in different areas of society. For example in her study about humour Baym (1995) shows how people create an online community to talk about soap operas. Rheingold (1994) provides too a rich picture of a online community devoted to share medical and family problems. There is also evidence of communities created for work purposes. An online community of journalists is studied by Millen and Dray (2000) and a group of public relations professionals was studied by Thomsen (1996). Elliott and Scacchi (2003) study a virtual organization of Free Software developers that successfully collaborate and resolve conflicts online. The members of this community share the same values about the use of free software and freedom of choice which are manifested in their online conversations and influence their practices.

Although that the heart of online communities is the people that are part of it, they would not exist without the technology that allow their members to meet online. The next section will briefly introduce the most popular software technologies used by online communities.

ONLINE COMMUNICATION TECHNOLOGIES

Many are the tools used for online communities to communicate and to hold their information. These online communication technologies have evolved through the years establishing new ways of online socialisation. In this section I will discuss what I believe are the three most popular: discussion forums, weblogs and social networking sites. These technologies have obvious implementation and visual differences. However the fundamental difference is the different social structures and dynamics that emerge from within them. Whereas discussion forums offer a single floor which is shared by everyone, interactions in weblogs happen at two levels: the authors' and the readers'. Social networking sites are a mixture, offering personal and group spaces. Additionally, interconnections between participants are inherent in the nature of the forums. When one member joins a forum it is assumed that he or she is now connected to the other members through that forum. In contrast, online members have to build their own connections in weblogs and social networking sites to not become isolated from the rest.

Discussion Forums

Discussion forums (or fora) have been around for quite a lot of time. They have changed shapes and names many times. You could associate them with the earlier mailing lists and Usenet or with discussion boards. Under the discussion forums heading I am including all these tools we use to converse between two or more people by sending or posting messages onto a public space, or by directing them to everyone else's emails. Discussion forums are asynchronous tools. That is, participants do not need to be present at the same time to be part of a conversation and the messages or posts stay on the public environment (or people's emails) for others to read for a long period of time. Usually discussion forums are structured by topic of conversation or interest. Discussions happen in threads of conversations started by any participant. Other participants can add their comments by replying to the original poster or by replying to any other replies. Posts tend to be short in length comprising a few lines. Depending on the design of the forum, it could be hard sometimes to follow a conversation, especially when many people have been participating. As in real life conversations, the focus of online conversations can change and two or more branches of that conversation can emerge. The discussion forums keep the logs of all these threads containing participants' thoughts and interactions between them. These are rich sources of information about the communities themselves and about the topics or areas of discussion in general. When a new member joins that community those threads are useful guidelines as well as the actual forum guidelines as to how to start participating in community activities.

Synchronous discussion or conversation tools exist as well. In those tools participants need to be present at the same time. However in most of them the content of the conversations and the conversations themselves only live for a short period of time. That is, logs of conversations are not saved anywhere. Therefore there is little, if any, evidence on the web of them ever happening. Example of synchronous discussion tools are instant messaging and chatrooms.

Weblogs

Weblogs or blogs are online journals containing posts written by one or more people. Most weblogs however are owned by only one person, usually called the blogger. Co-authored blogs are scarce but exist to represent the work or thoughts of a group of people, for example a project team. Weblogs provide means for personal expression and sharing information via the internet. (Allan, 2007) Blog posts are longer than discussion forum comments, perhaps like short articles. Writing styles, while informal, tend to be more elaborated than posts in discussion forums. Blog posts are ordered chronologically, the last one always being on top. Posts can be commented by the weblog readers. Therefore each post can become the starter of an online conversation, with readers commenting about the original post.

Bloggers are the hosts of their blogs. They are the ones who design the content and the layout. They can also decide whether to allow comments to their posts or not. When comments are allowed they can moderate them. Most common systems of moderation in weblogs are: placing a confirmation level whereby the blogger has to approve a comment before it can be published, or removing the comments which they do not find acceptable. Unlike discussion forums, where all participants share the same floor, bloggers are positioned in a centre stage within their weblogs, and their readers are positioned as their audience. However, where comments are allowed discussions can arise which can enrich the blogger's post by the addition of similar or opposite ideas. This is a useful source of information that can help the blogger to develop new ideas.

Bloggers can follow other weblogs, and be read by other bloggers. It is not that uncommon that beside their posts, bloggers add a list or blogroll

Figure 2. Blogroll (Source: Software, Software Development and other Atrocities http://clk0. blogspot.com)

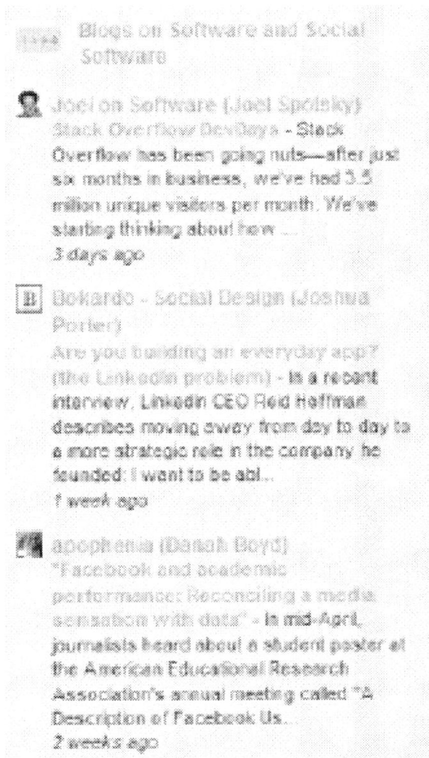

of weblogs they read or are interested in (see Figure 2). This list could be a simple inventory of links to other weblogs or RSS feeds containing short extracts from the latest posts in those blogs. Bloggers can also reference other bloggers' posts in their own posts. When that happens the original poster can keep track of those references by using LinkBack. Linkback is a method that allows authors to get notifications when another blogger inserts links to one of their posts in their posts. These notifications usually come in form of links which are attached next to the original post.

All these interconnections between blogs and bloggers create a network of blogs, also called Blogosphere. Some people could perceive the Blogosphere as a community of bloggers and readers, as they are interacting through their own postings and comments. Having blogs referenced in other media and vice versa – in online forums for example - enhances this perception.

Social Networking Sites

Social Networking sites host a variety of communication tools within them. Sites like MySpace, YouTube and Twitter are categorised as social networking services. Social networks gather users to interact around different themes. For example in MySpace members customise their own personal websites using html, YouTube is focused on sharing videos and Twitter is a micro-blogging environment. Members of these sites create a set of connections with other members who they can call contacts, friends or followers. All these connections form a network of contacts through which social interactions happen, hence the name social networking.

Social networking sites offer their members the spaces to create their own profiles. Through their profiles members can decide the way they can portray themselves to their network of contacts. Other tools available could be classified in two. First, tools which allow members to present an aspect of their persona or their lives, such as photo albums and personal music lists. Second, tools which allow members to interact with others such as discussion forums or commenting facilities. Traces of interactions are left in all these tools showing what has happened between users across time. Moderation of content occurs mostly at a personal level. So it is on everyone's hands to keep control of their own content and the content that is posted in their spaces. In extraordinary cases, profiles and content need to be removed when administrators of the social networking sites believe those belong to people who are trying to disrupt the harmony of the groups.

Concern exists among social networking site users who worry about the privacy of the information they upload in their profiles, most of which is of personal nature. However, in most of these sites users are free to choose who can see their information. So for example, they can choose to keep their data to themselves, show them to their contacts, to everyone in the site, or even to non members. There are also debates about the ownership of the data which some social networking sites would like to claim for themselves but which users believe is theirs.

DISCUSSION

This chapter has presented an overview of the nature of online communities and their dynamics. This overview has highlighted some aspects which concern the way online members organise themselves to keep their participations harmonious, and which keep the communities alive. Distinctions between different online environments show how various kinds of participation could be achieved depending on the tools and strategies used.

Being this a book on e-governance in this last section I would like to highlight some possible areas of intersection between the development of online communities and the efforts of governments at trying to provide better electronic services and get community engagement.

For a few years now, governments and local public institutions around the world have found the internet and other electronic networks useful tools to carry out some of their activities. In its beginnings e-government was associated with e-commerce and hence it had a focus on transaction and customer satisfaction. An example of this would be paying taxes online. However, with the advancements in technology and the emergence of new uses for that technology, public administrations have now the potential capacity to transform their processes by introducing e-governance. New web technologies – like web2.0 - allow people to interconnect and interact between them in different ways. Apart from the obvious benefits of modern electronic tools (saving money, increasing efficiencies, 24/7 access), there are other perhaps more important benefits to public administrations and their citizens brought by the use of the internet. Research in this area shows that the internet helps democratic institutions by increasing their transparency and accountability. (Torres et al., 2006) This brings e-government to a political plane (Calista and Melitski, 2007) with more citizen empowerment as citizens are able to see and have a say on what the authorities are doing for them.

UNESCO defines governance as the exercise of political, economic and administrative authority and e-governance as the performance of governance via electronic medium (UNESCO, 2005). In other words e-governance is about governments' efforts to encourage civil engagement via electronic tools. It is about allowing authorities interact with their citizens and vice versa; and thus establishing new channels of communications between them. By doing this, citizens are able to participate in processes such as decision making, development of new ideas, requesting changes or protesting. As these kinds of activities are typical of online communities, understanding their nature and dynamics could be useful for implementing governance in the online medium. An example of the use of online community strategies for governance purposes is shown in figure 3. This is an extract from the White House weblog where the writer (who reproduces the authorities' requests) offers readers (citizens) ways to contribute with

Figure 3. Authorities ask citizens for their opinions (Source: The White House weblog, www.whitehouse.gov/blog, date of post: 13th of May 2009)

- Have a suggestion for reforming food safety policy? Drop it in our comment form, or let the Working Group know on twitter (hashtag #WHsafefood) or at our Facebook page.

their comments on Food Safety issues. In this instance there are three possible ways for giving opinion, one being an online form, and the other two via social networking sites.

Figure 3 shows how authorities can develop initiatives by employing tools used by common citizens. With these media they become more accessible by placing themselves at citizens level but without abandoning their status as authorities.

Another important aspect that should be considered is that the online world is not only a technological world but a gateway through which people would like to contact other people. Having the most advanced technologies would not help if citizens do not get the impression that someone is listening and doing something about their concerns. Therefore what is needed is authorities or their representatives getting involved in online dynamics rather than portraying themselves as sophisticated electronic tools. An illustration of this is given in Figure 4. 10 Downing Street has read a post by a Twitter user and is replying to him through the same medium.

These last two illustrations (figure 3 and 4) are good examples of how governments can engage in online community participations by using media which is already established in the mind of citizens. Every common citizen can have a weblog or an account in Twitter, and indeed millions of them have one or more. Citizens like those would probably feel more indentified with a government's face with which they are familiar. By using these means of participation governments and citizens must of course play by the rules and norms that allow these communities to exist. Note how 10 Downing Street uses Twitter's conventions of communication by using the @ symbol to refer to another person or entity.

Rather than using top-down strategies e-government efforts can also focus on bottom-up approaches by giving citizens more freedom to shape their own ways at accessing their authorities. With this I do not mean that citizens should

Figure 4. 10 Authorities interact with citizens (Source: 10 Downing Street Twitter site, date of post: 15th May 2009)

design or own e-governance initiatives. That is in the hand of governments and authorities. However, I think that these initiatives could embrace some community success factors so as to incorporate all the richness from online dynamics and content. As I explained in the previous sections, these factors emerge from the online participations. For example, considering the reasons and motivations that will lead citizens to use a web site to access their authorities, or by being aware that such participations do not always happen on a personal basis but emerge from community interactions.

REFERENCES

Allan, B. (2007). *Blended Learning: tools for teaching and learning.* London: Facet Publishing.

Baym, N. K. (1995). The Performance of humor in computer-mediated communication. *Journal of Computer-Mediated Communication, 1*(2).

Calista, D. J., & Melitski, J. (2007). E-Government and E-Governance: converging constructs of public sector information and communication technologies. *Public Administration Qualterly, 31*(1), 87–120.

Coakes, E., & Clarke, S. (2006). The Concept of Communities of Practice. In E. Coakes & S. Clarke (Eds.), *Encyclopedia of Communities of Practice in Information and Knowledge Management* (pp. 92-95). Hershey, PA: Idea Group Reference.

Elliott, M. S., & Scacchi, W. (2003). Free software developers as an occupational community: resolving conflicts and fostering collaboration. In *Proceedings of the 2003 ACM SIGGROUP conference on supporting groupwork* (pp. 21-30). New York: ACM Press.

Greenfield, P. M., & Subrahmanyam, K. (2003). Online discourse in a teen chatroom: New codes and new modes of coherence in a visual medium. *Applied Developmental Psychology, 24*(6), 713–738. doi:10.1016/j.appdev.2003.09.005

Grefter, A. (2006, September 16). This is your space. *New Scientist*, 46–48. doi:10.1016/S0262-4079(06)60500-9

Jacobson, D. (1996). Contexts and Cues in Cyberspace: The Pragmatics of Naming in Text-Based Virtual Realities. *Journal of Anthropological Research, 52*(44), 461–479.

Kettinger, W. J., & Grover, V. (1997). The use of Computer-mediated Communication in an Interorganizational Context. *Decision Sciences, 28*(3), 513–555. doi:10.1111/j.1540-5915.1997.tb01321.x

Lee, F. S. L., Vogel, D., & Limayem, M. (2003). Virtual community informatics: A review and research agenda. *Journal of Information Technology Theory and Application, 5*(1), 47–61.

Marcoccia, M. (2004). On-line polylogues: conversation structure and participation framework in internet newsgroups. *Journal of Pragmatics, 36*(1), 115–145. doi:10.1016/S0378-2166(03)00038-9

McDaniel, S. E., Olson, G. M., & Magee, J. C. (1996). Identifying and Analyzing multiple threads in computer-mediated and face-to-face conversations. In M. S. Ackerman (Ed.). *Proceedings of the 1996 ACM conference on computer supported cooperative work* (pp. 39-47). New York: ACM Press.

Millen, D. R., & Dray, S. M. (2000). Information sharing in an online community of journalists. *Aslib Proceedings, 52*(5), 166–173. doi:10.1108/EUM0000000007011

Parks, M. R., & Floyd, K. (1996). Making friends in cyberspace. *Journal of Computer-Mediated Communication, 1*(4).

Rheingold, H. (1994). *The virtual community: finding connection in a computerized world.* London: Secker & Warburg.

Thomsen, S. R. (1996). @ work in cyberspace: Exploring practitioner use of the PRForum. *Public Relations Review*, *22*(2), 115–131.

Torres, L., Pina, V., & Acerete, B. (2006). E-Governance Developments in European Union Cities: Reshaping Government's Relationship with Citizens. *Governance: An International Journal of Policy, Administration, and Institutions*, *19*(2), 227–302.

UNESCO. (2005). *E-Governance capacity building*. Retrieved December 2008, from http://portal.unesco.org/ci/en/ev.php-URL_ID=4404&URL_DO=DO_TOPIC&URL_SECTION=201.html

Chapter 12

Ways of Citizen Learning: Political Deliberation on the Internet

María-Ángela Petrizzo-Páez
ENDOLOGICA, Open Informatics and Professional Services Cooperative, Venezuela
SPIRALIA C.A. Networking Learning Solutions, Venezuela

Francisco-Javier Palm-Rojas
University of Los Andes, Venezuela

ABSTRACT

Citizens talk, act, relate. In these processes, political deliberation emerges as a citizen exercise tool, and nowadays, also as a phenomenon of learning and training susceptible of being mediated through Web 2.0 tools. For the citizen to practice as such in full, he must, therefore, assume a political training process which contributes to his recognition in a broad and plural institutional context. What actions must be carried out by the institutions to account for the needs and expectations of citizen training? This paper pretends to reveal the pertinence of initiating the debate regarding the training needs of the citizens with the goal of understanding the use of ICT as a tool for deepening citizenship exercise.

INTRODUCTION: ABOUT THE CITIZEN SENSE OF BUILDING THE CITY

Western logic and common sense brings us near to the notion of a citizen defined by its normative character and which understands him as a receptacle of rights and duties that the State and institutions give to the individual. In this sense, a citizen is that individual that, referred to a determined territory and by means of laws, acquires a civil rights and duties package in the first place, then a political one and finally a social package consecrated like that in the document that orders the social and political endeavors in such territory. This classic vision of the citizen is, also restrictive and excluding on the very same rights that it pursues to guarantee, in the same way in which the genesis of the use of the term—the ancient Greek polis—also was.

This is a vision of the adjectived citizen: the citizen that it is understood as such if and only if there is a territory that determines it and if he can be ascribed as an individual in full use of his political rights. The citizen is of the territory, and not the opposite. In this perspective, not all the individuals that dwell in the same geographical space are citizens its citizens and besides, some may be citizens of

DOI: 10.4018/978-1-60566-860-4.ch012

several territories simultaneously. Thus, to be a citizen is a condition linked to the nationality. The citizen defines himself and in front of others as long as he enjoys a geographical reference more than a cultural one, although the latter is also present even if it is a little unclear. Without that reference, the citizen condition is not only lost but it becomes inapprehensive from an analytical and experience of life point of view. This notion of citizen concept results insufficient to account for the social processes of today which emerge marked by the prevalence of the global over the local and furthermore, it makes possible anonymity and passivity of groups socially excluded of key processes like political decision making (Bañez Tello, 2003).

There is, however, an emergent logic—although it is not recent—that brings us close to a more anthropological vision of the citizen in which any individual that enters into a conscious relationship with the surroundings in which he dwells (the city) is considered as such, without nationality mediating in that relationship as a pre–condition. From this vision, the citizen is understood as a unity which influences its surrounding and not as an additional element of it. We are talking about the vision of substantive citizen, that is to say, the individual that without nationality or territorial affiliation of the place he inhabits as pre-conditions (citizenship in normative terms), he is committed to the preservation of these spaces (social and environmental, for instance) and he does it in a more intense, active and constant manner than the adjectived citizen.

In essence, it is the same citizen who has been seen first from an exclusive perspective and strictly circumscribed to civil and political rights, and later under a holistic and integrating vision that, if in fact it does not completely spin off from the former, it has been able to surpass it and advance towards a more responsible and conscious idea regarding its role in the determination of the destiny of the social space (city, community) in which he dwells. The city becomes a physical and social place that is the subject of the collective construction by part of active citizens and it also becomes a space that influences directly in the citizen's endeavors. In one way or another, citizens that practice what in a another place has been called active citizenship (Bañez Tello, 2003; Petrizzo, 2005), defined by a set of processes that mark the citizen with the sense of belongingness and co-responsibility with the surroundings, seek to make the city a space in permanent construction in such a way that it allows holding and protecting that which is conceived as good for all the members of that community and which hereafter we will call: public good. It is, as Oliveira says (cited by Wainwright and others, 2007), an exercise of total autonomy in order to know how to decide, to decide and carry out the decisions with which he has a compromise. In short, it is about a process of democratic and social deepening, where participation and deliberation are not only activities of the individual but that it is supposed citizens conceive them as a subject of their participation in the social endeavors that they know well and of which they are main actors.

Several authors like Tomás Rodríguez Villasante (1994), insist on that justification of citizen participation as the active exercise of citizenship, considering it as something more than delegating in some specialists of politics or in some economic agents. It is an exercise of the free game of initiatives of different groups that feel responsible and give their own solutions, generating a dynamic and creative society. J. Ibáñez also insists on the significance of social participation in contributing to overcome exploitation in the following frames: in the economic frame with initiatives of social economy, with the democratization of the labor world, with cooperativism, etc.; in the ecological frame setting limits to growth in order to make it compatible with the ecosystem; and finally, in the cultural frame with respect of civil rights, participative democracy in the context of cultural multiplicity and with recognition of the right to be different (Bañez Tello, 2003).

The sense of being a citizen is, in essence, a social fact defined and seen in a progressive way through a number of tensions and exchanges among rights, duties, institutions and public policies, in short between the particular and the collective (Borja, 2002). We can advance towards the understanding of the nuance that it is imposed on this dialectics, in a scenario marked by the searching of an active citizen and by the presence of technologies in the social endeavors and in particular of information and communication technologies (ICTs), and also on ways in which their presence in government activities is going to be imposed on the citizens endeavors.

With the evident incorporation of the ICTs in the automation of the processes, the tasks and the functions of the State and the Government, its incidence has also been notorious in a progressive way in the forms by which citizens and institutions of the State are linked and converge in any phase of the process of public policies, not only offering them new channels of bidirectional communication with public institutions, but also revisiting the very sense of the idea of the citizen as actively responsible of his surrounding when he broadens the horizons of the formation of the citizen, of the tools available for that and of the very notion of the community and the surroundings. In the use of these new channels of linking of citizen-institutions-processes of public policies, a new category of analysis has emerged: the cybercitizen (e-citizen), understood as a label assigned to describe a new kind of citizen who uses technocratic aspects from the administrative apparatus to inter relate with public institutions. However, from a substantive vision of the sense of citizen endeavors, the exercise of this dialectics between citizen and institutions cannot be observed under the premise that it is the individual or collective demands market which inspires it, even if technology would seem to enclose this perversion. It is not the political market, understood from a simple instrumentalist position, what determines the emergence of citizen dialectics through the ways that the ICTs offer.

So, if we refer to citizen institution relations, the incorporation of ICTs offers new channels for linking them, but this is also true for the relations between citizens. In this sense the massification of the ICTs has led to the intensification and diversification of the information flows between people, but has also allowed the creation of spaces for conversation, action and social activism. Cyberactivism is the name that this set of tools and practices receive and which allow citizens ubiquitous and real time grouping around given causes. In some instances, these movements have shown effectiveness in the incidence in decisions and public policies, proof of that was the proposal in the European Parliament to create patents for the Internet and the course taken by the movement against that (La pastilla roja, 2006), most recently, transmission through internet, with real time updating in Web 2.0 platforms such as Twitter, Facebook and individual's and cyberactivist's Blogs of awful happenings such as the Mumbai terrorist attacks (Gonzalo, 2009), or significant situations such as elections or electoral debates (just to mention some examples, refer to Givtweets (hppt.//www.govtweets.ca/) or a page created recently to follow on the electoral promises of Barack Obama hppt:// politifact.com/truth-o-meter/).

Understanding, as stated before, the sense of being of the citizen as the construction of the city in which he inhabits, through the configuration of different spaces of convergence with his peers and with the institutions in the making of decisions and construction of public policies, the presence of means such as the ICTs so that this dialectic takes place doesn't imply the reconfiguration of the very sense of the citizen but its manifestation through other ways. This way, it is not wise to assume that the cybercitizen is something different fron the citizen. The best thing is to place the citizen as the creating and modifying entity of his surrounding and who assumes an active-deliberating role characteristic of open democratic surroundings, instead of a passive-approbatory role which is present in the majority of our traditional representative democracies. Hence, the idea of cybercitizen, in its last sense, is not very different to that of the sustantivated citizen that we have already looked at. It is not a person that branches off as a citizen and as a cybercitizen in his sociopolitical endeavors, it is an individual who exercises his condition of a citizen through different ways.

DIGITAL ENVIRONMENTS OF CITIZENSHIP EXERCISE

The concept of cybercitizen was mentioned before as an analysis category that accounts for the collective or individual action exercised by citizens through electronic media and therefore as the exercise of the responsibility in the construction of the surrounding of the individual (city or community) through other ways. Up to a few years ago, the direct and personal links were privileged in the exercise of conjugating wills around given causes. Today, however, the social causes or of citizen activism count every time more with answers in virtual spaces thanks to tools like blogs, cellular texts, profiles in virtual and microblogging platforms (through text, images or video) which not only allow keeping its members informed but also enable the very organization of the actions and recruiting new activists.

When Tim O´reilly and his team spoke for the first time in 2004, about Web 2.0, they tried to appoint a group of emerging web tools and how to distinguish them from what was done in that moment and which gave a static character to contents in the internet. Being Web 2.0 the set of

those utilities and services of Internet that are supported by a data base, which can be modified by users of the service, be it in its content (adding, changing or erasing information or associating data to the existent information), be it in the way of presenting them, or in content and form simultaneously (Ribes, 2007)

These tools finally conjugate, on the one hand, the opportunity of grouping users in communities and on the other hand the possibility of using tools like blogs, sms, wikis or social networks in exchange and customization of the information on which they convey a dynamic character regarding contents and flows among users. These features, together with sustained growth in the last five years of the Internet penetration and its number of users have had a direct impact on the progressive diffusion of the use of ICT tools in the type Web 2.0-social conversations and therefore in the generation of particular dynamics and synergies between users.

China is the territory with more users of Internet and represents 18% of the audience, according to data from ComScore which is estimated to be approximately 180 million surfers. Behind it, United States follows with 16,2% of users, Japan with 6%, Germany with 3,7% and the United Kingdom with 3,6%. In general, the region of the Asian Pacific is the world zone that comprises the greatest audience in the Network and it supposes 41% of it. After that, the European continent follows with a user percentage of 28%, North America with 18%, Latin America with 7% and the Middle East and Africa with 5%. The place that wins in visits is Google with no doubt, with a number of visitors of 777.9 millions in all its sites, followed by the virtual place of Microsoft with 647.9 million of visitors and Yahoo with the figure of 562.6 million. Finally, Facebook with 222 million users, the most famous social network in the world has experienced an overwhelming growth in its number of visits with 127% respect to last year and has become the seventh most popular property most popular in the world. Taken from (http://mouse.cl/detail.asp?story=1999/01/23/15/01/16), consultation made on 24 of January of 2009.

The dynamics of the evolution of services in the Internet has both a technical and a cultural component. In a way, it is evident that there has been a growth in the number of hosts from some 200 in 1980 to near 72.500.000 in 2000 and probably in the current days this figure has been surpassed more than 70 times, greatly due to the surge of cloud computing and Web 2.0. On the other hand there is also a cultural component that has to do directly with the contents available through these services and with the behavior of the internet users in terms of its use. The analysis of the contents in the internet in order to examine its growth can be understood by reviewing the statistics of languagesused in it (of the users according to their residence but also the languages of the contents) in face of statistics by geographic zones. With respect to the language and by what can be understood from the information supplied by the Internet World Stats (http://www.internetworldstats.com/stats7.htm), among the 1.463.632.361 of internet users, the English language continues to be the language used in a privileged way (with approximately 430.8 million users), followed by Chinese (with 276.2 millions) and Spanish in the third place (with 124.7 millions, 8.5% of the total number of users, which it does not represent even half of Brazil's population). This figure could vary drastically if accounting for the use of the internet, we add to the number of users that have English as their native language those who use English as a second language to communicate. This could raise to 750.1 million the number of internet users that communicate or that provide contents in English to the network. However, if we cross this information with the number of users per geographical area, we see that, for the case of Latin America, of the 139.009.209 estimated users in Latin America and the Caribbean, only 89.7% uses internet in Spanish, a quantity really low for a continent with only 51 million estimated users with languages different from Spanish.

Unfortunately the official statistics of internet use refer almost exclusively to the number of users connected, and little reference is found about integral and systematic analysis regarding trends in the use by the internauts in terms of most used applications, time of connection or contributions in contents by geographic regions. Nevertheless, something may be assumed with respect to internet contents and users who produce them especially with respect to its trends and attitudes in the use of the internet. Puyosa (2007) argues that Web 2.0 users are much more permeable to test innovations in Web 2.0 technological tools both in their links with other citizens (which, in turn tend to be greater every time), but also in their links with governmental activities and moreover, much easier to make them linkable to social movements and to become speakers of their own opinions and of those of others. These characteristics and others will undoubtedly affect the way in which individuals are using Web 2.0 tools to participate in them, interact with other individuals and hence to influence their environment. We will come back to this issue later.

The case of Wikipedia (http://www.wikipedia.org) is significant not only for the numbers but for the model in virtue of which the different articles go through an organization, revision and approval process previous to their publication and moreover, it is done by the same users of the tool to which they have free, open and not excluding access. While in Spanish, only a little more than 440,000 articles are available, in English there are more than 2,722,000. German is the second language used the most by users of Wikipedia with an estimated production of more than 860,000 although well below the English production. Certainly, these numbers do not correspond to geographic distributions but idiomatic in the first place and secondly a cultural distribution. More and more Wikipedia is the site of reference preferred by those who want to have the more basic ideas of whatever topic. Based on a wiki type platform (that allows the simultaneous co-edition and in real time of any page by any user), Wikipedia is nurtured on a daily basis with contributions of contents made by hundreds of users all over the world.

On the other hand, the revision of the use of networking, blogging and microblogging platforms (wordpress, Blogger, Hi5, Facebook, MySpace, Twitter, Plurk, Digg, Meneame or Pownce to only name a few), lead us to similar conclusions with regard to the characteristics of the users mentioned above. However, very functionalities of these platforms constitute them by themselves as open spaces to user conformation of groups and communities and therefore, in spaces exceptionally useful for the exercise of cyberactivism or citizen activism in the network, while its integration with institutional platforms of electronic government can also suppose an even deeper reach.

The use of ICTs in general and of Web 2.0 tools in particular, is mediated by learning and appropriation processes that generally happen guided by the hand of other users. Starting from the recommendation of one to another, users go on integrating themselves progressively with the use of any of these tools and, aided by the existing documentation and by trials and errors, they go on discovering its functionalities. This learning process guided by peers is based on mutual trust and recognition of the links around common interests that join the citizen to his peers.

Some state institutions, also in a progressive way, have incorporated ICTs in the automation of their processes and their administrative tasks which link them with other institutions, with enterprises and with the citizens as part of the so called Electronic Government. This is a public administration strategy that has been implemented since the 90´s in the XX century. The incorporation of the ICT to the governmental exercise has required, on occasions, simple operational changes in the tasks developed institutionally by Public Administrations and in other cases, deep modifications and revisions, both of the tasks and legal frameworks which regulate them and the entities that develop them. While this revision process of the public administration and the institutions that comprise it occurred, expectations of different intensities have been generated in the citizens and in the institutions. On the one hand, citizens hope to come near to real empowerment, which implies a revalorization of their role in the frame of the generation of the public policies agenda and also in the decision making processes. Institutions, in turn, expect a progressive decongestion of the bureaucratic apparatus and also a progressive increment in the values of efficiency and efficacy of services, by means of practices like outsourcing and management services and tasks, or the simplification and automation of transactions and processes.

If we look at it well, while the citizen hopes for a sustantivation in the exercise of his condition, it seems to be that from State institutions, the first intent is assisting the citizen in a client-supplier relationship that contravenes the expectations of empowerment by part of the users. The generation of this kind of dialectics on the sense of the citizen in the exercise of government is an important source of tension and a challenge for the processes of citizen's education launched by the initiative of the collectivity or of the institutions.

In this concert of tools, procedures, motivations and citizen causes, interactions among individuals linked to local activities also take place, taking advantage of Web 2.0 for collective action, be it local or not (surveillance tasks, popular initiatives, control and deliberation among others) and between these and public institutions (through public consultation of proposals for regulations or laws for instance). In virtue of this it weaves a web of ties to dialectical relations too which shape well differentiated dynamics between citizens and between them and the government institutions and which allow the construction of a sort of socio-political networks that incorporate, in addition to individuals, public institutions. The first ones have a conceptual reference in what has been advanced which is cyberactivism and they influence somehow the phenomenon of deliberation, and the second ones have their reference in the strategies of the e-government and should constitute a support for the emerging processes of political deliberation through the process of public policies. Examples of the first ones abound in the internet: from citizen journalism which keeps opening the participation of particulars in the day by day in the divulgation of local news with global or local impact (Castells, 1999), up to the surge of more subversive movements such as the so-called hacker-activism, those who seek the rein vindication of social rights for digital spaces acting against the regulation of the internet, those who support opening digital contents in free formats as part of the free knowledge (see the Devolución movement), and those that denounce activities against individuals, collectivities or the environment, among others. In the case of the second ones, the formulation of participative budgets in several parts of the world has in the ICTs evermore frequently a tool for support through the process.

We talked before about the constitution of the adjectived citizen as a permanent process and based on dialectic relations between institutions, norms and citizens. This process arises from the generation of learning dynamics based on the exchanges that occur in this web of dialectic relations and which lead to the exercise of political deliberation as a collective activity oriented to citizen participation in the construction of decisions of public character. This exercise assumes that the generation of citizen learning dynamics occur, in general, in non formal spaces and furthermore that a previous and progressive empowerment of the citizens exists parting from the principles of equity and amplitude during the process of deliberation as basic assumptions. Although the changes observed during this empowerment process occur to a rate lower than what is desired, there's no doubt that citizens that nowadays have access to the ICTs (a number very inferior to the minimum desired at the world level) have seen a refurbishing of their tools for participation and political deliberation.

The recurrent practice of political deliberation helps citizens and institutions walk forward towards the construction of the deliberative democracy, understood as

that in which the legitimacy of the political decisions depends on having paid the sane attention to the reasons of all of the citizens in processes of debate over issues of public matter (Bujanda, 2005).

Therefore, deliberation is a reflection and rational process that even though it depends on many factors for its achievement, it aims at making possible the creation of a space for exchange for the points of view known with respect to a course of action, and political deliberation is a participative process of confluence of citizens opinions and a convergence of these in the debate on what Kingdon (1995) called opportunity windows in public policies and by the same token in the conformation of the public or institutional agendas about decisions regarding the public good.

As a process linked to citizen exercise, political deliberation is increasingly considered as a fundamental principle of national political ordainment norms (supranational constitutions and norms). Its presence complements the idea of the participative democracy that at the same time already enhanced the reaches of the representative democracy, in such a way that the progression in political rights that the notion of citizenship has experienced is also being experienced by the models that have adopted the democratic exercise recently.

In the field of the dynamics of citizen learning, the exercise of deliberation brings with it, besides the natural learning about institutions, mechanisms and insertion procedures in the political endeavors of the proposals of the collectivities, one that is deeper referred to the mutual acknowledgement between citizens, social organizations and governmental institutions with a sui generis conversational process arising from this recognition which should lead to the evermore active participation of the citizen en decision processes, but also in his close ties with the elaboration and presentation of proposals to the collectivity and to a strengthening of the institutionalism and legitimacy of state entities.

This is how the deliberative process of public decision making becomes not only an instrument of citizen action, or exercise of active citizenship, but also a sponsor and support of the process of civic education. It is no coincidence therefore that content related to modes of deliberation and its requests, are ever more frequently a reference, explicitly or implicitly, in the courses of primary, secondary and tertiary levels and vocational courses.

If we assume that the public good is that which is good for everyone, that it belongs to everyone and that it is the responsibility of all without ever belonging specifically or being the exclusive responsibility of a particular interest and if on top of that we consider it a body of values and fundamentals directly identified with the social project that pursues its defense and cultivation, it will be a premise whose definition and attention must nor vary in function of the tool or space used for political deliberation and that must happen regardless of the range of action that the citizen, the officer or the public institution involved in such dynamics has.

The expectations of empowerment generated and cultivated by the citizens has influenced the opening of new spaces for public debate on decisions which opens up the possibility of exerting a higher citizen control on the decision processes. However, it is important to remember that in the frame of democratic exercise, respect for the idea of public good must be attended both by the citizen and by the public institutions, and this turns critical in spaces susceptible to drive the satisfaction of the individual needs like virtual space, mediated by the ICTs which make possible a direct contact between officer and citizen. In addition, when political deliberation occurs in the frame of its traditional mechanisms of exercise (for instance, open forums, popular consultations or a call for proposals presentations in local power spaces, among others), the idea of public good is much more clear than when this happens in spaces that we call virtual or digital in which, in general, the establishment of an order during the debates is a task somewhat more complicated, not always done in real time, with a systematization process of proposals generally more tedious and where, in general, in front of the citizen who goes to these spaces an officer is not found with a clear role in the public policy process but instead we find an officer with functions and attributions that are more specialized every time but with a limited range of action in the generation of operative changes by virtue of the permanent search for optimization of the bureaucratic apparatus.

In these contexts, preservation of the idea of public good as a political deliberation conductor axis becomes much more necessary as an antidote of the creation and strengthening of systematic vices in the bureaucratic apparatus that allow filtering particular interests hidden under the general interest and as an cornerstone of a debate to bind the public good with the most recent ideas of digital rights.

E-CITIZENSHIP, SOCIAL NETWORKS AND POLITICAL DELIBERATION

We have talked about e-citizenship above, in the way that citizens go on building frameworks of dialectic relations that make possible the formation of social and political networks and since this phenomenon occurs also enhanced by the incorporation of the ICTs to the social endeavors. A social network is a social structure comprised by individuals and by their ties. Social networks are weaved around interests, actions or motivations that are common to all members of the network and the flow of information, contacts and the power within it, through the links between individuals, allows maintaining the structure. The ties between the individuals occur with different intensity and this is a fundamental reason for them to have a different relevance in the network and for individuals with a better position than others.

As a relational structure and as social analysis category, social networks are prior to the recent emergence of the so called social networking sites. Furthermore, social networks support the creation of communities and maintain them through the relations that occur through interpersonal contacts, now mediated by ICT as a catalyst for the emergence of a sense of belonging to the group of individuals (Maya-Jariego, 2004). The surge of Web 2.0 tools has opened up the social networks to the utilization of social networking platforms facilitating a more rapid formation of the fabric of such relationships and also a greater intensity in the existing ties.

What makes social network sites unique is not that they allow individuals to meet strangers, but rather that they enable users to articulate and make visible their social networks. This can result in connections between individuals that would not otherwise be made, but that is often not the goal, and these meetings are frequently between "latent ties" (Haythornthwaite, 2005) who share some offline connection. On many of the large SNSs, participants are not necessarily "networking" or looking to meet new people; instead, they are primarily communicating with people who are already a part of their extended social network (Boyd y Ellison, 2007).

In this sense, social networking sites (or networking platforms such as Facebook, Twitter, Indenti.ca, Plurk and others), do not exclusively seek to create inexistent networks, but to serve as a tool to intensify ties in networks that, in some way, are already formed and that are developing in other spaces of the daily endeavor of the individuals. It is true that users come to meet new persons by means of these ICT tools, relating to others thanks to recommendations of third parties or with whom they perceive a certain degree of affinity with regards to their personal or collective interest, but this does not reveal itself as the central objective of the relationship of the individual with the social networking sites. In addition, these virtual spaces have made possible the verticalization of the social networks and the democratization of tools for its management: the intensification in the use of Web 2.0 has brought the end-user close to network management tools through drivers of contacts and creation of groups of users thus controlling the information that flows to each one.

It has been thanks to social networks supported by ICT that news, rumors and political opinions originated in their hearts and spread by their users have been able to have more space and receptivity in virtual and not virtual spaces generating an information exchange and divulgation dynamics of information which is interesting although it is not always contrasted with the original sources. Additionally, the emergence of what it is now called citizen journalism not only has stimulated the creation of blogs and other forms of digital logs in which every day more and more citizens display their opinions in face of the political options, but it has also driven the presence of informative sites on the Internet through alternative media of communication and information which increase the flow of information received by users. More and more, proposals for divulgation and debate of policies, news or initiatives have some repercussion on the internet, through discussion forums, blogs and fotologs, webpages of groups or events in spaces such as Facebook and Google Groups, and then poured into social networks. This way, surprising events have been covered such as terrorist attacks and accidents and events known as the most recent elections in the United States. Somehow the e-government strategies include within its functionalities the online connection with the citizens through forums, blogs and microblogging.

But, who have real access to the alternative digital communication media or to the use of Web 2.0 tools for the strengthening of communities or citizen action? A brief examination of the statistics presented in the first part of this chapter gives us an idea about the small number of persons that use the internet and by the same token Web 2.0 in comparison with the world population, all of that in spite of the notorious increment of subscribers who have tried tools like Twitter (more than 500% last year according to Compete Http://www.ojointernet.com/noticias/twitter-crecio-un752-en2008-segun-compete). This is a contradiction of bringing the telecommunications to the masses. The ICTs are working operatively in a good part of the provinces in some of our countries, but with important limitations to access for a good part the world population, in the first place because of issues of access to goods and services but also because of an issue in skills and abilities in their use and this results in a very small number of users accessing Web 2.0 tools and an even smaller number is that of the persons that uses these tools to deliberate, show their political tendency, practice citizen journalism – not just rumors – or to enrich digital spaces with dynamic content characteristic of the so called alternative communications media.

This panorama of excluding access to ICTs deserves, however, showing that such exclusion is not due exclusively to instrumentalist elements of access to computers or Internet connections. Internet access is then revealed as a complex issue which goes through reviewing the provision of equipment and connectivity, but also goes through the use citizens give to electronic media. Recently, Andrew Schrock (2009) published in The First Monday a study that reveals psychological characteristics of ICT users. Starting from the revision of the subscription in some so called technology clusters (Rogers, 1995), the author illustrates that the preference for use of these by part of the users obeys to some psychological factors such as extroversion, anxiety or skills for handling technology. Likewise, Zed digital (2008) in a study on social networks in Spain concludes that only 50% of internet users utilize Web 2.0 and that while the young use it to feel as a part of a group, adults close to 40 years old seek in its use an opportunity to follow the latest tendencies. A very high percentage of the people surveyed use Web 2.0 to download videos, movies or pictures, which may justify the inverse relation between Web 2.0 use and the consultation of other communication means such as television or the written press. Besides this, it is known that most of the so called internauts use between two and three Web 2.0 tools simultaneously and are subscribed to many more although they don't use them. Social networking sites such as Facebook however, keep enjoying user preference according to Compete, having at the beginning of 2009 close to 1,200 million visits.

Regarding the ties with the public administration and in terms of e-government little has been consulted on the preferences of citizens, neither are there clear observations about their expectations. There are values such as transparency, precision, clarity and integration of services that are appreciated both by citizens and by entrepreneur users of the services of public administrations, but even in those cases where electronic government strategies have advanced beyond the mere unidirectional transactions between administrators and citizens, it does not seem like a dialog or a debate has been established between users and public administrators regarding the necessary characteristics that platforms and applications should have in order to contribute to making process in the framework of mutual learning more dynamic, opening spaces towards the construction and enrichment the platforms of electronic government.

It is also probable that the lacking of spaces for dialog, deliberation and learning mediated by ICT between citizens and institutions can be understood by reviewing themes such as access, psychological factors of the users or the so called "digital gap". Eastin, M and Larose, R (200), state that the internet auto efficacy is a variable determined by ideas regarding control over the technique, indirect experience, verbal abilities and the psychological responses of the individuals. Complementing the idea of the digital gap with the so called auto efficacy of the internet allows, according to authors, counteracting the evident differences and exclusions not only through the democratizations of connections, access to the internet and computers (technical issues), but also paying attention to – with specified social interventions directed to identified population groups - elements related to attitudinal aspects pointed out before.

But there are two additional elements in the analysis of the use of the internet by part of the citizens. On the one hand, there is the development of tools, mashups and sites for social networking, whose spreading has also grown exponentially propping up the idea of integration between several social networking sites, the expansion of the services and functionality of the tools already available, and also expanding the range of available clients for the different Web 2.0 tools, but suffering sort of an extension of the problems of access experienced by the rest of the Web 2.0 tools. On the other hand, it is important to take up again the theme of the contents that the tools of Web 2.0 offer to its subscribers in terms of the democratization of the access to Web 2.0 and the democratization of the contents. To this respect, Jakcob Nielsen (2006) talks about the inequality participation, accounting for the not equal relation between those who create contents, those who edit them and reuse them and those who observe them. For the author, although figures vary according to the Web 2.0 tool we are talking about, the ratio is 1-9-90, with 90% of the information flows in the virtual communities in charge of 1% of its members. If we talk about blogs, for example, the ratio could be 1-5-95. In terms of the tools of electronic government, distribution of contents can vary significantly in favor of the creators of contents, making differences really huge.

The author points out the need to overcome some psychological aspects (fear, apprehension, and insecurity, for example) so that the users can revert these relations. For that it is essential to initiate and continue a learning process in users and towards the citizens. This takes us back again to the citizen education as a necessity and to the generation of dialogic learning processes as a way to move towards it and to reach levels of citizen appropriation and awareness which will drive the diminution of social and psychological barriers.

TO DELIBERATE AND DEBATE

Throughout this document we have tried to show how the citizen in his sustantivated conception, that is to say, compromised with the environment in which inhabits, faces the need to approach the formation and continuous learning aiming to initiate his own positioning tasks and recognition of himself and of the public organizations and institutions of his surroundings. Mead stated that the construction of his own meanings can only be made from the interaction with other persons and therefore, by means of conversations, debates and dialogs. This process of construction of meanings constitutes on its own, the key pillar for learning. By extension, the process of citizen formation and learning, requires the construction of his own meanings and the full and active exercise of citizenship with the participation of institutions and social organizations in a clear way in those dialogs and conversations.

The spreading of the Internet as a means of communications and ubiquitous data transmission, immediate and flexible, has presented new challenges to citizens and has intensified old challenges. Development of abilities and skills in technology management have emerged as challenges for all citizens, but especially for adults and medium – aged adults. The scarce deliberative capacities of the citizens and their little participation in decision-making processes, remains as pending matters in which ICTs can contribute.

With training for active citizenship as a pending task and the necessity to approach it not only by public institutions but also by the very citizens, this chapter has tried to reveal the challenge entailed by trying to show citizens the Web 2.0 as a tool for deepening their performance as citizens and not as a disruptive and dividing element in their reality.

We have pretended to show that citizen and deliberation cannot be seen in an isolated way and with a perception that puts out of focus the whole, favoring attention to any one of its parts. The greatest challenge, would seem to be, to attend immediately citizen formation mediated by ICT. The necessary generation of learning processes mediated or not by the ICTs and which consider these technologies useful tools in citizen exercise, must also aim towards the enrichment of the processes of political deliberation. Political deliberation may or may not be mediated by ICT and it has several manifestations: cyberactivism is, no doubt, part of this exercise as well as the occurrence of processes of public consultation of laws or presentation of concrete proposals in the face of certain problems.

ACKNOWLEDGMENT

This chapter was translated by Andrés Eloy Mendoza Rodríguez.

REFERENCES

Bañez, T. (2003). Ciudadanía y Participación. In M. J. Benruz & R. Susín (2003), *Ciudadanía: dinámicas de pertenencia y exclusión* (pp. 97-112). Universidad de la Rioja, Servicio de Publicaciones.

Borja, J. (2002). La ciudad y la nueva ciudadanía. *La Factoría,* Nº 17. Retrieved January 15, 2009, from http://www.lafactoriaweb.com/articulos/borja17.htm

Boyd, D., & Ellison, N. (2007). Social network sites: Definition, history, and scholarship. *Journal of Computer-Mediated Communication, 13*(1), Article 11. Retrieved January 2009, from http://jcmc.indiana.edu/vol13/issue1/boyd.ellison.html

Bujanda, M. (2005). *Ciudadanía y deliberación democrática en la escuela primaria con apoyo de las tecnologías digitales. Una experiencia de investigación y desarrollo curricular*. Paper presented at La Educación y la formación de una ciudadanía democrática, Harvard Graduate School of Education and Fundación Arias para la Paz y el Progreso Humano, Costa Rica, San José.

Castells, M. (1999). *La era de la información. Economía, Sociedad y Cultura*. Madrid, Spain: Alianza Editorial.

Eastin, M., & LaRose, R. (2000). Internet self-efficacy and the psychology of the Digital Divide. *Journal of Computer-Mediated Computation, 6*(1). Retrieved January 2009, from http://jcmc.indiana.edu/vol6/issue1/eastin.html

Folgueiras, P. (2005). *De la tolerancia al reconocimiento: programa de formación para una ciudadanía intercultural.* Unpublished doctoral dissertation, Universidad de Barcelona, Facultad de Pedagogía.

Gonzalo, P. (2008). *Los ataques de Mumbai, en la India, consolidan el periodismo ciudadano.* Retrieved from http://www.periodismociudadano.com/2008/11/28/los-ataques-de-mumbai-en-la-india-consolidan-la-eficacia-del-periodismo-ciudadano/

Kingdon, J. (1995). *Agenda, Alternatives and Public Policies*. New York: Harper Collins College Publishers

Lozada, M. (2004). El ciberciudadano: representaciones, redes y resistencias en Venezuela y América Latina. In D. Mato (Ed.), *Políticas de ciudadanía y sociedad civil en tiempos de globalización* (pp. 167-180). Caracas, FACES, UCV.

Martínez, G. (2004). Internet y ciudadanía global: procesos de producción de representaciones sociales de ciudadanía en tiempos de globalización. In D. Mato (Ed.), *Políticas de ciudadanía y sociedad civil en tiempos de globalización* (pp. 181-200). Caracas, FACES, UCV.

Maya-Jariego, I. (2004). Sentido de comunidad y potenciación comunitaria. *Apuntes de psicología, 22*(2), 187-211.Retrieved January 30, 2009, from http://www.cop.es/delegaci/andocci/files/contenidos/VOL22_2_3.pdf

Nielsen, J. (2006). *Participation inequality: encouraging more users to contribuite*. Retrieved February 2009, from http://www.useit.com/alertbox/participation_inequality

Nueva votación en el Parlamento Europeo sobre patentes. (2006). Retrieved February 2, 2007, from http://www.lapastillaroja.net/archives/001122.html

Osler, A. (1998). Conflicts, Controversy and Caring. Young people's attitudes towards children's rights. In C. Holdem & N. Clough (Eds.), *Children as citizens. Education for participation* (pp. 113-126). London: Jessica Kingsley Publishers.

Petrizzo, M. (2005). Participación ciudadana y tecnologías de la información y la comunicación: hacia una administración pública relacional. *Nueva Sociedad, 195*, 88–101.

Puyosa, I. (2007). *Cómo cosechar información sobre los mercados y la opinión pública en el campo de la Web 2.0*. Retrieved February 3, 2009, from http://www.scribd.com/doc/5579367/Como-cosechar-informacion-sobre-los-mercados-y-la-opinion-publica-en-el-campo-de-la-web-20-

Ribes, X. (2007). El valor de los metadatos y la inteligencia colectiva. In *Telos. Cuadernos de Comunicación e Innovación. de la Fundación Telefónica*. Retrieved from http://www.campusred.net/TELOS/articuloperspectiva.asp?idarticulo=2&rev=73

Robertson, R. (2003). Glocalización: tiempo-espacio y homogeneidad-heterogeneidad. In *Cansancio del Leviatán: problemas políticos de la mundialización*. Madrid: Trotta.

Rogers, E. (1995). *Diffusion of innovations* (4th ed.). New York: Free Press.

S/A. (2009, February 12). Facebook es la red social más utilizada. In *Google Noticias*. Retrieved February 2009, from http://www.google.com/hostednews/afp/article/ALeqM5jOlNcddVSFd-kJiJs_HArLf1D5JiQ

Schrock, A. (2009). Examining social media usage: technology clusters and social network sites membership. In *First Monday*, 14(1). Retrieved February 3, 2009, from http://firstmonday.org/htbin/cgiwrap/bin/ojs/index.php/fm/article/view/2242/2066

Wainwright, H., et al. (2007). *Networked politics. Rethinking political organisation in an age of movements and networks* (Work in progress). Amsterdam: The Transnational Institute. Retrieved from http://www.tni.org/docs/200701261425586570.pdf

Zarza, M. (2009). *Las redes sociales siguieron la amenaza de bomba y la explosión al minuto*. Retrieved February 2009, from http://www.abc.es/20090209/nacional-terrorismo/redes-sociales-siguieron-amenaza-200902090959.html

Zed Digital. (2008). *El fenómeno de las redes sociales. Percepción, usos y publicidad. Zed digital*. Retrieved January 2009, from http://www.zed-digital.es/Estudio de redes sociales_20_11_2008.pdf

Chapter 13
Articulated Planning

Jacinto Dávila
Universidad de los Andes, Venezuela

Ana Magaly Reyes
Freelance Consultant, Venezuela

ABSTRACT

The authors bring a notion called articulated planning that defies existing associations between centralised government approaches and socialism. With increasing availability of information and communication technologies (ICTs), the authors argue that governmental processes can be facilitated and develop more participative planning efforts towards developing e-governance in countries. The free software and simulation tools can contribute to generate appropriate spaces for citizens' participation in, definition and supervision of government plans.

INTRODUCTION

A "politically correct" argument, normally presented against socialist systems, is that central planning is an intractable problem, impossible to solve in any nation state with the minimal demands of modernity (Wikipedia, 2009). A complementary assumption states that central planning is the only possible choice for Socialism, from which one derives that socialist systems are doomed to intractability.

The argument (simplistically) explains that modern nations produce, by allowing freedom of choice and expression for its citizens, an extremely complex network of desires and needs, which is impossible to serve by means of a central entity. Sooner of later, sentences the argument, the bureaucracies, established by the central power to process citizens' requests, are overcome by that complexity or, even worse, by upheavals led by frustrated citizens who do not trust those inefficient filters in a clearly insufficient system.

Therefore, central planning, as required by a socialist state that tries to organize production to serve the needs of its people, is a futile exercise only conceivable by immature minds completely out of touch with the real nature of modern human society: individuals have infinite needs and only a free market economy, with all its lacks, can bal-

DOI: 10.4018/978-1-60566-860-4.ch013

ance those needs and individual aptitudes against available resources.

In this chapter, we aim to challenge that argument by attacking the premise that Socialism is equivalent to central planning, CP. We will defend another approach to national or community planning that we call articulated planning, AP, which is based on the intensive use of Information Technologies, IT. In AP, centralism and all its bottlenecks are avoided by implementing a distributed planning system that articulates regional contents by intensively using computers and telecommunication systems.

AP is not quite an actual reality. However, with the increasing availability of IT, specially under conditions of freedom, more and more steps are being taken to make AP commonplace. We want to report on some of those experiences and also speculate on others more elaborated possibilities with computational models and databases (Domingo, C. et al, 2006) to automate many fundamental tasks in any exercise to articulate massive collective contents and agreements.

We must start, however, by admitting that there is, indeed, a inherent complexity to Socialism, specially in its more recent developments which declare a strong commitment to democracy and adopt a bottom-up governance strategy. Given the size of human populations, when people want a real and egalitarian participatory access to government, a very efficient peer to peer, communicational system and a very well organized (an distributed) collective memory seem essential.

THE HISTORICAL PROBLEM OF PLANNING

Planning seems to be a quintessential human ability. Every big endeavor requires some previous thinking to guarantee success. But, when it is about planning a national economy, it both gets essential to any exercises that pretends rationality in the productions of goods, and complex, by the need to take a big number of actors and their intentions into account.

This complex human ability has evolved, in contemporary times at least, throughout a number of variants and points of view, all attempting to improve its results. At the beginning of the XXth Century, probably influenced by the mathematical and positivist tradition, planning became the purely technical process of calculating means to achieve certain, unproblematic, ends. It used to be a very authoritarian activity: a central planner produced a plan that everybody else just followed, on the ground that the planner was the only expert (Ahumada, Jorge, 1969). Later, the anti-positivism movement impacted all these practices and planners began to take into account other possibilities, most of those having in common a problematic set of goals, which created the need for some sorts of negotiation protocols to consider the preferences of all involved: strategic planning (Ansoff., I, 1965) was born at this stage. It also became clear at that time that planning must be an open process, to take into account uncertainty, risks and instability in a changing environment. In later times, interest have apparently shifted from the instrumental aspect of planning to consideration of motivational strategies, specially in enterprises that require the efforts of many people. We believe that the shift is only apparent, because those calculating capacities are now applied to manipulating the will of the persons involved with those motivational strategies. With the increase of computational capacity, it is ever more feasible to analyze preferences variants and produce a plan that consider more sophisticated scenarios in which actors try to maximize their own utilities in an always open and permanent process of planning and production (Mintzberg, H., 1994).

THE TECHNICAL PROBLEM OF PLANNING

A plan can be simply described as a set of actions to achieve certain goal. Producing a plan is, therefore, a matter of connecting that goal to those actions in some reasonable way. Reason is invoke in order to reduce the uncertainty of choosing actions at random, even though the very expedient of projecting actions into the future means that uncertainty cannot be completely eliminated.

Even reducing planning at that simply description is problematic from a mere technical point of view. As (Chapman, D., 1990) established, (that form of) planning is a mathematically intractable problem. However, a more realistic form of planning must account for the preferences of those involved, that may be many, and a dynamic of goal-changing that naturally arises as a result of many requests for action and many points of view about what to do next. As if all these were not enough, there is also the need to interleave planning and action. Continuous planning could be called the process wherein agents do planning while execute some actions. Those processes are not entirely simultaneous, because at acting the world may change and those changes could trigger new plans or, at least, a re-planning of the previous.

The technical problem of planning is, therefore, even more complex in any realistic setting, not to mention a number of psychological dynamics that could also take place among the agents involved. Any solution is bound to be of a complexity comparable with any other higher human endeavor. This could explain why many suggest that there is no real solution or that the solution is not to plan at all.

EMERGENT SOLUTIONS

Despite all those technical limitations, an endeavor as complex as national level planning (Alterman, Rachelle (Ed),2001) is a must for many national governments, even those clearly aligned to a free market economy (.ibid). It seems as if it has become clear that planning is fundamental for human communities although not necessarily related to population density or some other demographic variable (.ibid, pg 15) but instead it turns out that "the decision to adopt national level planning is a question of choice by decision-maker and voters".

Moreover, some national experiences tend to support the thesis that national level planning does not imply central planning or a centrally controlled economy, but just the opposite. Mastop and Priemous (cited .ibid, page 29) explain the preparation of the 2030 plan in the Netherlands: "the four major sectoral ministries, each of which has traditionally prepared its own sectoral plan, are on the verge of espousing a new role. The four ministries are Economic Affairs; Transport and Public Works; Water Management and Agriculture, Nature Management and Fisheries; and Housing, Spatial Planning and Environment. The latter is the ministry legally and traditionally in charge of land-use planning, and in all previous national plans it has taken the lead but coordinated poorly with the others. This time, each of the other three ministries has prepared a policy document which, in effect, is a comprehensive alternative conception to the national plan! One can therefore anticipate that the debates about the 2030 plan will be based on new grounds and with new dynamics. Instead of a debate between a lead agency in charge of comprehensive planning and sectoral agencies that in fact act as lobbies for their interests, the debate will now be between alternative comprehensive conceptions, from different loci. Thus, the notoriety of poor coordination among sectoral plans may have led the sectoral agencies to realise that the best way to effect comprehensive planning is by preparing it themselves!." (.ibid, pages 29-30). One direct outcome of that Dutch planning process is the National Environmental Policy Plan, or NEPP, now in its fourth realization articulated as one of the international Green Plans (RRI, 2009). These are "working models of sustainability in action. Unlike conventional approaches that address environmental problems in isolation of each other, green plans treat the environment as it really exists—a single, interconnected ecosystem that can only be safeguarded for future generations through a systemic, long-range plan of action". The three Green Plans (USA and New Zealand have their own linked plans) have apparently gather a lot of support: "More than 250,000 Dutch businesses are participating in the plan through performance-driven covenants with government. These legally binding agreements target emissions reductions, improved environmental quality of products, and energy conservation. Major reductions in pollution of air, water, and land have been achieved through the plan along with a streamlined regulatory process that is saving both time and money. The Dutch are reducing pressure on the environment even as their GDP climb." (.ibid).

One could say that the self-proclaimed first world has the advantage of a better educated and, therefore, very disciplined society. Once the politically inspired rejection to national planning has been overcome, those results as in the Netherlands must be expected (Notice, however, that the overcoming of that rejection is not a trivial move in a capitalist world). The third world, however, is learning fast. Venezuela, for instance, has adopted IT strategies to support planning in the very rigid and bureaucratic structure of its public sector, inherited from past attempts to establish a liberal, free-market oriented, society. National level planning and control is performed in Venezuela by means of the SIPEG Nueva Etapa (New Phase) system[1], which is used by public agencies all over the country to declare their plans and their achievements. There exists a Plan Operativo Anual Nacional, POAN, National, operative, annual plan for the whole country, made of POAIs, Institutional, operative, annual plans for each agency. Plans are organized by projects, defined in turn by a hierarchy of objectives, actions, subactions and tasks. An achievement is established by means of goals (metas), the name used to refer to, normally numerical, indicators or indexes of success or failure.

The process starts half a year ahead of the execution time, with each institution developing its POAI, which is then fed into the system to be verified. Plans are evaluated after been verified for structural correctness and then central government allocates resources to the whole POAN, which must be expended in the corresponding year. Multiyear projects are allowed, but must be approved explicitly and, nevertheless, financial resources are allocated and reported yearly.

The bottom up dynamic of the Venezuelan POAIs and the impulse of some, e-government initiatives[2] has, however, open the possibility for what we want to call Articulated Planning, AP. There is evidence that, with direct presidential encouragement[3], productive communities organize themselves for planning.

FROM E-GOVERNMENT TO SOCIALIST E-GOVERNANCE

Planning, however, is not the only aspect of public life that is being transformed and benefited by IT. As the UN states: "'Electronic government' – using modern information and communications technologies to facilitate governmental processes – is being used in many countries as part of reform initiatives"[4]. Those initiatives, however, normally fall short of transforming the actual relationship between government and citizens. The same report concedes: "Most electronic government initiatives, however, have been hampered by an over-reliance on technology, too little emphasis on building human capacity, insufficient intra-governmental collaboration and inadequate public consultation. Consequently, the focus has been shifting from electronic government to electronic governance – using technologies to improve interactions between governmental, non-governmental and civil society stakeholders" (.ibid)

Thomas Riley, in Canada, makes the case for the transition from e-government to e-governance (Riley, Thomas B., 2003). He starts by going back to the original concepts citing Ketll (2002): "Government is an institutional superstructure that society uses to translate politics into policies and legislation. Governance is the outcome of the interaction of government, the public service, and citizens throughout the political process, policy development, program design, and service delivery".

Thus, the transition from e-government to e-governance seems to be a movement from internal, self-serving provisioning of IT within the government to creating and sustaining a platform for the general public to actually participate in public affairs and political action. Borrowing again from Ketll, Riley summarizes the movement in Table 1.

Table 1. Movement from e-government to e-governance

Government	Governance
superstructure	functionality
decisions	processes
rules	goals
roles	performance
implementation	coordination
outputs	outcomes
E-Government	E-Governance
electronic service delivery	electronic consultation
electronic work flow	electronic controller ship
electronic voting	electronic engagement
electronic productivity	networked societal guidance

Riley goes on to explain that that view of government has been inherited (in all the western culture) from the views of sociologist Max Weber's theory of rule-governed bureaucracy (Weber, Max, 1920). According to Riley's reading of Weber, bureaucrats tend to believe that the most important aspect is to comfort to the rules that make the whole government enterprise a rational one. That is, bureaucrats are concerned with the "how". Citizens, on the other hand, are actually concerned with goals and outcomes much more than with how they are achieved.

This realization now justifies providing people not just with the "e-means" to interact with public services (electronic service delivery), but with means to participate (electronic engagement), supervise (electronic consultation) and control (electronic controller-ship and networked societal guidance based) governments.

This distinction between e-government and e-governance is crucially important to evaluate the evolution from a traditional, liberal government into a socialist and democratic one. Take again, for instance, the Bolivarian Republic of Venezuela. We have already mentioned some e-government initiatives[5]. But there are many more. There even exists projects to develop new free and open access tools[6]. Ministries (Government agencies and offices) have the own web services[7]. So does the Supreme Court (Tribunal Supremo de Justicia), publishing on line up to 250.000 sentences and resolutions and connecting, via email, the people involved[8]. Venezuelans can declare and pay their taxes on line[9]. Identification[10] and Social Health Services[11] are dealt with on line, at least partially. Country's statistics are also (partially) available on the web[12].

However, none of these systems actually open its corresponding institution to the scrutiny of its goals and evaluation of its performance and outcomes. There is only one, rather timid, effort to allow citizen to complain and follow a case for access to goods or services; originally a customer service agency (Indecu), Indepabis[13] is now becoming a more comprehensive organization for people's rights defense.

Moreover, despite the list of web-enabled institutions, there is almost no integration of online services. Perhaps one of the best examples of the situation of the country is the SIGECOF[14]

(Integrated System for Management and Control of Public Finances). Currently under the control of the national office for public accountancy, the system was originally designed in 1995, but was only release "for production" in 2001, and almost immediately abandoned, due to the lack of system's counterparts in other public offices and of value-adding applications. It was brought back to service in 2006 and currently processes up to 100000 requests per year from 12000 users. All its users, however, are public servants and many decentralized public agencies do not use the system or any other system connected to it, for that matter.

Thus, it is fair to say that a country like Venezuela can evolve in the direction of e-government without advancing in the direction of e-governance, and even less, socialist e-governance.

One could define socialist e-governance in such a way as to include that view of governance that put the actual power on people's hands by means of participatory mechanisms in the execution and supervision of public projects, i.e people's power. In a Socialism with a strong commitment to participatory democracy and, therefore, accountability, public servant's discretionary behavior should be constrained. This is to avoid that they concentrate on what they want (even if it is only following the rules of the bureaucratic state) "rather than what the citizenry want" (See Riley above).

So, again following Riley, "If templates for routinized decision-making could be developed for governance, and these could be loaded onto the Net, these aspects of governance could be built right into the social infrastructure and the arbitrary choices of government officials could be eliminated"(.ibid). A perfect memory of planning and execution of projects could serve to produce that "routinized decision-making" and will also address the issues of goals' supervision and outcomes analysis by the people. These are the reasons for the proposal of the following section.

There is, however, another aspect that could be hampering the progress of e-government and e-governance. Let us point it out with a simple test: we ran a search on Google Scholar® for the combined terms "e governance", "articulated" and "planning"[15] in the year 2009. Out of 22 references retrieved, only one was an open access document (a study of the Chilean case[16]). All the rest were reports, articles and books on sale for a charge, some of them not fully available on electronic form. This is, of course, the traditional dynamic for the distribution of academic knowledge. One could wonder, however, if all the insistence upon accountability and openness that appear on academic documents on e-government, should not be preceded by corresponding academic exercises transferring all that knowledge and experiences to the communities under conditions of freedom, i. e. that anybody could learn and use all the information without having to pay first.

EMPOWERING PLANNING: FREE SOFTWARE TOOLS FOR MICROPLANNING AND MICROEXECUTION

Many of the advances in e-government are due to increasing availability of supporting IT technologies. However, a real advantage for the adoption of these tools for a wider community is their availability under conditions of freedom. Free Software is a huge step in that direction. With Free (Libre, Open Source Software or FLOSS, as it is sometime called[17]), users see their rights increased as anybody has access to the "know how" of the planning tool and can adapt it to fit any specific circumstance. And, as we suggested earlier, adaptability to circumstances is a crucial requirement to make a planning process successful against a very complex system dynamics.

There exist many of those software tools[18]. We have had direct experience with three of them: Planner[19], GanttProject[20] and OpenProj[21]. The latter seems to be in a higher stage of development and it also has in its favour that it can decipher the file formats of another popular but proprietary tool: MS Project[22], making it easier to migrate. Stepping away for closed file formats (as those of MS Project) is fundamental for the sake of adaptability. However, we must acknowledge that proprietary formats have been the first to include consideration of cost and budget support, which are very important for planning purposes.

Nevertheless, the crucial effect of the availability of these software tools is that anybody can use them to plan activities at any level of detail. We can report an experience of what we call microplans and microexecutions construction, whereby an organization[23] is generating their global plan (actually, its POAI) by means of a combined top-down and bottom-up, iterative approach. Fundacite Mérida is trying a method of planning in which, 1st) in the usual manner, a number of high level objectives and goals are proposed, by the staff and following national guidelines, for the following year; 2nd) the central agency (in this case, the ministry) verified and approves those higher level plans; 3th) every person in the organization then unfolds that part of the higher level plan that has received as assignment, creating her/his microplan, 4th) microplans are combined into a big sheet (a.k.a. The planning savanna) for the whole institution, which is used by managers to do strategy planning and load balancing previous to execution and 5th) microplans, updated by new tasks and coordinations with the "savanna", are transformed into microexecution reports of the activities performed by each person in the organization during a fixed period of time. Microexecution reports then feed the personnel evaluation system of the organization. This strategy corresponds to so called management by objectives, as advanced by Peter Drucker[24], although here it is performed in a socialist context: Plans and execution reports end up being available for public, detailed scrutiny.

An important lesson from this experience is that higher level, collective, planning is relatively easy to do compared with the effort required by correct, complete and prompt microplanning. In general, advancing some interesting goals for the organization is a lot simpler than completing the whole plan in its full, including checking constrains on resources and "people-power" available. It is also clear that this kind of planning process is particularly sensitive to personnel motivation and commitment to the task. Any member of the staff can cause significant planning failures by just "dragging his legs" or by not reporting properly. This, however, may not be a weakness of this particular form of planning, but perhaps a challenge for any human collective endeavour. To address the matter, we believe we must keep exploring alternative forms of planning, one of which is articulated planning.

ARTICULATED PLANNING

We envisage Articulated Planning, at a first level, as a systematic exercise between a pair of organizations to cross relate their hierarchy of plans. A simple table with each organizations goals on each axis could be useful to guide the discussion and quickly inform those involve. A table like the one sketched in Illustration 1 would be sufficient.

The colour scheme, used to fill each cell as a traffic light, indicates whether goals are opposite, complementary, identical or unrelated. Thus, a real spreadsheet could be displayed as a tree like structure, like the one in figure 2:

At first sight, the sheet already displays useful information. Assuming that the goals are ordered from higher level to lower level, a predominant green on top (upper left corner) says that the organizations are very similar in intentions, even if they divert in the ways to achieve those intentions. Yellow everywhere indicates that the organizations constitute a suitable pair. And white or red on top with yellow and green towards the base of the tree indicate that the organizations are in a position to collaborate, even if they cannot cooperate (because they pursue different or unrelated higher goals).

Figure 1. Goals table

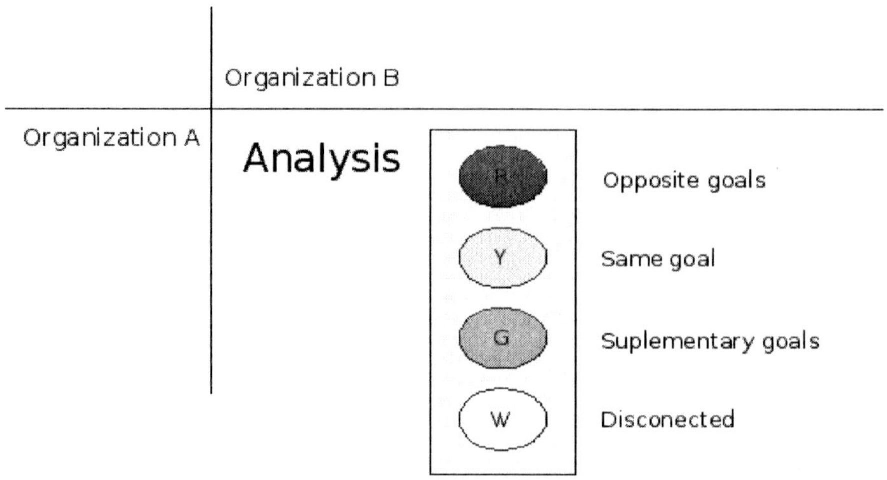

Thus, a very elementary spreadsheet of goals can serve as a articulating tool, providing useful information about the organizations involved. More tools, however, are already in stock to support the new forms of planning.

SCENARIO ANALYSIS FOR PLANNING WITH SIMULATIONS

Simulation is the realm of possible worlds. Computers are making it possible to foresee the evolution of a system, even before the actual system exists. One has to provide the computer with a model: a machine readable description of the system, which it uses to explain how the system progresses from one (initial) state to the following. Each system state is normally described as a list of variables with associated values, which change as (simulated) time passes. Some of those values for variables are defined by the modeler, before the simulation and aiming at configuring a particular family of states. Those families of states corresponds to alternative scenarios for the simulate system to ``live in".

Thus, simulation is the ideal tool to study possible scenarios in a system. A type of study which

Figure 2. Articulated planning spreadsheet

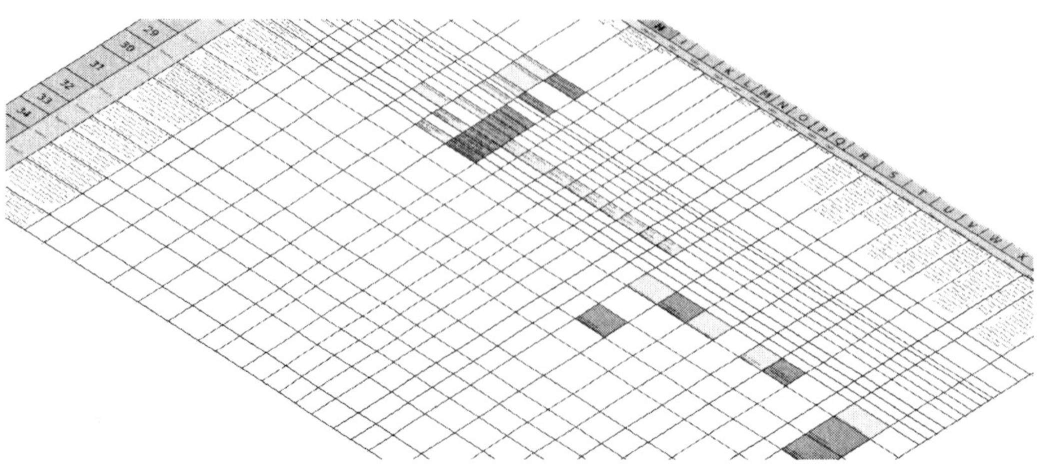

is specially useful for planning purposes. Let us illustrate this point with the following example of modelling and simulation.

The model described in what follows is an outcome of a international project entitled: Biocomplexity; Integrating Models of Natural and Human Dynamics in Forest Landscapes Across Scales and Cultures[25]. It is a model of land use and changes in vegetation cover as a consequence of human actions on a natural space (a forest reserve). As explained somewhere else (Acevedo et al, 2007), the exercise corresponds to a model of 1) the human dynamics, using the set of conceptual tools and data structures provided by an agent-oriented simulation platform[26], and of 2) the environmental dynamics, by integrating a cellular automaton (Quintero et al, 2004) library into the simulator of a forest reserve. The data structures of the simulator provide for the representation of the human agents' goals, beliefs and observations, and, also, for a very elementary reasoning engine to deduce actions for each agent, according to its circumstances.

In the actual system, settlers are people of limited economical resources that arrive at a forest reserve aiming to improve their economical status and to get the property of the land that they get to occupy. Initially they dedicated themselves to subsistence agriculture: they just try to maximize the benefits from their occupation of the area, without considering soil exhaustion due to poor management practices, and without much regard for ecological damage. After a few years, the land loses its fertility, and the settler must move to another available place (i.e. an area not under government supervision) or expand his farm by deforesting some adjacent land. The actual precise behaviour of the agents, however, depends of many variables and is, therefore, extremely hard to predict.

At this point a simulator makes all the different for planning and scenarios analysis. Simulation results for this system are portrayed as graphs (Illustration 3) that show the percentage of total forest area for each one of three policy scenarios (Agroforestry, Forestry, Hands-off) and maps that show the spatial distribution of land-use types obtained in each of the scenarios at each time step. Illustration 4 shows the final state of the Caparo Forest Reserve for each policy scenario.

Figure 3. Simulated scenarios

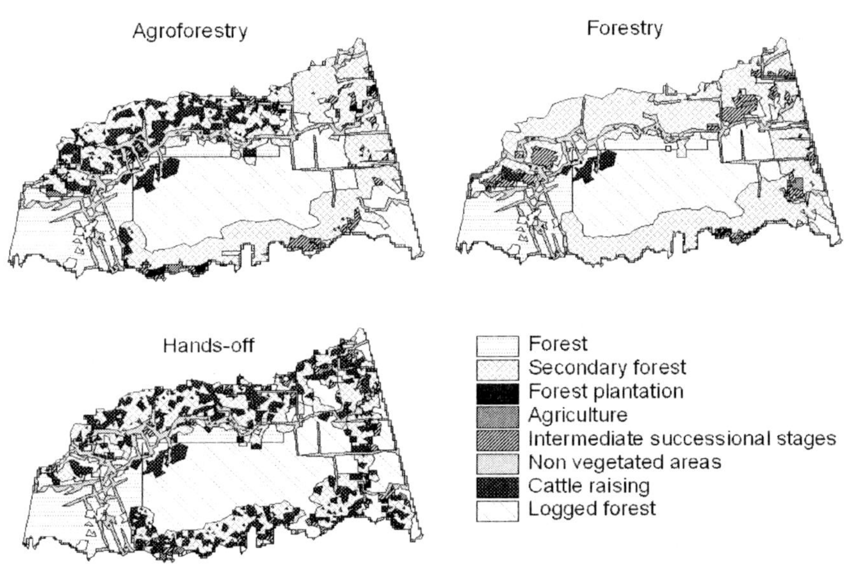

Figure 4. Final simulated state of Caparo Reserve

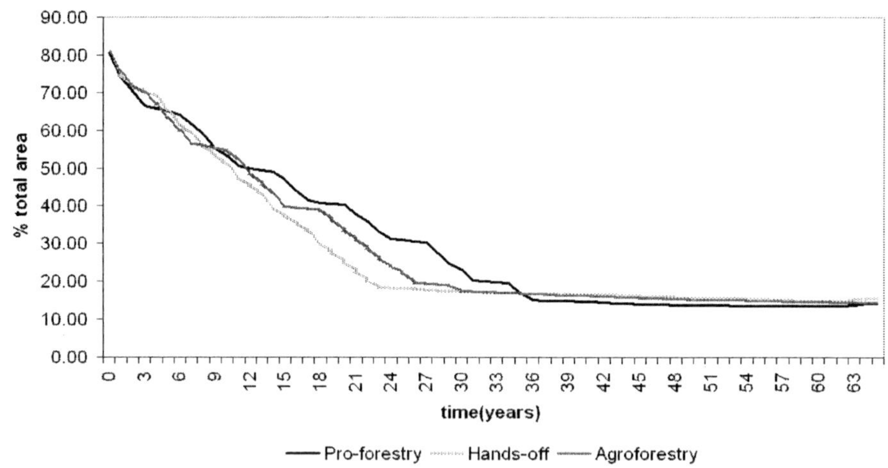

This example testifies that, once the fundamental elements on the description of a system have been represented, a simulation may exhibit non trivial features of the dynamics of that system. Thus, a planner with a simulator may explore unseen, future scenarios obtained whenever the intended actions are successfully executed. Choosing the best course of actions is, therefore, easier. This is the usual benefit of simulation for planning purposes.

Articulated Planning, however, introduces another element which has been discussed in the simulation communities for some time: the value of a model as a training and communicational tool. Models can play as discussion media for different organizations trying to articulate their plans. Alternatively, when no model exists, the process of creating one can also be useful for establishing a common language and, eventually, a common plan.

CONCLUSION

Throughout this chapter, we have been arguing that there exist new forms of planning that could suit a socialist system with a strong commitment to democracy and that do not reduce to central planning. Three new concepts arise in this context: socialist e-governance, articulated planning and microplanning. Socialist e-governance aims to support the participation of the people in governmental decision-making and validation of achievements. Articulated planning and microplanning are one step in that direction, by integrating public servant at all levels of responsibility (they are also part of the people, of course) in the planning process and providing a perfect memory for the general public to evaluate goals and achievements. Both latter concepts seems to required of two more elements to be successfully implemented: 1) a non-trivial information technology platform and 2) personnel trained and willing to use that platform. Fortunately, IT technologies are increasingly available under conditions of freedom, which means (is meaning) that anybody can learn of the tools and adapt them to suit specific situations. These abilities (a trained community of users and adaptability to specific conditions) could prove fundamental to overcome the limitations of planning in a complex and dynamic world. Furthermore, that planning platform could provide the base for a socialist state to function effectively, without having to sacrifice a strong commitment to the inclusion and participation of all.

REFERENCES

Acevedo, M., Callicot, B., Monticino, M., Lyons, D., Palomino, J., & Rosales, J. (2007). Models of Natural and Human Dynamics in Forest Landscapes: cross-site and cross-cultural synthesis. *Geoforum*. doi:.doi:10.1016/j.geoforum.2006.10.008

Acevedo, M., Rosales, J., Delgado, L., Ablan, M., Dávila, J., Callicot, J. B., & Monticino, M. (2007, September). Modelos de Interacción Humano-Ambiental: El Enfoque de la Biocomplejidad. *Ecosistemas, 16*(3). Retrieved April 2009, from http://www.revistaecosistemas.net/articulo.asp?Id=507

Ahumada, J. (1969). Notas para una teoría general de planificación. In *Cuadernos de la Sociedad Venezolana de Planificación Números 4 y 5*. Caracas: SVP.

Alterman, R. (Ed.). (2001). *National Level Planning in Democratic Countries. An International Comparison of City and Regional Policy-Making*. Liverpool, UK: Liverpool University Press

Ansoff, I. (1965). *Corporate Strategies*. New York: McGraw-Hill Education.

Chapman, D. (1990). Planning for conjunctive goals. In J. Allen, J. Hendler, & A. Tate (Eds.), *Readings in Planning* (pp. 537-558). San Mateo, CA: Morgan Kaufmann Publishers, Inc.

Domingo, C., Ramírez, V., Velásquez, A., & Zavarce, H. (2006). Use of Enterprises Data in Economic Planning. In *Proceedings of the Modelling, Simulation, and Optimization. MSO 2006*. Calgary, Canada: Acta Press.

Kettl, D. F. (2002). The Transformation of Governance. Baltimore, MD: John Hopkins University Press

Mintzberg, H. (1994). *Rethinking Strategic Planning*. Long Range Planning.

Quintero, R., Barros, R., Dávila, J., Moreno, N., Tonella, G., & Ablan, M. (2004). A Model of the Biocomplexity of Deforestation in Tropical Forest: Caparo Case Study. In C. Pahl, S. Schmidt, & T. Jakeman (Eds.), *Proceedings of the iEMSs 2004 International Congress: "Complexity and Integrated Resources Management,"* International Environmental Modelling and Software Society, Osnabrueck, Germany. Retrieved from http://www.iemss.org/iemss2004/proceedings

Resource Renewal Institute. (2009). *Green Plans in Action: Netherlands*. Retrieved July 20, 2009 from http://www.rri.org/greenplans_netherlands.html

Riley, T. B. (2003). *E-Government vs E-Governance: Examining the differences in a changing public sector climate. International tracking survey report '03. Number four*. Commonwealth Centre for E-Governance. Retrieved May 15, 2009, from http://www.electronicgov.net

Weber, M. (1920). *The Theory of Social and Economic Organization*. New York: The Free Press

Wikimedia Foundation. (2009). *Planned Economy, Wikipedia*. Retrieved July 20, 2009, from http://en.wikipedia.org/wiki/Planned_economy

ENDNOTES

1. http://sipegv3.mct.gob.ve/index.php?err=4, accessed in July 2009
2. http://www.gobiernoenlinea.ve/directorioestado/editorial.html, accessed in July 2009
3. http://www.minec.gob.ve/contenido.php?id=121, accessed in July 2009
4. http://update.unu.edu/archive/issue47_19.htm, accessed in July 2009
5. http://www.gobiernoenlinea.ve/directorioestado/experiencias.html, accessed in July 2009

6 http://forja.softwarelibre.gob.ve/http://forja.softwarelibre.gob.ve/projects/serviciosr-rhh/, accessed in July 2009. See alsohttp://sistemas.fsl.fundacite-merida.gob.ve/projects/said/, accessed in July 2009

7 The Ministry of Education has services on line for its employees (http://www.me.gov.ve/, accessed in July 2009) and for the general public (http://www.portaleducativo.edu.ve/, accessed in July 2009). The Finance Ministry has a portal to consult public investment (http://www.mf.gov.ve/, accessed in July 2009)

8 http://www.tsj.gov.ve/index.shtml, accessed in July 2009

9 http://www.seniat.gob.ve/, accessed in July 2009

10 http://pasaporte.onidex.gob.ve/, accessed in July 2009

11 http://www.ivss.gov.ve/, accessed in July 2009

12 http://www.ineinfo.ine.gob.ve/http://www.oncti.gob.ve/, accessed in July 2009

13 http://www.indecu.gob.ve/, accessed in July 2009

14 http://www.oncop.gov.ve/vista/boton.php?var=sigecofx, accessed in July 2009

15 http://scholar.google.co.ve/scholar?q=%22e+governance%22+articulated+planning&hl=es&lr=&scoring=r&as_ylo=2009&start=0&sa=N, accessed in July 2009

16 http://www.palgrave-journals.com/jit/journal/vaop/ncurrent/full/jit200826a.html, accessed in July 2009

17 http://flossimpact.eu/, accessed in July 2009

18 http://en.wikipedia.org/wiki/List_of_project_management_software, accessed in July 2009

19 http://live.gnome.org/Planner, accessed in July 2009

20 http://ganttproject.biz/, accessed in July 2009

21 http://openproj.org/openproj, accessed in July 2009

22 http://office.microsoft.com/en-us/project/default.aspx, accessed in July 2009

23 http://sistemas.fsl.fundacite-merida.gob.ve/projects/microplan/, accessed in July 2009

24 http://www.12manage.com/methods_smart_management_by_objectives.html, accessed in July 2009

25 http://www.geog.unt.edu/biocomplexity, accessed in July 2009

26 http://galatea.sourceforge.net/, accessed in July 2009

Section 4

E-Configurations, Citizens and the State

Chapter 14
Building the Digital Government in Regions:
Success Factors and Institutional Barriers in Spanish Comunidades Autónomas

J. Ignacio Criado
Universidad Autónoma de Madrid, Spain

ABSTRACT

The author traces the unfolding of modernisation efforts in the region of Valencia in Spain, showing a number of assemblages whose study requires developing varied employing complementary approaches and concepts. The inclusion of regional governments as important actors in modernization processes is put forward and explored in the case of Valencian Comunidad Autonómas (Autonomous Community). The chapter shows how the interplay of institutional practices, the development of new forms of organization to provide e-government services and the unfolding of regional policies within Spain and Europe all contribute to make e-government what is now in the region of study.

INTRODUCTION

This chapter tackles with some basic questions about *e*Government with a regional perspective: is it growing the importance of regional governments in the information age political environment? Is there a specific regional administrative modernization agenda? Which institutional factors shape regional electronic public administration developments? There is an increasing importance of regional government around the European Union, specially, in countries with strongly decentralised politico-administrative systems. This paper outlines a precise regional perspective to comprehend the global electronic government (*e*Government) phenomenon. In so doing, next pages deal with a country case (Spanish *Comunidades Autónomas*, in general, and *Comunidad Valenciana* region, in particular), with extensive regional competencies and socio-economic as well as political and legislative power. This study takes into account the significance of contextual institutional and historical factors, and internal institutional settings, that shape the evolution of administrative reforms, also in regional perspective.

DOI: 10.4018/978-1-60566-860-4.ch014

Internationally speaking, one can emphasize an extensive lack of analysis addressing the regional level of informatization and *e*Government. It is almost ironic that in this period of a global economy, an emerging network society, and ICTs, regions and regional government have been rediscovered again. A number of concepts such as regional development, regional government or regional governance have been a common place in the literature of geography, economics, political science or public administration during the last decades (Keating, 1985). However, '*there is a heightened interest in the region as a locus of political, economic and cultural activity*' (Tomaney & Ward, 2001: 1). In the last decade, this debate has turned into the concept of *new regionalism* and the idea that regions around the world are looking for new ideas and new competencies. Thus, the overall *objective* of this article implies providing evidences from a case study about success factors and institutional barriers for regional *e*Government development.

Methodologically, this chapter presents a detailed case study taking into account one of the Spanish *Comunidades Autónomas*. In brief, the interest on the regional level of government in Spain derives from the relatively successful process of devolution deployed during less than three decades. In that regard, this layer of regional governments and public administrations, together with modernization processes, grew up rapidly, resulting in case studies with interest for *e*Government scholars. In particular, the case of the *Comunidad Valenciana* is that of one of the most developed Spanish regions, in terms of informatization and ICTs diffusion, evolving under the monitoring of European Information Society regional strategies. In so doing, this region inaugurated links and connections with other pioneering regional cases in Europe.

Data for this study was gathered using different *data collection techniques*. First, thirty four semi-structured interviews with senior officials within the regional administration (i.e. successive general directors of ICTs and *e*Government central management unit, top civil servants governing ICTs in ministries and more technical profiles within the Valencian regional administration), as well as other actors from different context organizations with relevance for the case (i.e. managers and consultants from technology companies, academic consultants, etc.). Complementary sources of data derived from a documentary analysis of public strategies and internal documents about administrative reform and *e*Government, as well as statistical data from different public and private institutions. The fieldwork was deployed from 2004 to 2007, being funded by a research grant of the Spanish Ministry of Education and Science.

This chapter operates as follows. Firstly, it provides an approach to the theoretical framework, using different perspectives rooted in political science and public administration studies. Second, it delivers a description about the external and historical context for an electronic public administration in Spain. Third, it addresses in detail the analysis of the case study evolution. Then, a final discussion will be opened concerning the most salient aspects derived from previous sections, as a resume of results, unfolding final ideas about a future agenda of research with a regional focus.

THEORETICAL FRAMEWORK

This section draws together different theoretical perspectives to support the following analysis. During the last years, different aspects and dimensions of *e*Government have been studied using diverse objectives, methodologies, and theoretical frameworks. Danziger & Andersen (2002), Heeks & Bailur (2006), Ya & Brestschneider (2007), and Yildiz (2007) review articles have pointed out that variety of approaches to *e*Government phenomena, "*the first decade of academic e-government research was dominated by studies that focused on the observation and evaluation of the output of e-government initiatives* (...) *This study makes* *two methodological suggestions for the study of e-government. First, whereas examination of the outputs and outcomes of e-government efforts constitute the majority of the e-government literature, examining and explaining the processes of e-government development are also necessary. Careful analysis of the e-government policy processes is essential (...) The second suggestion is that certain methodological choices such as using primary data, triangulation of findings, and concepts with an analysis approach to produce theory are found rarely in the e-government literature*". (Yildiz, 2007: 659-661). Both types of suggestions fit in with this work. In so doing, three different theoretical foundations support this research: (1) *new institutionalism*, (2) *new public management and administrative modernization*, and (3) *ICTs in regional public administration*. Concisely, this section presents their most remarkable contributions for this research, basically, from a political science approach.

New Institutionalism in Public Administration

New institutionalism literature provides stimulating arguments to understand the constraints and drivers for change in public administration. Depending on the *new institutional* wing adopted (*historical, normative, rational choice, sociological*) institutional theory and institutions embody different analytical implications, because they underline distinctive facets of social/political facts (Peters, 2005). However, the conceptual dimension of *new institutionalism* provides key and broadly shared foundations for this study. First, we define *institution* as a group of *values, norms, rules, routines and processes* that structure the relationships among agents involved in social and political processes, giving the parameters of what is appropriate in the relationships among roles to follow up and defined situations (March and Olsen, 1989). This group of elements constitutes a *normative* and *regulative* framework, a cognitive outline that, with certain symbolic character, provides stability and meaning to the behaviour of agents, being developed from the organizational culture, the formal and informal structures, and the routines and processes (Scott, 1995). Following this definition, the role of institutions stands out as the 'rules of the game', that actors use to develop their strategies, decisions, and concrete actions.

Other specific features of institutional theory will be addressed using its conceptual potential. *Institutional isomorphism* relies on the fact that '*organizations are driven to incorporate the practices and procedures defined by prevailing rationalized concepts of organizational work and institutionalized in society. Organizations that do so increase their legitimacy and their survival prospects, independent of the immediate efficacy of the acquired practices and procedures*' (Meyer & Rowan, 1977: 357). This explains why some decisions are adopted even if apparently they are not positive for the organization from a rational point of view. On the other hand, *path dependence* notion implies the idea of adoption of lines of action that introduce restriction in future actions. In other words, past decisions shape in some extent the future available and acceptable options, making it impossible to choose others, including those rationally more efficient in the long term (Powell, 1991). In connection with this, *increasing returns* refers to self-reinforcing feedback mechanisms supporting *path dependence* processes. Using original approaches to technologies (Arthur, 1990; David, 1986), Powell (1991: 193) concluded that '*the more a technology is adopted, the more it is improved, and the greater its payoffs. When this occurs and adoption accumulates, the choice of the technology becomes structurally rigid and locked in (...) Path dependence models suggest that institutional arrangements are not likely to be flexible; they cannot change rapidly in response to perturbations in the environment. The self-reinforcing feedback mechanisms that support path-dependence processes make it difficult to explore alternative options. While these models have principally been applied to technology adoption, their significance is potentially broader*'. This not only may involve public sector reform policies, but also digital government building processes.

New Public Management and Administrative Modernization Agenda

Secondly, *new public management* (NPM) *and administrative modernization* literature frames in more public administration fashion the rationales of agendas for change in public sector organizations during the last decades. NPM describes the attempt to adopt a private management style in public administration (*managerialism*), emphasising economic theories to organise public services and introduce more rationality in public management (*new institutional economics*) (Barzelay, 2001; Hood, 1991, 1995; Hughes, 2003; Pollit, 1993). As Hood noticed (1991: 3) NPM was '*a shortland name for the set of broadly similar administrative doctrine which dominated the bureaucratic reform agenda in many of the OECD group of countries from the late 1970s*'. In that regard, during the last decades an impressive growing *corpus* of literature presented different studies and research related with NPM in different countries, addressing various types of levels of government and providing different kind of explanations about the adoption and diffusion of this new set of ideas in public administration. However, the approaches to NPM evolved as it did the extent of administrative reforms.

During more recent years, NPM has been under several criticisms. Previous optimistic approaches have been evaluated, providing a more systematic and moderate vision of administrative reforms (Pollit, 2002). Basically, evidences about results are less clear than one could be expected addressing the amount of publications including NPM in the title. On the other hand, where results have been identified, it is not clear that they had implied positive impacts for public administration resilience, relationships with citizens, or critical concepts such as efficiency, efficacy or economy (Pollit & Bouckaert, 2003). Besides, they were also underlined the important limitations of NPM diffusion, specially in politico-administrative regimes reluctant to market based theories and managerial instruments applied to public sector organizations, as European continental or Mediterranean cases (Capano, 2003; Pollit & Bouckaert, 2004; Sotiropoulos, 2004). In fact, last few years illustrated a shift within this modernization agenda, giving now priority to information technologies and communication networks adoption in public sector organizations. Some authors have highlighted that this new phase of ICTs in public administration can be considered as a second revolution in public management, another stage in public sector reforms that commenced in the late 1980s, or '*the modernization of public administration by other means*' (Hughes, 2003). In any case, further comprehension of NPM, administrative modernization and their relationship with ICTs is essential.

Informatization and ICTs in Regional Public Administration

Thirdly, prior *informatization* and later *ICTs in public administration* (or *e*Government) *corpus* of literature engender the specific ground for this study of strategies to consolidate regional digital government. Mainstream approaches to *e*Government have not concentrate in regional contexts. For example, the American group of academics under URBIS project rubric on ICTs deployed extensive scrutiny of local government using longitudinal data (Danziger *et al.*, 1982; Dutton & Kraemer, 1978; Kraemer & King, 2003). Even if the importance of States in American political life was also reflected within some studies, however they were much more recent and isolated, even if some of them deserve special mention (Gil-

García, 2004; Melistki, 2002; West, 2003; Ya & Bretschneider, 2007). On the other hand, the most part of studies addressing *e*Government in European countries provided very interesting results about national strategies, projects, and actions (Bellamy and Taylor, 1998; Chadwick & May, 2003; Dunleavy and Margetts, 2007; Margetts, 1998). Even in continental Europe, regional level has had less presence in the research agenda than central government, or local authorities (Frissen & Snellen, 1990; Lenk & Traunmüller, 2002; Snellen & van de Donk, 1998). Obviously, those contributions intellectually frame this paper, nonetheless regional governments across the European Union (EU) have not deserved parallel attention from the perspective of ICTs in public administration.

Taking into account the emerging significance of regional governments and public administrations across Europe, this chapter assumes that it is required a regional perspective on *e*Government. Recently, some contributions have increasingly underlined this regional focus within the EU context, even if it embodies very different politico-administrative regimes and types of territorial distribution of political powers. For instance, in recent European competitions to select the winners of *eEurope Awards for eGovernment* 2003, 2005, 2007, the largest number of applications in response to the call were received from local and regional authorities (Leitner, 2003; Leitner & Criado, 2006). In that regard, an author has noticed the fact that '*strong regions, with their extensive competences, have a real say when it comes to making the Information Society a reality in Europe*' (Heichlinger, 2004). At the same time, this perspective also implies the knowledge about the role of regional public administration in broader (national and European) *e*Government strategies and projects (Criado, 2006). Following sections aim to collaborate in developing that approach, regarding the case of Spanish regional tier of government.

AN OVERVIEW OF PUBLIC SECTOR MODERNIZATION AGENDA AND INFORMATIZATION IN SPANISH *COMUNIDADES AUTÓNOMAS*

Again, Spain became a Democracy after the francoist dictatorship in 1978, and now everyone accept the necessity for regional public administrations. The introduction of this regional tier of government in Spain had, at least, two main features determining the most recent administrative modernization processes (Aja, 2001). First, central government devolved political competencies to regional governments and, at the same time, its organizational, personnel, financial and technological resources became regional assets. Thus, during the last twenty-five years, Spanish regional public administrations have been expanded in capacity to manage growing amounts of public services, within a moment in which public sector in the context were under (opposite) NPM patterns and ideas (downsizing, privatization, and so on). Secondly, the organizational and functional structure adopted by regional government embodied almost a mimetic reproduction of central administration, including human resources and financial management techniques. In other words, regional public administrations tended to adopt the traditional/napoleonic model of functioning of Spanish central public administration: '*the hypothesis that we underline is that within the building process of regional public administrations have prevailed institutional isomorphistic and path dependence dynamics that mediated the final organizational design*' (Ramió y Salvador, 2003: 125). Both aspects shaped later opportunities to innovate in regional public sector.

Since the beginning of the 1990s, the process of administrative modernization in Spanish regional public administrations was surrounded by those general institutional arrangements defined in the previous paragraph. Different regional governments initiated processes of administrative reform, some years later than expected, because their previous efforts were devoted to consolidate political institutions derived from that recent territorial distribution of political powers (Arenilla, 2000: 24). In that context, regional public administrations required to adapt their adolescent organizational structures, management techniques, human capital, and technological resources, based on traditional and legalistic patterns, to growing levels of political autonomy, more mature social structures, and citizens demands of high service quality (Olías de Lima, 2002). Consciously or not, these administrative reforms in Spain shared NPM as intellectual administrative doctrine or operational toolkit, above all, with some ideas and instruments such as attention to the three 'e' (efficiency, efficacy, and economy), quality management (citizens charters, EFQM, awards...), one-stop shops, customer orientation, etc.

On the other hand, an administrative area did not share this general tendency through administrative reform harmonisation: *informatization*. Informatization policies had at their disposal broad margins for self-organization during the first moments of regional governments in Spain. From the beginning, they adopted their technological infrastructures following different organizational and functional models, and also developed independent informatization strategies during the process of administrative institutionalization. While harmonization exemplified the main results of politico-administrative devolution and later administrative reforms, in terms of informatization regional public administrations innovated in very different manners, within a general tendency to increase the amount of resources devoted to this area. *Table 1* provide comparative data from regional public administrations in Spain and

Table 1. Evolution of IT budget over total regional government budget (%)

	1991	1993	1995	1997	1999	2001
Total	**0,82**	**0,84**	**0,80**	**0,83**	**1,00**	**1,17**
Andalusia	0,48	0,52	0,57	0,99	1,40	1,21
Canary Islands	0,45	0,39	0,68	0,71	0,66	*
Catalonia	0,66	1,01	0,92	0,76	0,55	0,92
Galicia	0,65	0,36	0,23	0,32	0,60	0,78
Valencian Community	0,45	0,41	*	0,41	0,21	1,39
Aragon	0,35	0,45	0,82	*	*	0,64
Asturias	0,38	0,51	1,20	0,97	0,90	1,11
Balearic Islands	2,05	0,70	0,92	1,54	*	*
Cantabria	0,64	0,80	0,76	0,55	0,94	0,83
Castille & Leon	*	0,76	0,77	0,99	1,27	0,67
Castille-La Mancha	1,80	0,85	1,32	1,36	2,04	0,82
Extremadura	0,45	0,44	0,34	1,13	1,00	1,00
Madrid Community	1,85	1,52	1,38	1,62	1,87	1,25
Murcia	0,11	1,67	2,03	1,37	1,33	1,91
La Rioja	1,88	3,14	2,65	2,40	0,91	0,89
Navarre	1,07	1,15	1,05	*	1,05	*
Basque Country	2,45	2,27	2,12	0,90	1,87	2,54

Source: own elaboration from Ministry for Public Administrations.
* No data available

shows the important extent in which IT budget was growing as part of general budget in regional governments.

In the same vein, *Table 2* presents the increasing importance of civil servants with technological profiles in regional public administrations. However, even if general trends in both cases were generally oriented to expand resources and staff within IT areas, at the same time, it is not possible to identify any other general and shared direction on informatization patterns within Spanish regional public administration. For instance, in organizational terms, they develop very different models of IT exploitation (centralised, decentralised, externalization, even privatization...), political priorities evolved in different ways from case to case, etc.

More recently, the increasing importance of ICTs in Spanish regional level of public administration derived from the adoption of initiatives to develop within their territory Information Society and, in so doing, also *e*Government. In this sense, most of them have explicitly formulated strategic or director action plans to promote Information Society policies, and within them, even if they are named using different terms, it is usual to find out specific *e*Government strategies or actions plans to enhance this area in more detailed than informatization. In sum, this may imply an important shift in administrative reform policies: regional governments developed action plans to grow up themselves on the Internet, even if they did not implemented an extensive internal innovation in organizational and managerial terms, so it is required in depth analysis addressing different variables to understand those dynamics. This is the attempt in the subsequent section.

Table 2. Evolution of regional government IT employees

	1991	1993	1995	1997	1999	2001
Total	**2464**	**2885**	**3149**	**3005**	**3346**	**4043**
Andalusia	409	503	551	527	595	633
Canary Islands	82	82	234	230	283	*
Catalonia	593	715	748	731	284	526
Galicia	183	183	119	114	363	503
Valencian Community	248	194	158	162	172	392
Aragon	33	42	38	*	42	58
Asturias	20	20	45	62	61	67
Balearic Islands	19	19	19	28	21	35
Cantabria	17	19	17	21	26	37
Castille & Leon	*	214	247	244	310	359
Castille-La Mancha	70	54	125	173	247	295
Extremadura	66	73	9	123	127	211
Madrid Community	343	220	228	251	306	434
Murcia	5	76	76	94	114	166
La Rioja	21	37	37	38	6	6
Navarre	71	147	147	*	86	*
Basque Country	284	287	351	207	303	321

Source: own elaboration from Ministry for Public Administrations.
* No data available

BUILDING ELECTRONIC GOVERNMENT IN THE COMUNIDAD VALENCIANA

Generalitat Valenciana (GVA) is the term to name the executive (including government and public administration) located in the Valencian Community (*Comunidad Valenciana*) region. The Valencian Community is located in the Mediterranean coast of Spain, it is divided in three provinces (*Alicante*, *Valencia* and *Castellón*), being the capital city Valencia. GVA has actually more than 110.000 public employees, the annual budget is over 9.100€ millions and the amount of services offered to 4.5 million of citizens in the region sum more than 1.200. Some of those services are specifically oriented to the 541 municipalities of this region, above all, the group of less than 5.000 inhabitants. GVA is divided in departments (about 10) with a number of agencies and autonomous bodies, depending on regional administration with different legal status. During the last 25 years, political processes within this region can be separated into two broad periods: one from the beginning of the 1980s (origins of this region and its public administration) until 1995, and a second phase from that moment to date. They are rooted in political cycles: during the first (1983-1995) the socialist party (PSOE) ruled the government of this region, and during the second (1995-) the conservative party (PP) possessed this responsibility. In some extent, both periods can be identified with two phases in the process of adoption, use and diffusion of ICTs in public administration. This section assesses the success factors and institutional barriers to build up the digital government within this region, one of the most developed in *e*Government terms in Spain, addressing four dimensions, consecutively developed in the next pages: (1) evolution of institutional design for informatization prior to 1995; (2) public sector modernization strategies and political support; (3) funding, legal framework, and internal coordination; and (4) leadership and organizational actors.

Evolution of Institutional Design for Informatization and ICTs in Public Administration Before 1995

GVA actions related to *e*Government are in line with the policy path shaped in the first stages of informatization. During the first fifteen years of the GVA (until mid 1990s) the importance of informatization and IT made it possible a slow and regular adoption of ultimate technological innovations available in the market. It was also established an organizational structure for information systems management, that actually maintain some of its general features. At the same time, a stable group of senior civil servants started to serve within this administration, with technological profiles and competencies, preserving their employments (and organizational knowledge) in time, expanding their functions within the organization, and gradually improving their position in the formal structure and symbolic sphere of influence. Nonetheless, informatization remained as a peripheral political priority: even if this area grew up with assistance of civil servants with strong technological records, political willingness to provide ICTs wider significance in public administration then continued at low level.

In organizational terms, GVA was initially structured in functional departments, regional ministries (*consellerias*), and various public agencies and bodies (regional institutes, autonomous bodies, public enterprises, and so on), each of one with its own administrative competencies, resources, and budget. Historically, each functional department and the various public agencies and bodies, maintained their own *Data Processing Unit* (DPU), to support the provision of services of their own responsibility, designing and managing their information systems to reinforce internal processes. During that first period, the *Informatics Commission* (*Comisión de Informática*-CI) emerged with members from different departments to coordinate and expand some technological and organizational standards. Later, it was established a *General Directorate for Organization and Information Systems* (*Dirección General de Organización y Sistemas de Información*-DGOSI) (1985), within the regional Ministry for Public Administration, as strategic and central unit, but also with functions for operational coordination of DPUs, horizontal informatization development and standardarization, and control of technology procurement. Nonetheless, informatization within this case prior to 1995 presented a traditional lack of internal coordination, as well as a constrained political commitment with ICTs pervaded as a generally accepted conclusion for this early period.

Attributes of Public Sector Modernization Strategies

During the last years, the Valencian government has launched three strategies for public administrative reform, giving ICTs a growing role. At a glance, *Table 3* resumes the possibility to address different dimensions in each of the three strategic action plans, as to develop the following analysis of institutional factors shaping and being shaped by those programmes. This approach offers the opportunity to introduce an in-depth scrutiny regarding a long-period process of more than ten years. In that regard, timing in these strategies has been clearly attached to electoral cycles within this regional tier of government (1995-1999; 2000-2003; 2004-2007, etc.). Now, they are detailed the most salient attributes of each strategy, regarding different variables in connexion with the following pages.

The first *Strategic Plan for the Modernization of Valencian Administration* (*Plan Estratégico de Modernización de la Administración Valenciana*) (PEMAV) was inaugurated in 1995 as the first administrative reform strategic plan of the Valencian regional government. This general modernization strategy adopted four principles (GVA, 1996: 1): '*effective*, *responsible*, *networked*, and *decentralised*', as the most important for public sector in the future to promote a '*deep transformation in Valencian public administration*'. This initiative was based on 40 sectional projects, each of one focused on one of the principles mentioned before (*effective* (13), *responsible* (11), *networked* (11), and *decentralised* (5)). In the *networked* area, projects provided a key role to ICTs, in general, with actions directly oriented to internal management reform and diffusion of the corporative telecommunication network as platform for future external digital services. The majority of projects had a four years period of implementation, with a horizontal focus affecting critical aspects of the internal management of GVA, and addressing a broad group of units within the Valencian public administration. This embodied the starting point of project management to develop strategic plans, then adopted and later expanded in quantitative (total amount and dimension), but also qualitative (complexity, monitoring, etc.) dimensions.

The *Second Strategic Plan for the Modernization of Valencian Administration* (*II Plan Estratégico de Modernización de la Administración Valenciana*) (*Moderniza.com*) was launched in December 1999. By that moment, the conservative party had succeeded again in regional elections, so this strategic plan continued some of the basic lines defined within PEMAV. This general modernization strategy addressed three principles: *innovate*, *rationalize* and *transfer technology and knowledge* (GVA, 2000: 25). From this starting point, *Modeniza.com* provided a broader and ambitious vision of the modernization agenda, putting first ICTs (including the name *Moderniza.com*), and opening a new phase for *e*Government development. Within this strategy, orientation to the citizens was basic, at the same time, '*making it possible the emergence of the Information Society in the Valencian region, within an integrated society and innovating/transforming public administration*' (GVA, 2000: 29). The strategy leaders noticed the readiness of technologies and the GVA to boost its external capacities to interact digitally with citizens, thus misleading problems with internal management and organization structure.

Table 3. Strategic action plans for modernization & eGovernment in the Generalitat Valenciana

	Strategic Plan for the Modernization of Valencian Administration (PEMAV)	Second Strategic Plan for the Modernization of Valencian Administration (*Moderniza.com*)	Strategic Plan Advancing with ICTs (AVANTIC)
Period	1995-1999	2000-2004	2005-2010
President	Mr. Zaplana (Popular party)	Mr. Zaplana Mr. Olivas (Popular party)	Mr. Camps (Popular party)
Regional focus	Renovation of our politicians and public administrations	Valencian Community on the Net, as well as public administration and valencian people	Valencian Community as a point of reference in the international context. We are agents of the global Information Society
Administrative reform scope	Sectorial. Internal management and administrative reform	Intermediate. Electronic public service delivery	Intermediate. Electronic public service delivery (multi-channel and integrated)
eAdministration focus	Internal. First steps with innovative applications for internal management and development of technological infrastructure. Also information available on the Web	External, less internal projects. Development of the telematic channel to contact citizens, provision of digital information and services using the Internet	Completely external. Ubiquity (multi-channel access), Integration (middleware applications, and Interoperability (organizational, technological and semantic)
Political support	High	Medium	High
Leadership	Visionary. Without internal support	Emergent and Democratic. Internal support	Democratic. Stable with strong internal support
Funding	66€ millions	240€ millions	2.740€ millions (110€ millions i-Administration programme)
Organizational responsibility	General Directorate for Administrative Modernization (Ministry for Public Administration), Secretary for Administrative Modernization (Ministry of Treasure)	General Directorate for Administrative Modernization and Telecommunications (successive changes of department)	Autonomous Secretary for Telecommunications and Information Society (Ministry of Infrastructures and Transports)
Internal coordination	Office of PEMAV. CODESI	Office of Moderniza.com. CODESI	Office of AVANTIC. CIDIRTEL & CIDESTEC

Source: Own elaboration

The most recent action plan for the modernization of Valencian administration had a different, and more ambitious, scope, because it has been included under AVANTIC (*Avanzando con las TIC. Advancing with ICTs*), the regional strategy for infrastructure enhancement. AVANTIC was launched at the end of 2004 as '*the strategy to consolidate the advanced telecommunications and the technological knowledge society in the Valencian community*' (GVA, 2004: 15). The internal organization of AVANTIC was based on three blocks: two strategic plans, PEVTA (*Valencian Strategic Plan of Advanced Telecommunications-Plan Estratégico Valenciano de Telecomunicaciones Avanzadas*) and PETIC (*Strategic Plan for the Consolidation of the Technological and Knowledge Society-Plan Estratégico para la Consolidación de la Sociedad Tecnológica y del Conocimiento en la Comunidad Valenciana*). Plus horizontal or cross-sector programs, identified in and defined to support the former and give assistance to management structures. PETIC (Programme 3) was devoted to i-Government (*Intelligent Government*), a new term to overcome *e*Government with a different focus on citizen needs. This strategy was based on a very ambitious statement about the future of regional administration: '*Valencian government aspires to* *achieve leadership in the Intelligent Administration dimension to consolidate Valencian regional administration as a reference Administration in the XXI century*' (GVA, 2004: 214). Obviously, more concrete objectives for the concept i-Government has been outlined (*ubiquity*, *competitiveness*, *efficiency*, and *accessibility*), and the dimension of ICTs in public administration boosted digital public service delivery and cooperation between them (interoperability), specially, those (central administration bodies, local councils and other local entities) providing services to the Valencian citizenry.

E-Government Funding, Legal Framework, and Internal Structure

Understanding *e*Government strategies and choices require close attention to institutional settings. In general, they set up the organizational arrangements in which public organizations develop their work, and also they are the rules of the game that introduce some limitations in administrative reform policy formulation. *Funding, legal framework*, and *internal structure* are now presented briefly, as the institutional frame for this case analysis of the Valencian regional administration. *Funding* is the most inmediate translation of political commitment and leadership. The evolution of *funding* in different strategies has denoted the general significance of administrative reform and, specially, the relevance of *e*Government within the GVA. Broadly, the evolution of IT budgets within the GVA persisted more or less constant and moderate during the period of internal informatization. At the same time, an impressive change was fostered at the turn of the new century. From that moment, the magnitude of ICTs budgets was quantitatively different. In the first case, *Graphic 1* reflects the important growth within a long period of time in the GVA. This factor was akin with the adoption of *Moderniza.com*, making a diference in this respect. In consequence, *funding* for administrative reform (focused in management, without ICTs) was very precarious in comparison with magnitudes for *e*Government projects.

Secondly, during the last ten years different aspects related with the *legal framework* and the internal regulation of GVA have shaped administrative reforms and *e*Government programmes, addressing legalistic principles rooted in Mediterranean public administrations. In this regard, GVA faced two major problems: how to regulate internally the new ICTs in different management dimensions and, even more problematic, to ensure citizens rights in the digital sphere. For the first instance, the GVA adopted the *Decree 96/1998*, to regulate informatics functions (some general aspects), use of information systems (internal organization of ICTs) and design of electronic records (management of personal data). *Decree 96/1998* also combined some key aspects founded during the informatization phase, keeping almost unchanged the traditional approach to administrative problems. On the other hand, one of the concerns for public administrations in the last years has been how to establish a *legal framework* to guarantee digital interactions with citizens and businesses. In that regard, the majority of GVA *e*Government activities followed regional (but also national) norms. In regional terms, since 2000 (under the *Moderniza.com* strategic plan) important regulations should be mentioned, among others, the *Decree 87/2002* of *Advanced Electronic Signature* and the *Decree 18/2004* to create a *regional digital record* and regulate *digital registration and notifications*. In addition, GVA finally founded in 2002 its own *certification authority* (*Certification Authority of the Valencian Community-Autoridad de Certificación de la Comunidad Valenciana*, ACCV), to strengthen this renovated approach on the relational dimension of *e*Government with this own *regional legal framework*. In sum, the external orientation of this group of regulations changed the approach of informatization and *e*Government regulation, as defined in action plans.

The *internal structure* for ICTs management facilitates the understanding of institutional settings for *e*Government programmes within the GVA. In this case, the structure devoted to *e*Government issues laid on three different ambits: (1) a *central unit* (with different names and organizational responsibilities) devoted to strategic and co-ordination functions in different areas, above all, telecommunications; (2) DPUs or *sectional units* in the rest of departments and autonomous bodies of the GVA; and (3) specific *action plan offices* for successive modernization strategies. Also, other coordination bodies to integrate ICTs in the GVA worked together in project implementation under different plans, giving support to the rest of the structure to achieve their objectives. In the first instance, the traditional GVA *central unit* for informatics has been nodal in terms of orientation and monitoring of *e*Government, not only in formal, but also in more symbolic spheres. On the other hand, different *action plan offices* have served as an appendix of this *central unit*, not only as a consequence of their formal dependence, but also as a result of the type of activities deployed, for example, data centralization about levels of implementation or external communication with media. Finally, the role of *sectional units* or DPUs within *e*Government was confined to co-operate in design and implementation of specific projects, being excluded from different (and more executive) functions, as communication or diffusion of strategies within the organization. In consequence, decoupling between formal coordination structures and real functioning institutions also deserved attention to disseminate *e*Government within the whole GVA.

Organizational Actors: General Service Managers vs. IT Managers

This final dimension is essential to understand administrative reform and *e*Government in the GVA. Here, the author focuses on differences about perceptions and interests of those with managerial responsibilities regarding technological functions and the group with other general functional competencies in public services, units, and agencies. The interaction of ICTs managers with general service managers is one of the challenges for public administration, a key aspect to comprehend the extent of decisions about *e*Government policies and strategies. *Figure 1* provides a view about both groups, with politicians playing a role as referees in some cases.

Figure 1. Evolution of ICTs budget in the Generalitat Valenciana (Source: Own elaboration from CODESI data (Technical Committee for Development of Information Systems in the Generalitat Valenciana-Comité Técnico para el Desarrollo de los Sistemas de Información de la Administración de la Generalitat Valenciana) and Ministry for Public Administrations)

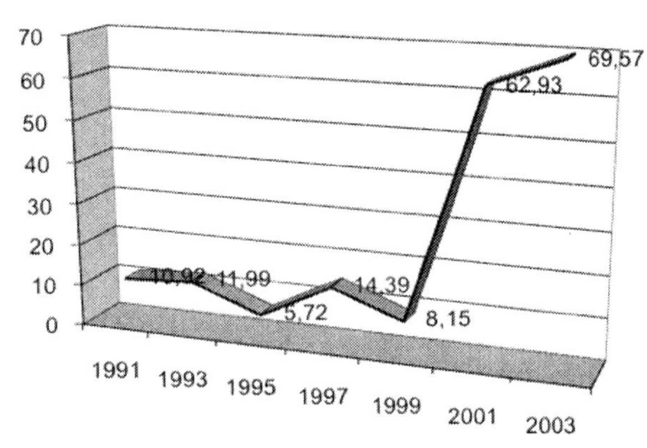

Figure 2. ICTs triangle of perceptions, interests, and decisions (Source: Own elaboration from Heeks 1999)

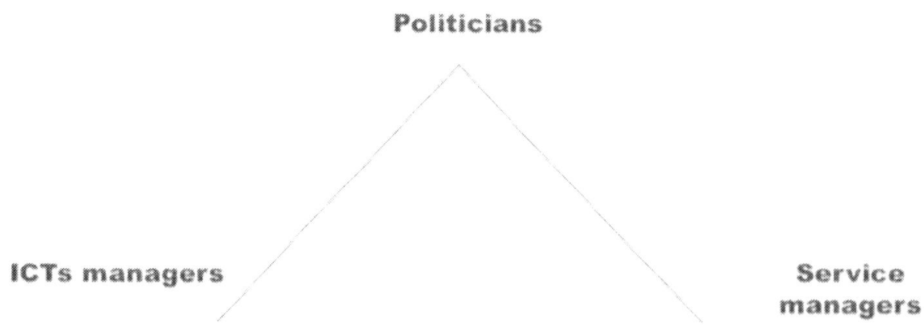

In general, the analysis of relationships between these two types of managers requires to emphasize an important lack of communication, based on divergent interests and perceptions about the potentials of ICTs for personal and organizational objectives. As Heeks recently has stereotyped (Heeks, 2006: 21): '*senior public managers have a limited understanding of IT or its possible uses in their agency. They may resist any centralised approaches they fear they cannot understand or control. IT professionals do not understand the public organization's businesses and are only really interested on the technology. This prevents centralised approaches being business-led, and leaves them being technology-led. Such approaches to e-government, if developed at all, therefore address only IT, not data or information systems or business objectives*'. The study of the GVA has confirmed some of these general statements, however it is important to develop other specific aspects of this case analysis.

The institutional context of the GVA has drawn a positive picture for public managers with technological profiles, even if their capacity to influence the political level was minimum. A *decentralised model* of information systems development and management provided the possibility to make use of them within DPUs in ministerial departments and, in that regard, being closer to specific requirements of service managers served in different agencies and units. In spite of formal institutional arrangements, personal interviews have demonstrated the persistence of basic divergences between both types of top managers, not only in day by day relationships, but also during implementation of *e*Government projects. In this case, one of the inhibitors to complete satisfactory the process of public service digitalization laid on the continuous misunderstandings between service managers and ICTs managers. On the one hand, some services managers, experts in general management, did not have skills related with technologies, so they did not understand vocabulary, capacities, and expected results from their services digitalisation. On the other hand, ICTs managers used to encompass problems when understanding business processes in specific services involved, so they did not integrate digitalisation with organizacional and human dimensions. At the end, politicians facilitated equilibrium using *e*Government policies, strategies, and decisions to promote their own interests, even if they did not appreciate the significance of ICTs for governance, as well as perils and problems they may embody in public administration.

FINAL DISCUSSION

The analysis of Valencian Community public administration (*Generalitat Valenciana*-GVA) now provides the opportunity to open up this final debate about different aspects underlined above. Here, the chapter provides a set of preliminary answers addressing the introductory questions, but also its theoretical foundations and empirical evidences, for a better comprehension of issues related with drivers and constraints for innovation in public sector modernization agenda, in general, and *e*Government strategies and orientations, in particular. Finally, it was expected from the analysis of this case study to contribute with an assessment of the regional dimension of *e*Government. Also, the approach of this paper developed more general conclusions about the relevance of ICTs in public administration.

Institutional Settings and Innovation in Administrative Reform

One of the core assumptions of *new instituionalism* is that choices and political processes are shaped by institutional arrangements. In that regard, this paper has pinpointed a limited set of institutional factors affecting administrative reform in public sector organizations, not being the most important technologies. Attending the case of the GVA, timing and context of this process kept fundamental to understand the later progresses of this experience, above all, its origins and initial institutional choices. In that regard, the first steps in administrative reform design occurred simultaneously with a favourable context for regions and regional influence in the international context, as well as within the EU, where Valencian Community governments became an active agent, above all, regarding programmes for regional Information Society development. At the same time, the political decentralization process in Spain from mid 1990s facilitated the formulation of this first package of administrative reform initiatives within regional governments.

The contextual (and historical) evolution of the GVA confirmed some of the conclusions about institutional constraints for public sector modernization. Mimetic reproduction in regions of napoleonic/legalistic model embodied in Spanish central public administration, within the process of devolution, originated problems and introduced noticeable limitations for later public sector innovation strategies. In the case of the GVA, this administration inaugurated its first administrative reform strategy in 1995, and even if its initial objectives and expectations transpired optimism, next opportunities for organizational change kept much more limited, because they were rooted on institutional arrangements adopted during the previous decentralisation political processes. Thus, dynamics of *increasing returns* emerged in the GVA, derived from self-reinforcing processes within this legalistic/napoleonic administrative model. It extraordinary hindered the (a) competence and creation of learning mechanisms; (b) long term periods of time for action and programme implementation; and (c) opportunity to change the *status quo* in public administration, as public service and policy, and institutions are designed to prevent radical changes (Pierson, 2000).

Consequently, the case of the GVA made clear that political processes of administrative reform are founded on specific organizational settings in which they are both rooted and implemented. In general, changes in public administration are more incremental than radical, so even different institutional forces have been in operation, they have also provided the opportunity for limited innovations in the long term. In other words, durable processes will be crucial in creating the pre-conditions for institutional change in public administration derived from ICTs (Fountain, 2001). These processes might operate mainly to gradually diminish the benefits (*increasing returns*) associated with a particular institutional arrangement (*path dependence*), so it remains plausible that extended periods of disappointing results attributed to a particular institutional arrangement may lead internal and external actors to reassess the benefits derived from the *statu quo* (Pierson, 2004). However, this was not the only theoretical perspective addressed in this study.

Modernization Agenda, NPM, and ICTs

Another important aspect underlined in this chapter is the evolution of the modernization reform agenda during the period of analysis in the GVA. In general, there was a shift from the fisrt moments of administrative reform, almost exclusively oriented to internal management aspects, to the most recent technology driven modernization agenda, almost monopolised by ICTs. This is not to say that changes have been radical. Before the turn of the century, administrative reform also gave importance to reinforce the internal dimension of technologies in public administration (management and organization), because public administration was not ready for developing its external dimension (public service delivery). Since then, some authors have postulated the end/death of NPM, due to increasing predominance of *e*Government issues for agendas of public administration modernization (Dunleavy *et al*., 2006). Onward attention was given to the connections between public service delivery and ICTs. However, the critical point derived from the fact that this process was parallel with the existence of unchanged institutional arrangements.

In the case of GVA, with institutional settings founded on public law, this strategic shift became even more evident. In the last years, technologies within the GVA have turned into the core issue of administrative reform strategies, and even an issue of interest for politicians and senior officials. However, to understand the scope of this technology driven modernization agenda, one of the external senior advisors of the GVA underlined the next words during an interview:

If you look at eGovernment and people working on this, after more than fifteen years working for the GVA for different governments, I would distinguish two levels: who is doing what and, in second place, how is it done. Addressing the first, I have lots of doubts about the possibilities to transform the decisional and policy-making process in public administration, even using ICTs, even more within bureaucratic contexts like the Spanish. Competencies about what to do and who is doing that are very clear, and the real possibilities for change derived from the political level, which confronts the opposition of senior managers to transform the rules of the game. On the other hand, how is it done, it is accessible to change with ICTs. How things are done implies that public administrations produce and provide public services and then offer them to the citizens, so without alter the line of power and the competencies of different units and people involved, it is possible to adopt new technologies in these processes to improve them (Interview C5).

These preceding words indicate the external role conferred to ICTs in public administration. In other words, that external focus given to ICTs in the modernization agenda of reform changed the interest from the internal (*back office*) to the external (*front office*) dimension of public administration. Besides, it is remarkable that this external approach to the citizens presents them as mere consumers of electronic public services, who '*a) should be empowered so that she is able, more than before, to act as a 'homo economicus' who actually has a choice, which b) can obtain more client-friendlier and more cost efficient services, or even c) can participate as co-producer in the way services should be provided in order to strengthen the responsiveness and need-orientation of the public service delivery process*' (Bekkers & Korteland, 2005: 22). In consequence, even if the existence of an electronic supermarket of public services deserves serious attention, it is even more provocative the existence of administrative structures (physical and digital) working in parallel.

Our analysis of human capital focused on relationships between services managers and IT managers, however the third part at the top of the triangle (politicians) needs to be also underlined here. Politicians have an even '*more limited understanding of ICTs, and want quick and visible results. They may resist large-scale centralized approaches and prefer smaller, citizen-oriented initiatives that have a lower risk and shorter time-scale*' (Heeks, 2006: 21). It seems that modernization agendas limited to external *e*Government projects and concerns fit well with politicians' interests. Besides, they confront different forces that will incline them to maintain this general approach, at least in the short term: internally, they do not envisage demands to change structural or organizational elements of *statu quo* from both groups of actors addressed before, so they will be inclined to maintain in-house stability. Externally, they are under strong *isomorphistic pressures* from *e*Government reform markets, with different private agents in action, such as, ICTs companies, consultancies, but also other public administrations and general businesses and professional groups (lawyers, architects, doctors, etc.). In general, preserving this narrow path for administrative reform implies following orientations given by private sector companies and e-commerce service delivery focus. Here, ICTs companies dominate extensively and they have rapidly conformed different types of *e*Government products and packages (CRMs, e-signature, and so on) supporting electronic public service delivery. Nonetheless, they did not address more politically and sensitive issues, such as the extension of citizens rights or political participation in the cyberspace.

Building the Digital Government: A Regional Focus?

The final dimension of this discussion became the first in the initial pages of this chapter. Even after general conclusions of previous dimensions of this debate, it could be unclear if it is possible to confirm (or not) the need for an *e*Government regional focus. At the beginning, we formulated different questions about the existence of autonomous regional administrative modernization agendas, specific institutional factors shaping regional electronic public administration developments, or the capacity of regional governments to shape more general electronic government reforms. From that first moment, this chapter has shown the growing importance of regional governments in the information age political landscape, because '*this informational paradigm bring capital, people, institutions, information and technologies into more intensely motivated by and stimulated through localized geographical agglomeration and spatial clustering*' (Castells, 1996). In more political science fashion, it is also a political decision about where democracy properly lies, about who should have a say in what sort of decisions.

From the point of view of political scientists, regions tend to represent sub-national administrative units, and have become a synonymous with a sphere of political power or an area over which someone has the capacity to govern. Thus, this debate is about whether or not states need a level of government intermediate to central and local tiers with powers devolved from the centre or local authorities and, if so, what form it should take (Keating and Rhodes, 1979). Regional tier of government facilitates governance in two senses of the word. On the one hand, regional government may express the cultural/national aspirations of cultural/national self-identified communities. On the other hand, there is a vision of improving service delivery and policy and decision-making process participation. GVA focused on both dimensions using ICTs and defining *e*Government policies. In consequence, and within a narrow point of view, it is possible to answer *yes* to the question about the existence of a regional approach to *e*Government. However, the problem is about the limits of this affirmative answer. The case study has shown the emergent significance of political powers within a young European region, in a decentralised political system, and its close relationship with the capacity to design its own regional administrative modernization agenda, even if they were also shaped by different institutional factors.

In fact, the existence of this *e*Government regional focus derives from the assumption about the capacity of this tier of government to facilitate governance, improving service delivery and policy and decision-making process participation. In general, developments on *e*Government also may support regional economic development, social welfare of the region's citizens, and enhance regional democracy and political autonomy. In the concrete sphere of public service delivery and internal public management, this should imply more proximity to the demand side, more integration of regional perceptions and interests, and more support to local government. All of them based on accurate knowledge of politico-administrative regional implementation forces.

On the other hand, the symbolic dimension deserves specific attention here. Even in little scale, regional modernization strategies of reform prioritise a regional/symbolic factor, in other words, they communicate something to the citizens of the region. In the case of the GVA, different official documents and publications epitomized that rhetoric: '*renovation of our politicians and public administrations*; *Valencian Community on the Net, as well as public administration and Valencian people*; and *Valencian Community as a point of reference in the international context. We are agents of the global Information Society*'. This symbolic dimension also contributes to reinforce the modernization discourse, once it is attached to regional self-identification based on cultural, idiomatic, or historical affinity.

Finally, future developments within this regional approach will contribute decisively to enrich *e*Government studies. First, it may be probably, at least in the mid-term, that regional governments in Europe will comprise more budget and public services and policies than other tiers of government. Thus, it is feasible to think they will probably develop even more influential *e*Government strategies and projects, so deserving more interest from scholars in the field. Second, to complete a general picture about global progress in *e*Government it will be required to overcome the traditional dearth of this type of regional-based studies, which make interesting to foster comparative research among different countries. Third, future growth of digital governance will likely rely on relationships among different tiers of government, in political science terms, intergovernmental relations, and in *e*Government words, interoperability. The intermediate position of regional government, between local and central/national authorities, perhaps will provide them a nodal role in political systems of this new digital age.

REFERENCES

Aja, E. (2001). Spain: Nation, Nationalities, and Regions. In J. Loughlin (Ed.), *Subnational Democracy in the European Union* (pp. 229-253). Oxford, UK: Oxford University Press.

Arenilla, M. (2000). Introduction (Introducción). In M. Arenilla (Ed.), *Modernization Processes in Public Administrations. La Rioja Case* (El Proceso de Modernización en las Administraciones Públicas. El Caso de La Rioja) (pp. 23-34). Logroño, Spain: Gobierno de La Rioja - Consejería de Desarrollo Autonómico y Administraciones Públicas.

Arthur, B. W. (1994). *Increasing Returns and Path Dependence in the Economy*. Ann Arbor, MI: University of Michigan Press.

Barzelay, M. (2001). *The New Public Management. Improving Research and Policy Dialogue*. Berkeley, CA: University of California Press.

Bekkers, V. J. J. M., & Korteland, E. (2005). *Governance, ICT and the Modernization Agenda of Public Administration: a Comparison of some European Policy Initiatives*. Paper presented at European Group of Public Administration Annual Conference, 1Bern, Zwitserland.

Bellamy, Ch., & Taylor, J. A. (1998). *Governing in the Information Age*. Buckingham, UK: Open University Press.

Capano, G. (2003). Administrative Traditions and Policy Change: when Policy Paradigms Matter. The Case of Italian Administrative Reform during the 1990s. *Public Administration, 81*, 781–801. doi:10.1111/j.0033-3298.2003.00371.x

Castells, M. (1996). *The Information Age: Economy, Society and Culture*. Oxford, UK: Blackwell Publishers.

Chadwick, A., & May, Ch. (2003). Interaction between States and Citizens in the Age of Internet: "e-Government" in the United States, Britain, and the European Union. *Governance: An International Journal of Policy and Administration, 16*(2), 271–300.

CODESI. (2003). *Report of the Technical Committee for Development of Information Systems in the Generalitat Valenciana* (Memoria del Comité Técnico para el Desarrollo de los Sistemas de Información de la Administración de la Generalitat Valenciana. Año 2002). Valencia, Spain: Generalitat Valenciana.

Criado, J. I. (2006). *Regional Electronic Public Administration and Modernisation Agenda in Spain*. Paper presented at European Group of Public Administration Annual Conference, Milano, Italy.

Criado, J. I., Hughes, O. E., & Teicher, J. (2002). *e-Government in Australia and Spain: a Study of Contrasts and Similarities*. Paper presented at VI International Research Symposium on Public Management, Edinburgh, UK.

Criado, J. I., & Ramilo, M. C. (2007). Local Public Administration in the Age of the Web. The Case of Spain. In P. Trevorrow, D. Griffin, & E. Halpin (Eds.), *Developments in e-Government. A Critical Analysis* (pp. 49-72). Amsterdam: IOS Press.

Danziger, J. N., & Andersen, K. V. (2002). Impacts of Information Technology on Public Administration: an Analysis of Empirical Research from the "Golden Age" of Transformation. *International Journal of Public Administration, 25*(5), 591–627. doi:10.1081/PAD-120003292

Danziger, J. N., Dutton, W. H., Kling, R., & Kraemer, K. L. (Eds.). (1982). *Computers and Politics. High Techology in American Local Governments*. New York: Columbia University Press.

David, P. (1986). Understanding the Economics of QWERTY: the Necessity of History. In W. Parker (Ed.), *Economic History and the Modern Historian* (pp. 30-49). London: Blackwell.

Dunleavy, P., & Margetts, H. (2007). *Government on the Internet. Progress in Delivering Information and Services Online*. London: National Audit Office.

Dunleavy, P., Margetts, H., Bastow, S. H., & Tinkler, J. (2006). *Digital Era Governance: IT Corporations, the State and E-government*. Oxford, UK: Oxford University Press.

Dutton, W. H., & Kraemer, K. L. (1978). Management Utilization of Computers in American Local Governments. *Communications of the ACM, 21*(3), 206–218. doi:10.1145/359361.359364

Fountain, J. E. (2001). *Building the Virtual State. Information Technology and Institutional Change*. Washington, D.C: Brookings Institution Press.

Frissen, P. H. A., & Snellen, I. Th. M. (Eds.). (1990). *Informatization Strategies in Public Administration*. Amsterdam: Elsevier Science Publishers.

Gil-García, J. R. (2004). Information Technology Policies and Standards: a Comparative Review of the States. *Journal of Government Information, 30*(5-6), 548–560. doi:10.1016/j.jgi.2004.10.001

GVA. (1996). *Strategic Action Plan for Valencian Administration Modernization* (Ejes del Plan Estratégico de Modernización de la Administración Valenciana (PEMAV)). Valencia: Generalitat Valenciana.

GVA. (2000). *Moderniza.com. 2nd Action for Valencian Administration Modernization.* (Plan Moderniza.com. 2° Plan de Modernización de la Comunidad Valenciana. La Administración de Todos). Valencia: Generalitat Valenciana.

GVA. (2004). *Avantic 2004-2010*. Valencia: Generalitat Valenciana.

Heeks, R. (1999). Reinventing Government in the Information Age. In R. Heeks (Ed.), *Reinventing Government in the Information Age. International Practice in IT-Enable Public Sector Reform* (pp. 9-21). London: Routledge.

Heeks, R. (2006). *Implementing and Managing e-Government. An International Text*. London: Sage.

Heeks, R., & Bailur, S. (2006). Analyzing e-Government Research. Perspectives, Philosophies, Theories, Methods, and Practice. *Government Information Quarterly, 24*(2), 243–265. doi:10.1016/j.giq.2006.06.005

Heichlinger, A. (2004). *eGovernment in Europe Regions. Approaches and Progress in IST Strategy, Organisation and Services, and the Role of Regional Actors*. Maastricht, The Netherlands: European Institute of Public Administration.

Hood, Ch. (1991). A Public Management for All Seasons? *Public Administration, 69*(1), 3–19. doi:10.1111/j.1467-9299.1991.tb00779.x

Hood, Ch. (1995). Emerging Issues in Public Administration. *Public Administration, 73*(2), 165–183. doi:10.1111/j.1467-9299.1995.tb00822.x

Hughes, O. E. (2003). *Public Management and Administration* (3rd ed.). London: Palgrave Macmillan.

Keating, M. (1985). Regional Autonomy in the Changing State Order: A framework for analysis. *Regional Politics and Policy, 2*(3), 111–132.

Keating, M., & Rhodes, R.A.W. (1979). Is there a regional level of government in England? *University of Strathclyde Studies in Public Policy* (No. 49).

Kraemer, K. L., & King, J. L. (2003). *Information Technology and Administrative Reform: Will the Time After e-Government Be Different?* Paper presented at Heinrich Reinermann Schrift Fest, Post Graduate School of Administration, Speyer.

Leitner, Ch. (2003). *eGovernment in Europe: the State of Affairs*. Maastricht, The Netherlands: European Institute of Public Administration.

Leitner, Ch., & Criado, J. I. (2006). eGovernment in Europe: Towars a Focus Based on Benchlearning (El Desarrollo de la e-Administración Europea. Hacia un Enfoque Basado en el Benchlearning). In L. Cotino (Ed.), *Freedom, Democracy and Electronic Government* (Libertades, Democracia y Gobierno Electrónicos) (pp. 263-299). Grenada: Comares.

Lenk, K., & Traunmüller, R. (2002). *Electronic Government: Where Are We Heading?* Paper presented at Electronic Government, I International Conference, EGOV 2002, Aix-en-Provence, France.

March, J. G., & Olsen, J. P. (1989). *Rediscovering Institutions: the Organizational Basis of Politics*. New York: Free Press.

March, J. G., & Olsen, J. P. (1996). Institutional Perspectives on Political Institutions. *Governance: An International Journal of Policy and Administration, 9*, 247–264.

Margetts, H. (1999). *Information Technology in Government: Britain and America*. London: Routledge.

Melistki, J. (2002). *The Adoption and Implementation of e-Government: the Case of e-Government in New Jersey*. Unpublished doctoral dissertation, Graduate School Newark Rutgers, The State University of New Jersey, New Jersey.

Meyer, J. W., & Rowan, B. (1977). Institutionalized Organizations: Formal Structure as Mith and Ceremony. *American Journal of Sociology, 83*(2), 340–363. doi:10.1086/226550

OECD. (2003). *The e-Government Imperative*. Paris: Organisation for Economic Co-Operation and Development.

Olías de Lima, B. (2002). Spanish Public Administration: Assessment and Perspectives (Las Administraciones Públicas Españolas: Balance y Perspectivas). In J. Crespo & G. Pastor (Eds.), *Spanish Public Administration* (Administraciones Públicas Españolas) (pp. 367-393). Madrid, Spain: McGraw-Hill.

Peters, B. G. (2005). *The New Institutionalism. Institutional Theory in Political Science* (2nd ed.). London: Continuum.

Pierson, P. (2000). The Limits of Design: Explaining Institutional Origins and Change. *Governance: An International Journal of Policy and Administration, 13*(4), 475–499.

Pierson, P. (2004). *Politics in Time: History, Institutions, and Social Analysis*. Princeton, NJ: Princeton University Press.

Pollit, Ch. (1993). *Managerialism and the Public Services: the Anglo-American Experience* (2nd ed.). Oxford, UK: Blackwell.

Pollit, Ch. (2002). Clarifying Convergence: Striking Similarities and Durable Differences in Public Management Reform. *Public Management Review, 4*(1), 471–492.

Pollit, Ch., & Bouckaert, G. (2003). Evaluating Public Management Reforms: an International Perspective. In H. Wollmann (Ed.), *Evaluation in Public Sector Reform. Concepts and Practice in International Perspective* (pp. 12-35). Cheltenham, UK: Edward Elgar.

Pollit, Ch., & Bouckaert, G. (2004). *Public Management Reform. A Comparative Analysis* (2nd ed.). Oxford, UK: Oxford University Press.

Powell, W. W. (1991). Expanding the Scope of Institutional Analysis. In W. W. Powell & P. J. DiMaggio (Eds.), *The New Institutionalism in Organizational Analysis* (pp. 183-203). Chicago: University of Chicago Press.

Powell, W. W., & DiMaggio, P. J. (Eds.). (1991). *The New Institutionalism in Organizational Analysis*. Chicago: University of Chicago Press.

Ramió, C., & Salvador, M. (2002). The Configuration Public Administrations in Comunidades Autónomas (La Configuración de las Administraciones de las Comunidades Autónomas). In J. Subirats & R. Gallego (Eds.), *Twenty Years of Autonomies in Spain* (Veinte Años de Autonomías en España) (pp. 99-133). Madrid, Spain: Centro de Investigaciones Sociológicas, Scott, W. R. (1995). *Institutions and Organizations*. London: Sage.

Snellen, I. Th. M., & van de Donk, W. (Eds.). (1998). *Public Administration in an Information Age. A Handbook*. Amsterdam: IOS Press.

Sotiropoulos, D. A. (2004). Southern European Public Bureaucracies in Comparative Perspective. *West European Politics, 27*(3), 405–422. doi:10.1080/0140238042000228077

Tomaney, J., & Ward, N. (2001). Locating the region. An introduction. In J. Tomaney & N. Ward (Eds.), *A Region in Transition. North East England at the Millennium* (pp. 1-22). Aldershot, UK: Ashgate.

West, D. M. (2003). *State and Federal e-Government in the United States, 2003*. Providence, RI: Center for Public Policy, Brown University.

Ya, A. N., & Bretschneider, S. (2007). The Decision to Contract Out: A Study of Contracting for E-Government Services in State Governments. *Public Administration Review*, *67*(3), 531–544. doi:10.1111/j.1540-6210.2007.00735.x

Yildiz, M. (2007). E-government Research. Reviewing the Literature, Limitations, and Ways Forward. *Government Information Quarterly*, *24*(3), 646–665. doi:10.1016/j.giq.2007.01.002

Chapter 15
The Process of E-Government Public Policy Inclusion in the Governmental Agenda:
A Framework for Assessment and Case Study

Federico Monteverde
Electronic Government Researcher, Uruguay

ABSTRACT

The incorporation of information systems in government agencies is a process that started several decades ago, but it has sped up its pace due to the evolution of information technology and communications. This process could be characterized as being responsive to specific or by-sector requirements without policies or guiding principles. To this date, national and local government agencies obtain, record and process information in order to accomplish their duties. Their ways illustrate a fragmented scheme, noticeably based on the criteria of a self-referenced, bureaucratic state. Accordingly, the state has a myriad of information systems which are frequently disconnected. In spite of this, their remarkable evolution is a valuable asset that policy makers should consider. When the concept of electronic government emerged in the late nineties, it came to bridge the gap in the creation of public policies regarding the use of state information technologies and communications. However, electronic government it is not just limited to technology applications; it also introduces an innovative view on state modernization. The latter is presented from an external point of view that adopts the citizen's perspective and emphasizes the need for coordination and integration in inter-agency processes. The development of electronic government has two basic approaches: electronic government initiatives and electronic government policies. The first approach resembles the traditional incorporation of information systems in government agencies,

DOI: 10.4018/978-1-60566-860-4.ch015

where multiple electronic government initiatives respond to specific vertical requirements. The second alternative incorporates a holistic view intended to build a global model of State through an electronic government public policy. This chapter focuses on the political process leading to the development of an electronic government policy, and it is illustrated by the analysis of the political process that led to the development of electronic government policy in Uruguay. The analysis uses John W. Kingdon's (1995) multiple streams model, a conceptual approach that is intended to explain the mainstreaming of policies in the governmental agenda. The chapter further proposes that the multiple streams model could be extremely useful as a framework to be used on an ex ante basis for advocates of electronic government policy making.

INTRODUCTION

This chapter focuses on Electronic Government as a form of public policy and its development as a political process. It proposes an analytical framework that allows for assessment of advocates of Electronic Government during primary stages of this process. Bearing this roadmap in mind, the journey starts with the shaping of the distinction between Electronic Government *initiatives* and *policies*.

Electronic Government initiatives are projects whose deployment are either subject to by-sector requirements, or intended to solve a particular problem of the Public Administration. Depending on their own nature they can be classified as applications in the fields of government to citizen (G2C), government to business (G2B), government to employee (G2E) or government to government (G2G). In contrast, *Electronic Government policies* are intended to build a global model of State, introducing integrated contents and providing a new level of cooperation among government agencies, supported by a networking infrastructure. The resulting benefits can be measured in the increased efficiency and improved quality of public services, thus increasing transparency and improving government communication with citizens and businesses (Valenti et al., 2003).

Having stated the public policy aspect of Electronic Government, let us review the chapters' structure. The first section discusses the importance of mapping out the policy context, so as to carry out an effective policy assessment. It provides support to conceptual policy making acting as a non-linear dynamic process, which involves multiple actor networks with varied interests and resources, sometimes influenced by competing policy standpoints. It introduces the multiple streams conceptual model developed by John W. Kingdon (1995) as an analytical framework, with the intention of explaining the flow of the policymaking process in context. Over time, this flow determines the chances of success or failure of a certain policy initiative.

The second section draws on the political process that led to the development of Electronic Government policy in Uruguay, using the conceptual framework introduced in the first section. This has the merit of being a practical application of the framework and configuring an interesting case study that allows for the critical analysis of the success of the ongoing Uruguayan Electronic Government policy.

Extensive literature exists regarding the implementation phases of Electronic Government (Araya, 2004; Reilly, 2004; PRYME, 2006; Stamoulis et al., 2001; Sware & Deane, 2003). Hence, this chapter attempts to reflect upon previous phases that are mostly related to the setting of the agenda, the specification of alternatives and the decision-making processes.

ANALYTICAL FRAMEWORK

Why a Framework?

The complex nature of public policies requires an integrated framework that can explain causal links among the elements participating in the policy process. Dye (1992) argues that frameworks and theoretical models in public policy analysis are useful to simplify and clarify the existing knowledge on public policies, identify the highlights of political problems, increase the awareness in public policies in order to facilitate the distinction between important and accessory elements, and explain public polices in order to foresee their consequences. Such a framework constitutes a theoretical model from which hypothesis can be derived and tested (Ostrom, 1999).

The Public Policymaking Process

Harold Lasswell´s (1951) early work established the traditional approach to the study of policymaking. His basic metaphor of a stages model consists of a linear and discrete process, used to break down policymaking, allowing for the analysis of each component part sequentially. Analytic units are treated as temporally and functionally distinct; therefore, each of the stages in the policy process involves distinct periods of time, political institutions and policy actors. Both original and revised versions of Lasswell's stages model (Jones, 1970; Anderson, 1975; Brewer and Deleon, 1983), also known as the stages heuristic or textbook approach (Nakamura, 1987), include the following basic stages: (1) the identification of policy problems; (2) agenda setting; (3) the formulation of policy proposals; (4) the adoption and legitimization of policies; (5) the implementation of policies and, (6) the evaluation of a policy.

The stages model has two major conceptual strengths. First, by focusing on the process stream, it surpasses the boundaries of specific institutions; secondly, by reducing the complexity of policy making to smaller, manageable analytical units, it has allowed new researchers the benefit of a discrete set of events and ideas to focus on (Pressman and Wildavsky, 1973; Kingdon, 1984).

The basic criticism of the stages framework is that it does not allow for a whole explanation of the policy making process, because the stages-focused approach tends to analyze it in a separated and disconnected way. Another source of criticism is its close relation to a rational decision making model. As a matter of fact, Lasswell first perceived it in that sense, which simply does not describe policymaking realities very well (Braybrooke y Lindblom, 1963:23.31; Lindblom, 1979:518). Finally, Sabatier (1999) argues that the assumption of a single policy cycle regarding policymaking oversimplifies the process of multiple, interacting cycles involving numerous policy proposals and actors at multiple levels of government. Initial emphasis on policymaking as a rational process shifted to a new approach of policymaking as a social interaction process between various actors in a certain context.

These shortages were addressed by Kingdon, who adapted the garbage can model of choice proposed by Cohen, March and Olsen (1972). The garbage can model was intended to explain decision-making in organizations, a process depicted as "organized anarchies where ambiguity is rampant" (Cohen, March and Olsen, 1972:1) by the authors. Kingdon extended its usage to the government, with the aim of policy formation analysis, involving two pre-decision processes: agenda setting and alternative specification (Zahariadis, 1999). Kingdon's model deals with policymaking under conditions of ambiguity, defined as "a state of having many ways of thinking about the circumstances or phenomena." (Feldman, 1989:5) Ambiguity is a fact of policymaking, but there are constraints and inhibitions to ambiguity in policymaking as well, that is where Kingdon's model comes in to explain the complex, unordered process of policymaking (Zahariadis, 1995).

Therefore, public policy making can be considered as a set of processes including at least "the setting of the agenda, the specification of alternatives from which a choice is to be made, an authoritative choice among those specified alternatives, and the implementation of the decision" (Kingdon, 1995: 3-4). The first two stages of the public policy making constitute the pre-decision phase, which is largely related to the setting of the agenda and the specification of alternatives. The adoption of a policy is a well defined, observable conclusion to the policy formulation or pre-decision stage, which needs to be legitimized by means of a political or administrative authoritative body. Finally, the implementation phase is the deployment of the public policy in the real world.

In these pages, the focus will be set on the pre-decision processes, because they are the key stages to include an Electronic Government public policy in the agenda and succeed to further achieve a positive adoption.

So far, our discussion has presented a series of descriptions of the policy process, in which the agenda plays a central role. In order to further clarify these concepts, it is worth reviewing the definitions and characteristics of the agenda.

Cobb and Elder (1983) refer to the policy agenda in terms of the *systemic* agenda and the *institutional* agenda. The *systemic* agenda is formed by issues that the members of the political community consider worthy of public attention. The *institutional* agenda, on the other hand, contains explicit items that the government is actively and seriously considering (Cobb & Elder, 1983: 85-86). Kingdon further differentiates the institutional agenda in the *governmental* agenda and the *decision* agenda. The *governmental* agenda includes subjects "to which people in and around government are paying serious attention", while the *decision* agenda includes those subjects "moving into position for some sort of authoritative decision" (Kingdon, 1995: 142).

These concepts add to the understanding of public policy making as a political process, a new dimension where the interaction of several stakeholders takes place (Klijn, 1995; Lindblom & Cohen, 1983), therefore emphasizing the importance of the context of the political process.

The process of setting the agenda is a dynamic and complex phenomenon. Many different factors should be simultaneously considered in order to explain why some issues are included or excluded from the agenda, allowing for the dynamics of policy change. Kingdon's Multiple Streams Model aims at shedding some light on the way pieces fit in this puzzle of agendas, alternatives and public policies.

Kingdon's Multiple Streams Model

John W. Kingdon (1995) developed an analytical model with the aim of deepening the study of the stages in setting the agenda, and the system that leads government actors to reach a decision. His model, called "multiple streams or window of opportunity", focuses on the study of the actors who influence and build policy proposals, as well as the processes involved in the formulation of the agenda and identifying alternatives.

The model describes the policymaking process as the convergence of three streams: the *problem stream*, which is the process of emergence and social awareness of various issues whose solution may require governmental action; the *policy stream*, which embodies the process of developing possible solutions to a problem and the generation of policy proposals, and the *politics stream*, which represents the political context determining the chances for the selection of a policy proposal and a decision to be made.

These three streams have their own logic and flow independently; however, the actors involved can be affected by the ongoing processes of any stream. Eventually, the three streams converge at a given point, and then a *window of opportunity* opens, creating a unique moment for changes in policy. An open policy window is an opportunity for policy advocates to bring their solutions to the decision-making agenda with big chances of success.

In contrast with a classical stages model (Lasswell, 1951), this approach stands for a non-linear policymaking process. It also accounts for multiple actor networks, inside and outside the government, with varying interests, resources and power. These processes and actors are the building blocks of Kingdon's analytical framework.

THE POLITICAL PROCESS OF E-GOVERNMENT IN URUGUAY

Actors and processes are the basic elements of Kingdon's blueprint, clearly illustrating the conception of policy networks (Jordan & Schubert, 1992; Klijn, 1995; Rhodes, 1997) on which his theoretical development was built. The following paragraphs are intended to identify and describe the involved actors, as well as the policy process towards Electronic Government in Uruguay, studied in three separate processes: the problem stream, the policy stream and the political stream.

Identifying the Actors

The analysis starts with the identification of the actors participating in the political process. These actors are classified according to certain criteria that will assist in the determination of their capacity to influence the agenda or generate alternatives, also reviewing the available resources and the sphere in which they operate (Table 1).

Problem Stream

International experience shows that the incorporation of information systems into the Public Administration is a process that started many de-

Table 1. Actors and capacities

	Actors	Influence	Resources
Governmental Sphere	President of the Republic	Agenda	Presidential powers established in the Constitution of the Republic.
	Office of Planning and Budget (*OPP*)	Agenda Create alternatives.	Advising the Executive Branch in budget matters. It has legal powers to design plans and develop programs.
	Ministry of Economy and Finance	Agenda Create alternatives.	Financing and Budget control sources. Some executive units show great technological development (Internal Tax Revenue Division, National Accounting Division, National Customs Division).
	Honorary IT Advisory Group (*GATI*)	Create alternatives.	Directly advising the President of the Republic in matters of Information Technology.
	Pro-Government legislators	Agenda	Legal powers granted to the Legislative Branch by the Constitution of the Republic. The pro-government sector represents the absolute majority in both legislative chambers.
	Honorary Directive Committee of AGESIC	Create alternatives.	Technical resources for the creation of proposals.
Non-Governmental Sphere	Business Chambers	Create alternatives.	Economic resources.
	International bodies	Create alternatives.	Financial capacity. Academic and research resources.
	Universities and Academic Institutions	Create alternatives.	Academic resources, research.
	Technology Companies	Create alternatives.	Technological resources. Economic resources.

cades ago and that has increased its pace to match the progress of communications and information technology (Margetts, 1995). This process typically responds to concrete or by-sector needs, with no guiding principles or policies. The national and local public agencies collect, register and process information in order to fulfill their objectives, and their dynamics openly reflect the fragmentation criteria arising from a self-referenced, bureaucratic State model. As a consequence, the State has at its disposal a myriad of information systems that, although frequently disconnected from each other, are remarkably evolved. This great asset should be taken into account by policy makers.

The situation in Uruguay shows similar characteristics. The development of information systems in the Public Administration originates in by-sector initiatives, showing sharp asymmetries in the relative development achieved by different public agencies. The agencies that have self-generated resources, or obtain them in their capacity as tax collection agencies, are the ones that show a higher level of development. Therefore, commercial and industrial state companies, such as the electric power company (*UTE*) and telecommunications company (*ANTEL*), tax collection agencies such as the Internal Tax Revenue Division (*DGI*) and the Social Security Bank (*BPS*), budget control executive agencies, such as the National Accounting Division (*CGN*), or locally, the Montevideo City Council, have been forerunners in the incorporation of information systems. It is worth mentioning that all of them share a common trait: the objective of all the systems incorporated is to solve issues related to their internal administration.

Paradoxically, the technological development in the Public Administration reflects the features of a development model guided by the market. Some of these traits include the absence of public policies defined by the State, and the deregulation and lack of coordinated action. In this sense, during the first computerization phase, decisions were liberally made, providing technology to those agencies and institutions with the economic capacity to afford the associated costs.

State modernization and reform processes take place during the period in which the development of information and communications makes its most remarkable progress. It is interesting to note that during this process, both developments start to operate on a horizontal level, and their application contributes to the modernization of the Public Administration. In facilitating an efficient and consistent administration, they become a central element in the State's modernization process (Rivera, 2003). As stated by Prats (2005) "technologies themselves do not transform the roles, the structure and the functions of the Public Administration; however, in a democratic system, they open significant possibilities for the agents of change to foster genuine reforms." (p. 21).

Awareness of the volume of assets that the State administers in the form of information and technology infrastructure, the evident need for State modernization and reform, and the certainty of these two concepts being closely related, are the main factors in the identification of the problem and of the emergence of a need to act through the creation of public policies.

Policy Stream

When the model of New Public Management incorporated the values of the economy, efficacy and efficiency of the Public Administration into the political agenda, the use of technology acquired a special functional prominence in relation to this State modernization processes. At this point--the early and mid nineties--technology entered the framework of State modernization policies.

In its initial stage, the model proposed by the New Public Management tried to attain these values through the incorporation of techniques and tools pertaining to the sphere of private administration to the public sector; this constitutes one of the New Public Management´s guiding principles (López, 1998; Lynn, 1998; Peters and Savoie,

1998). Mendoza (1990) arranges these administration techniques and tools into the following categories: (i) general management techniques of the organization (ii) the techniques that respond to the function of a company, and (iii) managerial capacity developing techniques. The importance of information technologies and communications in order to fulfill many of these techniques is undeniable (Bonina & Lepore, 2005).

Another guiding principle of the New Public Management is its strong focus on the citizen, who starts to gain a central role in the Public Administration activity, as opposed to the bureaucratic, self-referenced model. This focus on the citizen, who is even described by some authors as a "client-citizen" (López, 1998), together with the technological evolution and the popularization of the Internet, constitutes a turning point in the role of information and communication technologies in the Public Administration field. These technologies start to be perceived as economical, effective and efficient means to bring public services closer to the citizen while improving the services rendered. The concept of Electronic Government, centered in the citizen and in the digital rendering of public services, starts to be shaped.

As the concept of an Electronic Government evolves, it acquires new applications: government to citizen (G2C), government to business (G2B), government to employee (G2E) and government to government (G2G) (Campanegra Vallé, 2002; Hilbert, 2001; Valenti et al., 2003). These new dimensions suggest new application fields and needs--organizational changes that can be fulfilled by the New Public Management (Schedler & Scharf, 2001).

It is worth mentioning that many of the preexisting information systems of the Public Administration can be considered Electronic Government initiatives. Their "rediscovery" leads us to make the distinction between Electronic Government *policies* and *initiatives*.

The *Electronic Government initiatives* are projects whose development responds to a by-sector perspective or tries to solve a particular issue of the Administration. Due to their character they can be classified as applications in the spheres of government to citizen (G2C), government to business (G2B), government to employee (G2E), or government to government (G2G).

Electronic Government policies, on the other hand, aim to the creation of a global State model, presenting integrated contents and facilitating cooperation between public agencies based on a network infrastructure. The benefits sought are an increase in administrative efficiency, the improvement of the quality of public services, an increase in administrative transparency, and the improvement of communications with citizens and businesses.

The level of development and conceptual evolution attained by the Electronic Government through its innovative vision--the modernization of the State from an external angle--established the foundations for the creation of Electronic Government policies, which bring solutions to the above-mentioned problems.

Politics Stream

The first sign that could be interpreted as a State attempt to create central policies in the field of information systems in the Public Administration was perceived in 1994. The enacted Decree 271/1994 stated that the establishment of a national policy in matters of IT is reserved for the Executive Branch, and that the National IT Commission (*CONADI*) is the Executive Branch's advising agency in IT matters. The latter operates in the sphere of the Planning and Budget Division (*OPP*), which is dependent on the Office of the President of the Republic. However, *CONADI* neither met these existing expectations, nor did it create IT policies; it limited its intervention to public requests for proposals within the State sphere.

In 1995 the Executive Committee for State Reform (*CEPRE*) was created. Its main goals were State reform and modernization, but it also approached aspects related to the Public Administration information systems, and even acted as an advocate and developer of vertically-integrated systems. Some examples are the State Purchases Integrated System, the Remuneration System and the Uruguayan State Portal. The *CEPRE* was a good attempt to render public services electronically, but it lacked an integrative approach, failing to create comprehensive Electronic Government policies, and it focused on fostering Electronic Government initiatives, whose functions were related to the State reform processes carried out during its administration.

In August 2000, the National Committee for an Information Society (*CNSI*) was created in compliance with Decree 225/2000. Its aim was to define and establish a national strategy that allowed for the development of the Information Society in Uruguay. The central themes created for the development of the strategy were: a) a literacy campaign in information and communication technologies; b) development of services for citizens and businesses based on information and communication technologies; c) modernization of the Public Administration; d) promotion of the telecommunications and Internet market, and e) development of the software industry sector in Uruguay. The work carried out by *CNSI* in order to achieve its goals was disrupted by the severe economic crisis suffered by the country in 2002.

Although there were some subsequent initiatives, they all aimed to cut spending and to raise tax revenue. With this purpose, Decree 15/2003 created the Committee for the Application of the Information Exchange System, which seeked to improve the efficiency of the administration of the main tax collection agencies through the streamlining of information systems. In this line, other measures were implemented in order to strengthen controls over State purchases, using the State Purchases Integrated System proposed by the *CEPRE*.

When the new administration took power in 2005, the situation of the Public Administration information systems remained unchanged, and it even showed signs of regression, due to the economic contraction, which virtually froze State investments in technology. On March 18, 2005, the President of the Republic resolved to create the Honorary IT Advisory Group of the Presidency (*GATI*). This group was formed by renowned professionals in the field, and its purpose was to advise the Presidency of the Republic in matters of information, communication and education technologies. The President´s declared purpose was to create a strategy for the elaboration of a digital agenda that provided a guideline for the country´s technological development.

The emphasis by the new government on matters of information and communication technologies clearly indicates that the politics stream flows in the right direction.

GATI recommended the creation of an agency for the Electronic Government. The creation of such agency was not included in the Executive Branch Budget Act, but it was incorporated during the legislative process through the agreement reached by the pro-government sectors, the *GATI*, the *OPP* and the Ministry of Economy and Finance.

Finally, section 72 of the National Budget Act 17.930 of December 23, 2005 established the creation of the Electronic Government Development Agency. The Agency worked within the Presidency of the Republic, and it assimilated the goals previously established for *CONADI* and *CNSI*, together with the administration of the technology projects that used to belong to the *CEPRE*. It was also provided with new goals related to "the improvement of the services rendered to the citizens through the utilization of the possibilities provided by information and communication technologies."

The organizational structure of the agency was completed through subsequent regulations. It was provided with new budget allocations, new complementary goals, and an Executive Director. Its name changed to Agency for the Development

Figure 1. Kingdon's three streams model

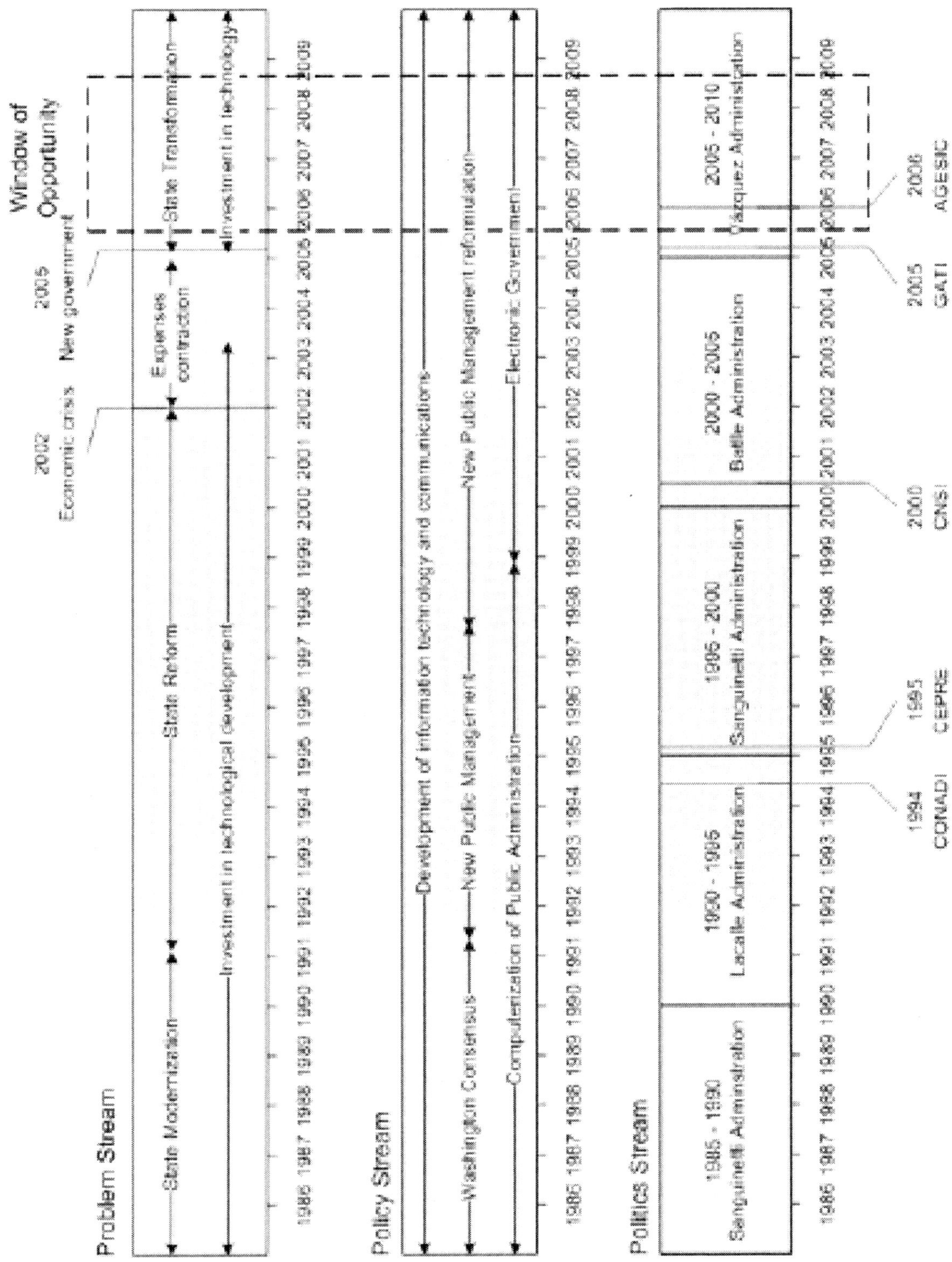

of the Electronic Administration of Government and the Society of Information and Knowledge (*AGESIC*).

FINAL CONSIDERATIONS

A central element in the *Kingdonian* model is that the opening of the window of opportunity is determined by the convergence of the three streams at a given time. This landmark indicates the right time for the incorporation of a policy to the decision-making agenda. However, it also has a very important temporal component: if for any reason the momentary conditions are ignored, the window of opportunity may close, and we would consequently have to wait for the renewal of these conditions or political cycle.

For this reason, the role played by certain actors, who Kingdon calls *political entrepreneurs* (Kingdon, 1995), is very important. These actors participate in the processes taking place within the sphere of one or more streams, but they also acquire an essential role when the window of opportunity opens, working politically to reach decisions and to position them within the agenda.

The Honorary Directive Council of the *AGESIC*, the directors of the *OPP*, the Ministry of Economy and Finance and the *GATI* played a crucial role in the consolidation of the political process. They were the political entrepreneurs (Kingdon, 1995) that made possible the incorporation of the Electronic Government policy to the decision-making agenda.

The following diagram provides a summary of the political process analyzed in Kingdon´s framework.

The three stream model makes it possible to explain the possible reasons why an Electronic Government policy was not encouraged during the four previous administrations (between 1985 and 2005), but it was during President Vázquez Administration. This is described in the following analysis.

Previous Administrations (1985 – 2005)

Problem Stream

This stage is governed by the perception of the State modernization and reform as a problem to be solved. Information technologies are not part of the problem; they are in fact considered part of the solution since the new paradigm of the New Public Management reveals the need for coordination between the agencies. This need arises from the processes of decentralization and the delegation of capacities, as well as from the citizen-oriented perspective, which requires the implementation of new services. Information systems start to be perceived as an essential tool for progress.

Policy Stream

No satisfactory proposals for the instrumentation of State technology streamlining policies were provided during the Public Administration IT development phase. However, the following development and conceptual readiness of the Electronic Government by the end of the nineties provides new possible solutions.

Politics Stream

It was not flowing in the right direction, and the politics stream did not provide the political actors with solid foundations. By the time the first favorable sign manifested—the creation of the *CNSI*—the deep economic crisis that struck the country forced the government to prioritize crisis-abating policies over all others.

Vázquez Administration (2005 – 2010)

On 2005, on the other hand, the three streams underwent substantial changes, thus creating an opportunity for the new government to progress in the design of a new Electronic Government policy, whose beginnings materialized in the Budget Act of 2005.

Problem Stream

The program of the new government includes a relevant reform package that stems from a process of State transformation. The transformation process was conceived in relation to the need of improving Public Administration, citizen participation and State intervention in those fields deemed relevant for the improvement of the citizens′ quality of life. Some examples were the tax system reform, the health care reform and the reform of the State itself, all of which were focused on the information exchange between agencies and services rendered to the citizen.

Policy Stream

Conceptual readiness is not the only given asset for the Electronic Government: there were also widespread international experiences that show successful scenarios, practices and application models.

Politics Stream

The President of the Republic's inauguration address, which expressed his determination in fostering the country's technological development, together with the prompt creation of the *GATI*, a few days after taking office, are clear signs that the politics stream flowed in the appropriate direction.

In Kingdon′s words, the intersection of the three streams made it possible to position the Electronic Government policy in the decision-making agenda of the new government. Its materialization was the creation of *AGESIC*, a promoting and executive agent of such policies in Uruguay.

CONCLUSION

Government, in our networked society, is faced with controversies over complex problems that cut across the traditional jurisdictions of organizations. Involved actors disagree about the solutions, but even about the problems. Traditional methods of dealing with these complex problems no longer suffice, because actors are forced to operate in the context of networks that are new to them and in which their traditional operating procedures are no longer adequate (Koppenjan and Klijn, 2004). Policy networks have become a key concept to analyze policymaking in multi-actor settings (Klijn, 1996). Complex issues like public administration modernization and State reform, incorporation of information systems in government agencies and Electronic Government meet these characteristics.

This chapter has its roots in Electronic Government conceived as a public policy. Therefore, it has dealt with the policymaking process with the aim of contributing to Electronic Government advocates a conceptual framework, standing from the policy networks perspective, to be used in a networked society.

The close relationship between public administration modernization, State reform and information technologies advances has been analyzed in the context of Kingdon's framework. One key observation is that its mixture has played different roles along this timeframe. They have been part of the problem and part of the solution as well, depending on the prevailing wave of State modernization, ranging from the incorporation of information technologies into public administration to Electronic Government. Consequently, its analysis is spread among the streams in Kingdon's model.

Although every case study of the Electronic Government is unique, and relies on the country's particular conditions, and although the existing experiences cannot be fully transferred and applied to third countries, it is important to highlight the fact that the application of the *Kingdonian* model does not necessarily have to be established *ex post*.

It is therefore clear that the application of the model *ex ante* can be extremely useful as a framework to assess the possibilities of success, and the conditions that should precede the promotion of an Electronic Government policy in any country.

REFERENCES

Anderson, J. (1975). *Public Policy-Making*. New York: Praeger.

Araya, R. (2004). Tres perspectivas para observar el gobierno electrónico. In R. Araya & M. Porrúa (Eds.), *América Latina Puntogob. Casos y Tendencias en Gobierno Electrónico* (pp. 23-42). Chile: FLACSO Chile – OEA – AICD.

Bonina, C. M., & Lepore, W. (2005). Desafíos y oportunidades del gobierno electrónico en el marco de la nueva gestión pública: el caso de Argentina. In *Proceedings of the X Congreso Internacional del CLAD sobre la Reforma del Estado y de la Administración Pública*, Santiago, Chile.

Brewer, G., & Deleon, P. (1983). *The Foundations of Policy Analysis*. Monterey, CA: Brooks/Cole.

Capanegra Vallé, H. (2002). El Gobierno Electrónico: Hacia una verdadera Reforma del Estado. In *Proceedings of the XVI Concurso de Ensayos y Monografías del CLAD sobre Reforma del Estado y Modernización de la Administración Pública "Gobierno Electrónico"*.

Cobb, R. W., & Elder, C. D. (1983). *Participation in American Politics: The Dynamics of Agenda-building*. Baltimore, MD: Johns Hopkins University Press.

Hilbert, M. R. (2001). *Latin America on its path into the digital age: where are we?* Division of Production, Productivity and Management of the Economic Commission for Latin America and the Caribbean of the United Nations, ECLAC.

Jones, C. (1970). *An Introduction to the Study of Public Policy*. Belmont, CA: Wadsworth.

Jordan, G., & Schubert, K. (1992). A Preliminar/ Ordering of Policy Network Labels. *European Journal of Political Research, 21*(1-2), 9–10. doi:10.1111/j.1475-6765.1992.tb00286.x

Kingdon, J. W. (1995). *Agendas, Alternatives, and Public Policies* (2nd ed.). New York: Harper Collins.

Klijn, E. H. (1995). Policy Networks: An Overview. In W. J. M. Kickert & J. F. Koppenjan (Eds.), *Managing Complex Networks*. London: Sage.

Klijn, E. H. (1996). Analyzing and Managing Policy Processes in Complex Networks. *Administration & Society, 28*, 90–119. doi:10.1177/009539979602800104

Koppenjan, J. F. M., & Klijn, E. H. (2004). *Managing Uncertainties in Networks*. London: Routledge.

Lasswell, H. D. (1951). The Policy Orientation. In D. Lerner & H. D. Lasswell (Eds.), *The Policy Sciences*. Stanford, CA: Stanford University Press.

Lindblom, C. E., & Cohen, D. K. (1979). Usable knowledge: Social science and social problem solving New Haven, CT: Yale University Press.

López, A. (1998). *La Nueva Gestión Pública: Algunas Precisiones para su Abordaje Conceptual*. INAP. Serie I: Desarrollo Institucional y Reforma del Estado. Documento Nr. 68.

Lynn, L. (1998). The New Public Management: How to Transform a Theme into a Legacy. *Public Administration Review, 58*(3), 231–237. doi:10.2307/976563

Margetts, H. (1995). The Automated State. *Public Policy and Administration, 10*(2), 88–103. doi:10.1177/095207679501000207

Mendoza, X. (1990). Técnicas Gerenciales y Modernización de la Administración Pública en España. In *Documentación Administrativa* (Nr. 223). Madrid.

Nakamura, R. (1987). The Textbook Policy Process and Implementation Research. *Policy Studies Review, 7*(1), 142–154. doi:10.1111/j.1541-1338.1987.tb00034.x

Peters, G., & Savoie, D. (1998). *Taking Stock: Assessing Public Sector Reforms*. Montreal, Canada: Canadian Centre for Management Development and McGill-Queen's University Press.

Prats, J. (2005). *De la Burocracia al Management, del Management a la Gobernanza*. Madrid: INAP.

Pressman, J., & Wildavsky, A. (1973). *Implementation*. Berkeley, CA: University of California Press.

PRYME. (2006). *Gobierno Electrónico en Chile 2000-2005: Estado del Arte II*. Ministerio Secretaría General de la Presidencia. Proyecto de Reforma y Modernización del Estado.

Reilly, K. (2004). Estrategias de e-gobierno en ocho países de América Latina. In R. Araya & M. Porrúa (Eds.), *América Latina Puntogob. Casos y Tendencias en Gobierno Electrónico* (pp. 71-110). Chile: FLACSO Chile – OEA – AICD.

Rhodes, R. A. W. (1997). *Understanding Governance: policy Networks, Governance Reflexivity and Accountability*. Buckingham, UK: Open University Press.

Rivera, E. (2003). *Nueva Economía, Gobierno Electrónico y Reforma del Estado*. FLACSO, Editorial Universitaria.

Ruiloba, J. M. (2007). Un marco para el análisis de las políticas públicas: características y elementos definitorios del modelo de las corrientes múltiples de J. W. Kingdon. *Revista Ciencias Jurídicas y Sociales, 3*.

Schedler, K., & Scharf, M. C. (2001). *Exploring the Interrelations Between Electronic Government And the New Public Management. A Managerial Framework for Electronic Government*. Switzerland: Institute for Public Services and Tourism at the University of St. Gallen.

Schware, R., & Deane, A. (2003). Deploying e-government programs: the strategic importance of "I" before "E". *info, 5*(4), 10-19.

Stamoulis, D., Gouscos, D., Georgiadis, P., & Martakos, D. (2001). Revisiting public information management for effective e-government services. *Journal of Information Management & Computer Security, 9*(4), 146–153. doi:10.1108/09685220110400327

Valenti, P., Anta, R., & Bendersky, M. (2003). *Manual.Gob. Estrategias de Gobierno Electrónico en los países de la Región 1: La definición de un modelo de análisis y estudio de casos*. Inter-American Development Bank, Washington.

Zahariadis, N. (1995). *Markets, States, and Public Policy. Privatization in Britain and France*. Ann Arbor, MI: University of Michigan Press.

Zahariadis, N. (1999). Ambiguity, time and multiple streams. In P. A. Sabatier (Ed.), *Theories of the policy process* (pp. 73-93). Boulder, CO: Westview Press.

Chapter 16
A Historical Perspective of the Development of E-Gov in Brazil

Alexandre F. Barbosa
Brazilian Internet Steering Committee, Brazil

Álvaro Junqueira
Secretariat of Finance of State of São Paulo, Brazil

Eduardo H. Diniz
FGV-EAESP, Brazil

Otávio Prado
FGV-EAESP, Brazil

ABSTRACT

Initiatives of reform and modernisation of the public sector in Brazil have intensified, not only as a consequence of the fiscal crisis that occurred during the 1980s, but also as a result of the exhaustion and collapse of bureaucratic management and government interventionist models. The development of the Brazilian e-gov program officially started in 2000 and today is in its third phase after two different government administrations. It plays an important role in the modernisation of the public sector in the country and faces the challenge of giving proper support to public policies to universalise opportunities that promote economic and social development. The currently available literature analysing the history of the Brazilian e-gov program focuses too narrowly on the identification of relevant technological aspects affecting its designing and development, failing to use a theoretical framework that takes into consideration social actors and other intervening factors. This chapter presents and employs an approach that takes into consideration the relevant social groups involved in each phase of the construction process of e-gov programs, the intervening factors impacting this construction, and the strength of the relationships among them. Based on qualitative empirical research, this chapter discusses the validity of the proposed framework for studying the history of e-gov in Brazil. As a result of an in-depth documental analysis and interviews with key actors throughout the different administration tenures, this chapter identifies important intervening factors that guided successful and unsuccessful government initiatives in the e-gov program.

DOI: 10.4018/978-1-60566-860-4.ch016

INTRODUCTION

The idea of electronic government (e-gov) is associated with the intensive and strategic adoption of ICT (Information and Communication Technology) in the public sector. E-gov is often associated with modernisation of public administration, with improvement of the efficiency of governments' internal processes (Osbourne 1997; Agune and Carlos 2005) and with the use of the internet as a channel for delivering electronic public services (Ronaghan 2002; Grant and Chau, 2005).

The reason for the intensive and strategic adoption of ICT by governments lies in the fact that citizens and companies have been resorting more and more to technology, giving preference to online transactional services, as well as the universalising of the internet itself (Holmes, 2002; O'Looney, 2002). In addition, Fountain (2001) and Heeks (2003) also indicate forces deriving from the state's reform and modernising movement as reasons for adopting ICT in the public sector. The use of technology in transforming the government's internal processes, in modernising management and control tools, and in delivering public services has made imperative a change of paradigm in assessing public sector efficiency (Fountain 2001; Pavlichev and Garson 2004). This has driven a substantial increase of governments' use of ICT (Heeks 2003), causing technologies to play a vital role in the modernisation process of public management. These technologies have been acknowledged as agents of change and as tools that make it possible to implement reforms in the public sector (Fountain 2001; Heeks 2003; Holmes 2002).

It is possible to affirm that the development of the public sector has become dependent on the use of ICT and, particularly, of e-gov programs (Fountain 2001; Klitgaard and Light 2005). E-gov programs are imperative in order for the public sector to attain new levels of efficiency so as to transform itself into a high-performance organisation. The analysis of factors of success and failure in e-gov initiatives makes it possible to identify different aspects associated with the political, social and technological dimensions that affect the performance of these programs.

The existing literature analysing the history of Brazilian e-gov focuses mostly on identifying the relevant technological aspects influencing its creation and development. There is a gap in the literature concerning the adoption of structured models or frameworks in a historical analysis considering the social actors and their relations, as well as other intervening factors, such as those of a political and social nature. This paper proposes a structured framework of analysis considering the social actors involved in each construction phase of the e-gov, the intervening factors influencing this construction and the strength of the relationships among them. Thus, the proposed model seeks to identify social, political, institutional and technological factors that significantly influence conception, implementation and evaluation phases of the e-gov programs. We start from the assumption that the nature and strength of the social relations among social actors influence each phase of this construction process.

E-GOV IN THE BRAZILIAN FEDERAL PUBLIC ADMINISTRATION

The initiatives of reform and modernisation of Brazil's public administration have intensified, not only as a consequence of the fiscal crisis of the 1980s, but also as a result of the exhaustion of the bureaucratic management model and of the interventionist state model in economic activities. The movement known as New Public Management (NPM) had as its crux principle to seek to guide and optimise the public services in terms of citizens' needs. This movement was based on management principles concerning results, efficiency and governance, guiding the public management towards market management practices. Behn (1998) defines this movement, which was characterised as New Public Management, as a group of concepts applied in public administration, consisting of several interrelated components, the use of technology being one of the necessary factors achieving results of high performance in the government's internal processes.

The strategic use of ICT as an enabling element of a new model of public management in Brazil has evolved into what is now called e-gov (Agune and Carlos 2005). In addition to being one of the most important forms of state modernisation, e-gov is strongly supported by a new vision of the use of technologies in delivering public services, changing the way the government interacts with citizens, companies and other governments. E-gov is not limited to automation of processes, making public services available by means of online services on the internet (Abramson and Means 2001), but represents a change in the way the government reaches its objectives, by means of the use of ICT, so as to fulfil its role based on the governance principles: transparency, democratic participation and accountability (Heeks 2003).

Information and Communication Technologies are important enablers in many modernization and transformation initiatives in Public Administration, and its adoption and use has significant internal and external impacts. From the Public Administration's perspective, at its core, e-gov plays an important role in governance, process rationalisation, cost control, efficiency, productivity and service quality. From a citizen perspective, e-gov is about genuinely transforming the way public services are delivered to citizens to meet their growing expectations in the new ICT-enabled world.

The Brazilian e-gov program was formally constituted through a specific legal framework in the year 2000. Its efforts aimed to fulfil three strategic areas aligned with the Information Society Brazilian Program: Services Universalisation; Government Accessible to All; and Advanced Technological Infrastructure (Brasil 2002a). This program prioritized incorporating ICT into administrative processes and providing electronic public services to citizens.

In a historical perspective, in many countries e-gov is associated with the evolution of the computerisation of the public sector. In the historical context of Brazil, Reinhard and Dias (2005) propose a historical analysis of e-gov from the identification of four phases associated with the evolution of the computerisation process in the public sector and with the relevant technological factors in each phase:

- **Pioneering:** Early 1950s through mid-1960s
- **Centralising:** Mid-1960s through late 1970s
- **Outsourcing ICT services:** From early 1980s on
- **E-gov services:** From early 1990s on

However, Diniz (2005) proposes a historical analysis focusing on the technological applications developed in three separate phases:

- **Internal management:** From 1970 to 1992
- **Services and information to the citizen:** From 1993 to 1998
- **Internet service delivery:** From 1999 on

Although, in both analytic propositions, the division into phases contributes to a historical analysis of the construction process of e-gov in Brazil, an excessive emphasis on the technological factors can be identified (Figure 1). Other influencing factors in the process, such as context, the nature of relationships among social actors, and institutional and political factors, are not always present.

Figure 1. Historical analysis of e-gov in Brazil (sources: Reinhard and Dias 2005; Diniz 2005)

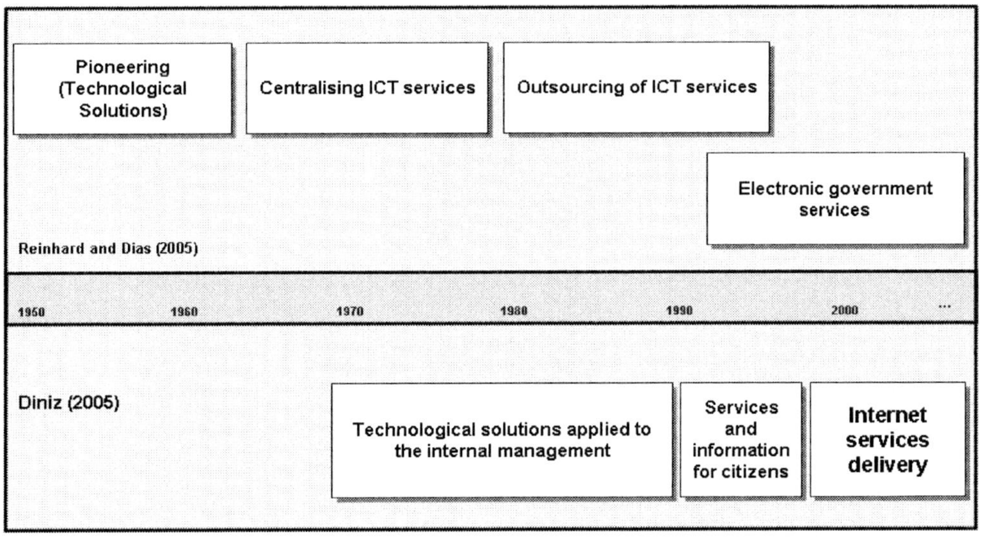

STRUCTURED FRAMEWORK FOR HISTORICAL ANALYSIS OF E-GOV IN BRAZIL

The framework proposed in this paper aims to be useful as a methodological support for the study and analysis of the factors contributing to the development, consolidation and institutionalization of e-gov programs. Although the framework emerged from in-depth analysis of data collected during interviews and documentation of the Brazilian e-gov program, its elements could also be extended to any national e-gov program historical analysis. Fountain (2001) states that the e-gov construction process is associated with the process of institutionalising ICT use in the public sector. Therefore, its historical analysis requires consideration of different aspects linked to the process of its construction and institutionalisation. In this context, one must consider the social actors involved and the influencing factors in its construction, such as: existing institutional arrangements, legal framework and public policies, and technological structure. In this context, the role of institutional arrangements and public policies concerning use of technology in the public sector take on a fundamental importance in the development of e-gov. The public policies concerning development of e-gov programs are important components of these institutional arrangements.

Frey (2000) and Peters (1996) discuss the development of public policies based on a process perspective divided into separate phases that range from conception to monitoring and evaluation of public policies. The proposed framework assumes that the construction of the e-gov is a result of a process of development of public policies and is influenced by the action of social actors and their specific contexts. It also embodies the notion of e-gov construction phases and considers the influence of the action of social actors in each phase. We consider that the strength and the nature of the relationships deriving from interactions among social actors and influencing factors throughout the various phases of construction of the e-gov affect the development, consolidation and institutionalisation of an e-gov program.

Figure 2 below presents the concepts adopted in the framework, showing the possible relations among actors and influencing factors in the five phases observed in the e-gov construction process.

Figure 2. Structured framework for historical analysis of e-Gov (source: elaborated by the authors)

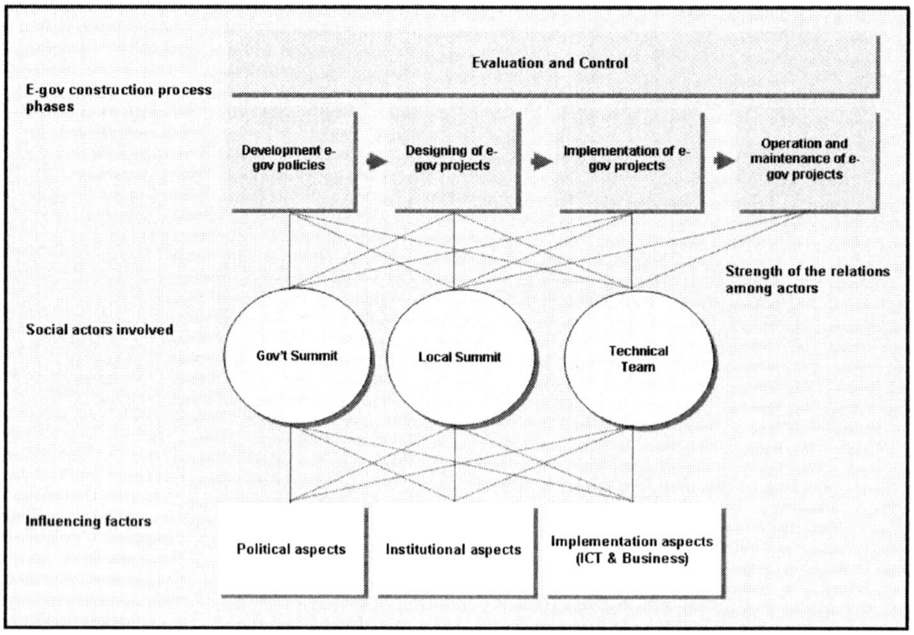

E-Gov Programs Construction Process Phase

The construction of an e-gov program can be understood as a process composed by various phases, where each phase incorporates political, technical and organizational aspects, starting with a designing phase and ending with implementation and evaluation phases.

Albeit specific, the construction phases of an e-gov program generally follow along the lines of the described phases of the public policy cycle, as pointed out by Frey (2000) and Peters (1996), as shown in Table 1. These authors established a series of steps for the formulation, implementation and evaluation of public policies, varying slightly the number of phases and the focus given to each one of them.

Considering that the e-gov construction process is a result of a specific public policy, it can be divided into five specific phases: i) development of e-gov public policies; ii) designing of e-gov projects; iii) implementation of e-gov projects; iv) evaluation and control; and v) operation and maintenance of e-gov programs.

The "development of e-gov public policies" phase consists of elaborating the directions that will provide guidance for policies development concerning use of technology in the public sector. These policies are preferably conveyed to the internal and external environments by means of formal communication tools available in the public administration. This phase comprises activities performed by political actors at the summit of governments, whose objective is to align interests among the middle areas, responsible for the support functions, and the end areas, responsible for providing public services to society.

The "designing of e-gov projects" phase involves creating new projects or converging existing technological projects with public policies concerning technology. The scope of the project, its budgetary source and its main interrelations with other projects and resources are critical in this phase.

Table 1. E-gov and public policies construction process phases

Phases proposed by Peters (1996) – public policy cycle	Phases proposed by Frey (2000) - public policy cycle	Phases in the proposed framework
Defining the problem Agenda Setting	Perceiving and defining problems Agenda Setting	Development of e-gov public policies
Formulating Legitimating Connecting to resources	Elaborating programs and decision	Designing of e-gov projects
Implementation	Implementing policies	Implementation of e-gov projects
		Evaluation and control
Evaluation	Evaluating policies and making corrections	Operation and maintenance of e-gov programs

The "implementation of e-gov projects" phase involves the execution of the previously defined projects. It requires the technical detailing of projects, which may lead to a redefinition of its scope or of its dismemberment in subprojects. According to Gil-Garcia and Pardo (2005), the greatest challenges and critical factors for success when implementing an e-gov project are transversal technological compatibility and integration, budget continuity, and bureaucratic rigidity.

The "operation and maintenance of e-gov projects" phase refers to the e-gov project's operation and maintenance processes. These involve the updating of information, continuity in service providing, and tuning and refining processes or systems. Specific performance indicators for each project must allow performance measurement and evaluation.

The "evaluation and control" phase encompasses definition of the evaluation and control mechanisms that will be used for managing the e-gov project. Such mechanisms aim to define and manage specific performance and reliability levels, considering the established scope, costs and deadlines.

Social Actors Involved

The social actors considered in the proposed framework were those who have a significant influence on the various phases of the e-gov construction process. A reduced group of actors was chosen, aiming to facilitate understanding of their roles and of the political, institutional and technological influencing factors affecting the construction of the e-gov. The social actors considered in each phase described in the framework are: government summit, local summit and implementation technical team. The scheme of actors here presented is a simplification of the reality. As Checchi et al (2003) point out, other social actors are usually present and should be taken into account as well, such as citizens, private and non-governmental organisations, academic institutions and any other relevant actors that may be of a particular interest in the analysis. The groups of actors accounted for by the framework are limited to:

- **Government summit:** Group of actors participating at the government's strategic level. They are responsible for formulating strategic directions and public policies. They are highly representative actors and have a relevant projection within government with great political power.
- **Local summit:** Group of actors composed of middle-level public managers in governmental agencies linked to public administration's end and middle areas. They are usually involved with the formulation and monitoring of the e-gov projects and they represent the intermediate hierarchical level of public managers.

- **Implementation technical team:** Comprised of specialist technicians in the data processing public companies or in the governmental agencies themselves. They are responsible for detailing and implementing businesses and technological projects.

Influencing Factors

Political, socio-economic and technological factors influence social actors' actions when performing their roles in each phase of the elaboration of public policies and, consequently, the construction of e-gov programs (Frey 2000; Peters 1996). The proposed framework used for historical analysis considers three main groups of influencing factors:

- **Political aspects:** Regarding both implicit and explicit political demands of the stakeholders in the e-gov construction process, directly or indirectly involved with the technological projects
- **Institutional aspects:** Regarding institutional arrangements influencing e-gov projects or policies, such as organisational structure, legal framework and public policies, power of the actors, service-hiring process, ICT services outsourcing model, organisational bureaucracy, etc
- **Implementation aspects:** Regarding technological impacts of the e-gov projects, such as technological standards in use, complexity of technological requirements of projects, interoperability, technological infrastructure, development methodology, etc

It is important to note that other influencing factors such as costs and efficiency of the government actions along the e-gov history could also be explicitly taken into account in the framework as important elements of analysis. Nevertheless, costs and efficiency can be considered as part of the influencing factor " implementation aspects".

Strength of Relations Among Actors

The strength of relations among actors and influencing factors in each phase varies according to specific contexts. In the so-called political networks, in which the interactions of the social actors at upper executive and legislative levels involve generative discussion of public policies, the intensity of relations in the public policy conception phase is stronger. However, those very same actors present low-intensity relations in the operation and maintenance phase of technological projects. Table 2 below shows assumptions regarding the strength of actors' relations in different phases of e-gov construction.

FRAMEWORK VALIDATION METHODOLOGY

Aiming to validate the proposed framework, qualitative research techniques and an exploratory

Table 2. Actors' strength of relation in different phases (source: elaborated by the authors)

Actors	Phases of the process				
	Development of e-gov policies	Designing of e-gov projects	Implementation of e-gov projects	Operation and maintenance of e-gov projects	Evaluation and control
Government Summit	*Strong*	*Medium*	*Weak*	*Weak*	*Strong*
Local Summit	*Medium*	*Strong*	*Medium*	*Weak*	*Strong*
Technical Team	*Weak*	*Medium*	*Strong*	*Strong*	*Medium*

case study research strategy using primary and secondary data were applied. This strategy is in accordance with the understanding presented by Babbie (1999), which indicates the case study as a technique to broadly describe a given phenomenon, based on the interrelations of the various components being studied.

The collected primary data correspond to the gathering of documents on e-gov in the Brazilian Federal Government and of the contents of semi-structured interviews with key actors conducted by the researchers. In order to understand the strength and nature of the relations among the social actors throughout the e-gov construction process, technicians and politicians directly involved with this process were interviewed.

An extensive analysis of documents and of a reduced group of interviews with key actors was conducted, aiming to identify:

- The strength of relations among the social actors in each phase
- The influencing factors, such as coordination structure, legislation and norms, program financing, projects agenda (per sector or transversal), aspects of the implementation of the program (leadership, inter-bureaucratic coordination and communication), and, finally, articulation with the society

Document Analysis

For the document analysis, we used academic publications concerning the theme, as well as available data related to the technical documentation of projects, laws, decrees, news and other documents related to e-gov initiatives in Brazil, collected directly from the electronic pages of governmental agencies of the federal public administration.

The history of e-gov in Brazil is directly linked to the development of public computerised services offered by state data processing companies since the early 1960s. The first demands for computer applications in the Brazilian public sector were related to fiscal information and data associated with government's revenue and expense control. Agune and Carlos (2005) point out that, starting from that demand, computing in public management has evolved through the years towards delivery of services to citizens through various access channels.

The most noticeable aspects in this analysis are the ones related to the development of the e-gov policy in April 2000. The great advance in Brazilian e-gov policy was the proposal of directions and norms related to the new electronic means of government/society interaction. The creation of an organisational structure to politically coordinate the Brazilian e-gov program was another important aspect identified (Brasil 2001). This structure made it possible to institutionalise some important initiatives, such as promoting the effective universalisation of access to information technologies and services, which was done aiming to facilitate development of the project (Brasil 2002a, 2002b). Another relevant factor consisted of the creation of the Electronic Government Executive Committee (CEGE) in 2003, and the institutional and political support given to the e-gov program by the Planning, Budget and Management Ministry, by means of the Information Technology and Logistics Bureau (SLTI). This institutional arrangement secured the technical-administrative support needed for the CEGE to function. Still in 2003, the Presidency of the Republic published a decree instituting eight technical committees in CEGE: Digital Inclusion; Online Services and Websites Management; Free Software Implementation; Systems Integration; Networks Infrastructure; Strategic Information and Knowledge Management; Government to Government Relationship (G2G); and Legacy Systems and Software Licenses.

According to Pinto and Fernandes (2005), despite the success of the e-gov program's institutionalisation, it has been losing governmental priority since the tenure transition in 2003. According to those authors, this is due to four main factors:

- **Political leadership change:** The role performed by the Head of the Civil House, Pedro Parente, was not performed with the same dexterity by the new representatives in Lula's tenure
- **Lack of intra-bureaucratic coordination:** There are no instances of e-gov program coordination within each ministry, which makes it difficult to run integrated actions
- **Difficulty in articulating with society:** Legal restrictions on the public sector still generate little stimulus for the establishment of partnerships with private initiatives, worsening the uncertainty about the model of relationship among ICT service providing companies and the government
- **Lack of resources:** E-Gov projects have scarce resources and are routinely subjected to budget rationing. Successful initiatives tend usually to be per sector, separate from a strategic solution by the government.

The documental analysis also made it possible to identify influencing factors leading to difficulties in coordinating the Brazilian e-gov program, such as lack of monitoring of e-gov initiatives, lack of performance indicators, existence of multiple governmental internet portals (making it harder to manage the contents), and loss of political leadership and of CEGE's power.

Analysis of the Interviews with Key Actors

Interviews were conducted with key actors who participated in the e-gov program construction process in the administration of President Fernando Henrique Cardoso (1999-2002) and in the administration of President Luís Inácio "Lula" da Silva (2003-2006). The data, collected from semi-structured interviews, were rich in strategic information, complementing the document analysis.

These interviews were conducted and analysed by four different researchers. The results of the individual analyses converged into identification of important elements in the validation of the proposed framework, especially those related to phases of the e-gov construction process and to the relevance of considering social actors and influencing factors in this process. In addition, the interviews brought out facts that were not documented, as well as some relevant concepts for the research. The analysis of the collected data in this phase revealed eight new influencing factors in the e-gov construction process in Brazil that were not registered in the available documents (Table 3). The assumptions related to the intensity of the actors' relations, as presented in item 3.4, were also validated by analysis of the interview content.

The interviews were organised according to the proposed framework and were conducted during the second semester of 2005 and the first semester of 2006. The interviewees were: a high governmental executive at the ministerial level; a local executive, responsible for the e-gov executive committee in Brazil during the period of President Fernando Henrique Cardoso's administration (1995-2002); and a local executive, responsible for the Brazilian e-gov program management in the first administration of President Luís Inácio "Lula" da Silva's tenure (2003-2006).

The first influencing factor disclosed by the interviewees was that all the nationwide initiatives and all resources allocated by SLTI to address the "millennium bug – Y2K" triggered the action plan for the construction of an e-gov program in Brazil. The "millennium bug – Y2K" was considered a strategic project of the Brazilian government, managed directly by the Civil House and coordinated operationally by SLTI, and it played a very important role in the Brazilian e-gov history. The

Table 3. Influencing factors on e-gov in Brazil (source: elaborated by the authors)

Influencing factor	Description
1. The Millennium Bug	SLTI was the agency that articulated the actions regarding the updating of the ICT resources of the Federal Government so as to deal with Millennium Bug issues. The institutional and operational structure created was used as a structuring mechanism for the e-gov program in Brazil.
2. Pedro Parente's political strength.	Minister Pedro Parente's determination and political strength (initially in the Budget and Management Ministry and, later, in the Civil House) were decisive factors in overcoming the obstacles to successful conducting of the e-gov program, on what concerned the conducting of CEGE's activities.
3. Brazilian electricity crisis	The electricity outage crisis in Brazil brought about a priority change to the e-gov project. Pedro Parente, who had been supervising CEGE's work very closely, ended up assuming the Chamber of Management of the Crisis in May 2001. E-gov project was given less priority in the government, reflected by the absence of the Civil House in CEGE's meetings, resulted in a decrease in the pace of e-gov initiatives.
4. Electoral process	With the electricity crisis in Cardoso's tenure and the strengthening of Lula, who was then running for the presidency of Brazil, the governmental priority was more devoted to a government with macro-economical stability. With that, the possibility of continuity loss, due to the electoral process in which the government candidate was less likely to win, brought about a lack of motivation among teams for long-term work.
5. Cardoso/Lula Transition	The year 2003 was marked by a large discontinuity in CEGE's works, which were only resumed in July 2004, when the decree creating the E-gov Department in the SLTI was published.
6. Mismatches in e-gov actions	Research among citizens, companies and public institutions in ten Brazilian capital cities revealed the mismatch between supply and demand in electronic public services.
7. Service quality evaluation	It wasn't until December 2004 that a study concerning e-gov services performance evaluation began. This study aimed to define performance indicators and goals for e-gov services. In 2006, the implementation of a pilot project started, covering seven public services at the federal, state and local levels of Brazilian government.
8. Difficulty in articulating and prioritising	The capability of overcoming obstacles and prioritising activities is identified as a problem at the end of Cardoso's tenure, as well as during Lula's tenure. One of the reasons pointed out is SLTI's lack of articulation power with the other ministries. It is necessary to institutionalise a forum with administrative and political power to manage the e-gov program.

skepticism related to the capacity of coordination of large IT project by SLTI was minimized by the success of the Y2K project:

The success of this project has shown that the SLTI's IT team is able to coordinate large IT projects and nationwide actions. It was a very good example of the capacity of organization, planning and implementation. This was fundamental to the design of a nationwide e-gov project. (Interviewee nr. 1).

Moreover, this influencing factor has also revealed within the political context, the importance of IT to the government modernization process, moving the IT concept from a operational to a strategic perspective. The resources allocated to the "millennium bug" project were used to support structured actions of the e-gov project by the CEGE.

The political strength and sponsorship of the Minister of Budget and Management – Mr. Pedro Parente, revealed also as very relevant influencing factors. He was the sponsor for the "millennium bug" project and was also responsible for the initial e-gov action plan. Despite the fact that he was assigned as the Minister of Civil House in 1999, he maintained SLTI as the coordination agency for the e-gov project and has formally assigned the CEGE as the coordinator of the project in 2000:

Pedro Parente and Marcos Tavares decided to created a decree in October 2000 to establish a formal coordination for the Brazilian e-Gov Program since the basis for this coordination already existed. The basis was the team set up for the "millennium bug". This was how the CEGE was created. (Interviewee nr. 1).

The strategy of creation of the CEGE expressed the political will of Pedro Parente and it was one of the most decisive factors to overcome the obstacles of priority and resources allocation. This strategy reduced the barriers and resistance against the program.

SLTI itself hadn't the required political force necessary to intervene at the Federal Government level mobilizing all the Ministers as sponsors of the e-gov program. It was why we created the CEGE – the electronic government program steering committee – managed by the minister of the Civil House. This committee was responsible to monitor all the e-gov initiatives, the resources allocation and the inventory of the public services offered by agencies through the e-gov program. (Interviewee nr. 2).

The Brazilian society started to support the e-gov program as the first results started to appear. (Interviewee nr. 1).

Another influencing factor was the electricity crisis that has started in May 2001 brought about a priority change set by the Civil House. Mr. Pedro Parente who was very close to the coordination of CEGE took the crisis management bureau resulting in a loss of focus and the speed of actions of CEGE.

It is important to mention that although the electricity outage crisis has not caused a disengagement of the e-gov program it has slowed down the actions of CEGE. (Interviewee nr. 1).

The government tenure transition from Cardoso to Lula has caused a great discontinuity of the e-gov program during 2003.

During the 10 first months of Lula's tenure we hadn't a sponsor for the e-gov program. All the actions planned to develop new electronic services were stopped and a lack of coordination and leadership was very visible, although key-employees were kept after the transition. (Interviewee nr. 3).

During the end of Cardoso's tenure and at the beginning of Lula's tenure a lack of legitimacy of SLTI was observed. The government did not recognize the SLTI as an agency able to centralize and coordinate the e-gov program actions. This has resulted in a total lack of coordination of the governmental actions. Institutional actions were to be taken to motivate and promote engagement of all agencies.

A sensitization movement was designed and put in place in order to promote engagement from the agencies. We discussed the priorities, the benefits of the program, the redesign of processes and services. (Interviewee nr. 3).

Table 3 summarizes eight key influencing factors identified during the interview process with key social actors.

A consolidation of the main historical breakpoints observed in the analysis of the documents and of the interviews is shown in Figure 3.

CONCLUSION

The structured framework of historical analyses of e-gov programs proposed in this article makes it possible to identify the main social actors and influencing factors that affected the initiatives of the electronic government program in Brazil.

The crucial concepts of the theoretical framework that support the proposed framework were validated by means of a structured analysis of a large volume of existing information and of interviews with key actors. Those concepts are: vision in the e-gov construction process phases, the nature and force of the social actors' relations, and influencing factors. Although the framework's validation was performed with relevant social ac-

Figure 3. Historical perspective of e-gov in Brazil from the application of the structured framework used of historical analysis of e-gov (source: elaborated by the authors)

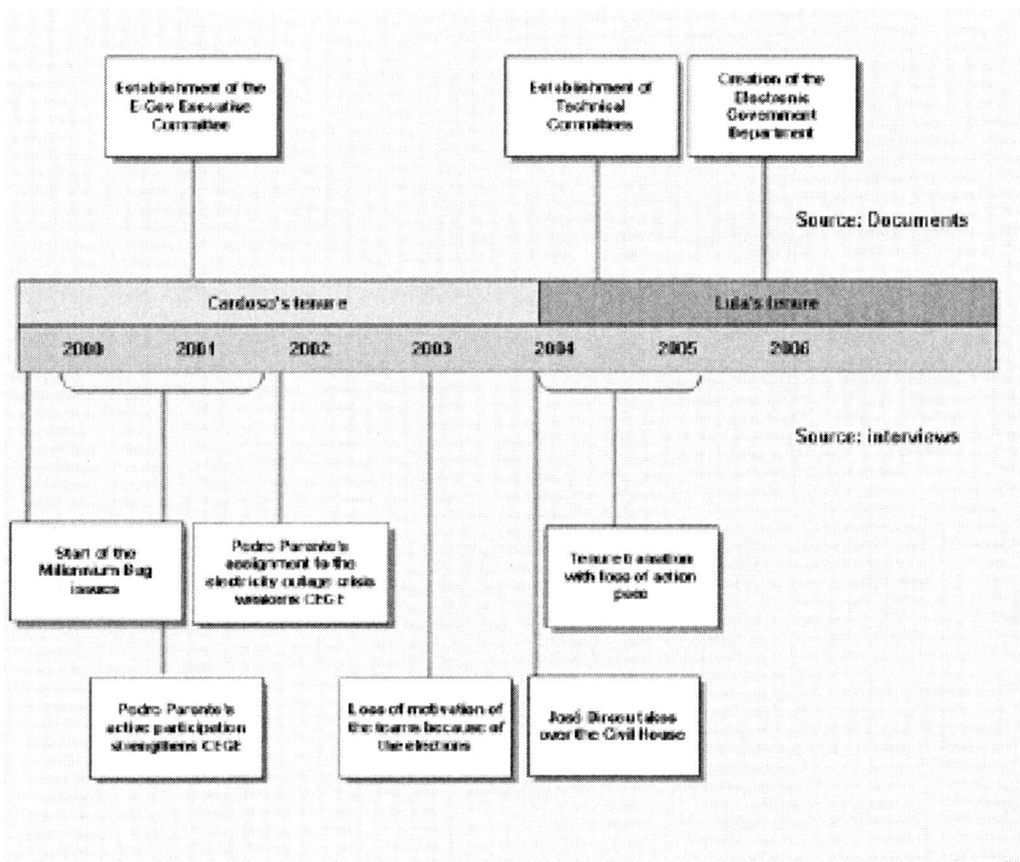

tors in the process during two Brazilian Federal Government administrations, this paper could account for a greater number of social actors as a hypothesis for its greater precision, including the group of implementers of ICT solutions, public managers who were not directly involved with public policy process, and representatives of the civil society.

The application of the proposed framework identified, for instance, that the Brazilian e-gov program was officially begun and became technically viable because of the influence of a previous initiative which previously had not been clearly mentioned in the literature, involving the Millennium Bug. The overall mobilisation to deal with the bug ended up promoting the necessary structures for the SLTI to take up the role of e-gov program coordinator, and for another specific institutional arrangement to be created so as to institutionalise strategic initiatives concerning use of ICT in the federal public administration. Political and institutional aspects were also clearly identified as influencing factors: the force and leadership of Minister Pedro Parente, the Brazilian electrical crisis, the electoral process and governing transition. Institutional aspects linked to the Brazilian e-gov program's coordination and operation structure also emerged as relevant influencing factors.

Thus, this paper's contribution is in the presentation of a structured framework of analysis for studying the history of the e-gov program.

This framework does not intend to describe the reality in all its complexity, but it can show relevant aspects and dimensions which should be more widely analysed and understood given the intention of evaluating the historical evolution of an e-gov program.

REFERENCES

Abranson, M., & Means, G. E. (2001). *e-Government 2001 – IBM Endowment for The Business of Government*. Oxford, UK: Rowman & Littlefield Publishers, Inc.

Agune, R., & Carlos, J. (2005). Governo eletrônico e novos processos de trabalho. In E. Levy & P. Drago (Eds.), *Gestão Pública no Brasil Contemporâneo*. São Paulo, Brazil: Fundap.

Babbie, E. (1999). *Métodos de pesquisa de Survey*. Belo Horizonte, Brazil: Editora UFMG.

Behn, R. D. (1998). O novo paradigma da gestão pública e a busca da accountability democrática. Revista do Serviço Público, Brasília, DF, 49*(4).*

Brasil. (2001). *A política de governo eletrônico no Brasil* [online]. Comitê Executivo do Governo Eletrônico, Brasília. Retrieved from http://www.governoeletronico.gov.br/governoeletronico/index.html

Brasil. (2002a). *2 anos de governo eletrônico: balanço de realizações e desafios futuro* [online]. Comitê Executivo do Governo Eletrônico, Brasília. Retrieved from http://www.governoeletronico.gov.br/arquivos/Balanco_2_Anos_Egov.pdf

Brasil. (2002b). *Levantamento das ações de governo eletrônico* [online]. Comitê Executivo do Governo Eletrônico, Brasília. Retrieved from http://www.governoeletronico.gov.br/arquivos/Anexo_Balanco_Inventario.pdf

Checci, R. M., Hsieh, J., & Straub, D. (2003). Public IT Policies in Less Development Countries: a critical assessment of the literature and a reference framework. *Journal of Global Information Technology Management*, *6*, 45.

Diniz, V. (2005). *A história do uso da tecnologia da informação na gestão pública brasileira através do CONIP*. Paper presented at X Congreso Internacional del CLAD sobre la Reforma del Estado y de la Administración Pública, Santiago, Chile.

Fountain, J. (2001). *Building the Virtual State – Information Technology and Institutional Change*. Washington, DC: Brooking Institution Press.

Frey, K. (2000). Políticas públicas: um debate conceitual e reflexões referentes à prática da análise de políticas públicas no Brasil. [IPEA]. *Planejamento e Políticas Públicas*, *21*, 211–259.

Gil-Garcia, J., & Pardo, T. (2005). e-Government Sucess Factors: mapping practical tools to theoretical foundations. *Government Information Quarterly*, *22*, 187–216. doi:10.1016/j.giq.2005.02.001

Heeks, R. (2003). *Reinventing Government in the Information Age – International practice in IT-enabled public sector reform*. London: Routledge.

Holmes, D. (2002). *e.gov – eBusiness Strategies for Government*. London: Nicholas Brealey Publishing.

Klitgaard, R., & Light, P. C. (2005). *High-performance Government – Structure, Leadership, Incentives*. New York: RAND Corporation.

O'looney. J. A. (2002). *Wiring Governments – Challenges and Possibilities for Public Managers*. New York: Quorum Books.

Osborne, D. (1997). *Banishing bureaucracy: the five strategies for reinventing government*. New York: Plume.

Pavlichev, A., & Garson, G. D. (2004). *Digital Government: Principles and Best Practices*. Hershey, PA: Idea Group Publishing.

Peters, B. G. (1996). *The Policy Capacity of Government* (Research Paper, No. 18). Ottawa, Canada: Canadian Centre for Management Development.

Pinto, S., & Fernandes, C. (2005). *Institucionalização do Governo Eletrônico no Brasil*. Paper presented at X Congreso Internacional del CLAD sobre la Reforma del Estado y de la Administración Pública. Santiago, Chile.

Reinhard, N., & Dias, I. (2005). *Categorization of e-Gov initiatives: a comparison of three perspectives*. Paper presented at X Congreso Internacional del CLAD sobre la Reforma del Estado y de la Administración Pública, Santiago, Chile.

Ronaghan, S. (2002). *Benchmarking e-Government: a global perspective. Assessing the progress of the UN member states*. United Nations Division for Public Economics and Public Administration & American Society for Public Administration.

Chapter 17
E-Government and Its Evaluation in South East Europe: The Case of Albania

Endrit Kromidha
Military University "Skenderbej" Tirana, Albania

José-Rodrigo Córdoba-Pachón
University of London, UK

ABSTRACT

E-government (electronic government) has spread globally as a phenomenon that involves the use of information technologies supports delivery of public services. There are different actors and organisations being involved in e-government implementation and use, three of which are government organisations, businesses and citizens. In the case of government institutions, a common goal is to provide an appropriate service environment for both businesses and citizens. Citizens want to maximize their benefits received when for instance they pay taxes, and businesses have a goal to maximize profits. How to assess and improve e-government services is still a challenge in many countries. This chapter will offer some insights into the e-government phenomenon which could inform its future evaluation. The objective is to review e-government in the light of new public management (Barzelay, 2001), a term used in the last 30 years to describe new themes, styles, and patterns of public service management. This will lead us to consider alternative ways to evaluate e-government services (e-services). One of these is a technology acceptance model (TAM) to describe e-government related reforms in Albania and the Balkans region towards building a better understanding of how evaluation could be conducted. The insights from the Albanian case open up a number of opportunities that could be taken forward by policy makers, developers and other stakeholders. For those researching on e-government evaluation, the chapter offers a view of why we need to consider different aspects in the evaluation process.

DOI: 10.4018/978-1-60566-860-4.ch017

INTRODUCTION

The South-East Europe, or better the Balkans, is often mentioned for its communist past, for its problems but also for its present huge reforms. Surprisingly the region has the lowest negative economic growth predicted in Europe for 2009, the year of the world crisis, due to reforms and economic development. In Albania, according to the International Monetary Fund (IMF), the situation looks even more promising, with positive instead of negative growth being predicted for 2009, which has turned it into one of the most interesting European countries in terms of investments.

With current integration developments of the Balkan to the Euro-Atlantic structures like North Atlantic Treaty Organization (NATO) and European Union (EU), e-government services to improve relationships between government institutions, businesses and citizens are crucial. These services can help improve the quality of the country's electronic networks of information, communication and transaction systems. Despite a perceived degree of progress, some key questions remain: To what extent are e-government services being accepted specifically in Albania? What are the e-government systems' drawbacks? How to adapt and encourage the use of these systems by different actors?

In this chapter we will review the Albanian case to give some answers to these questions. Throughout the discussion, we will use ideas of new public management (NPM) to contextualise the definition of existing policies and strategies of the Albanian government and private sector to improve e-government systems. We will suggest ways to evaluate systems by considering a variety of actors and e-government configurations that have emerged throughout. Our chapter begins by situating e-government as a worldwide and emergent phenomenon.

A CONTEXT FOR THE EMERGENCE OF E-GOVERNMENT

In a so-called era of technological progress, a new kind of rationalization has been introduced in the public sector worldwide by the use of modern information and communication technologies (ICTs) (Dunleavy, Margetts, Bastow, & Tinkler, 2006). Increasingly the use of ICT tools and applications is leading to transformational shifts in public policy, processes and functions of e-government. Such transformations are not only aiming to provide citizen-centred services but to increase the efficiency of public sector organisations, as well as improving cost efficient transparency and accountability in government functions. These transformations are being labelled under the banner of e-government.

Transformations are also shifting traditional roles. Governments are becoming partners rather than dictators of rules and regulations. Citizen groups have come to expect a 24/7 convenient service user interface provided by the government with ease of use, in a language the user understands and which is tailored to individual needs. This is the case in countries like Albania or any other so called 'developing' one. In order to be able to implement this 24/7 type of vision, government institutions are acknowledging that they need to integrate their government systems (UN survey on E-government for 2008), as well as accompany this integration with more fundamental (legal) reforms, training and infrastructure. As a result, we now have in governments complex arrays of organizational projects across government institutions, some of which aim to connect diverse technological platforms across them as well as automate delivery with the help of internet or intranets. Profound changes are envisaged within government institutions and citizens are becoming more aware of services they can use online.

Within the variety of projects labeled under the banner of e-government worldwide, something can be distinguished as common: the use of ICTs to improve public service delivery at the administrative (internal) level as well as at the relational (citizen) level (Bannister, 2007; Carter & Bélanger, 2005; Grimsley & Meehan, 2007). Joining this common aspect is a widely accepted perspective on e-government development (see also introductory chapter in this book) that proposes an evolution of e-government systems from purely administrative ones (e.g. electronic billboards of web pages) to fully interactive transactional and relational, in which citizens can exert accountability over the performance of services (West, 2004). To many, the technological component is a fundamental part of what electronic government (e-government) systems development is and should be about. To others, e-government is not about technology as an end, but about strengthening democracy and decision making aspects (Heichlinger, 2004). This latter aspect means that citizens could have some degree of power to shape the content and nature of such services (e-services). With increasing demands and expectations on government institutions and electronic government systems, it is worth asking if we can provide more comprehensive (e.g. holistic) ways to understand the dynamics of e-government implementation and use in diverse cultural and socio-economical contexts. The dynamics are multifarious (Dunleavy et al., 2006), and lead us to consider the importance of e-government services (e-services) evaluation to link both systems implementation with users (citizens, businesses, government organizations), their perceptions about government, and their degree of acceptance to proposed transformations in service delivery.

New Public Management Theory and E-Government

As a form of 'technology' (involving also procedures, norms and ways of communicating), e-government inherits many ideas already being proposed by advocates of **New Public Management (NPM)** (Barzelay, 2001). According to Thomas Gordon (2002), new public management (NPM) is a management theory about how to reform government by undertaking a number of changes that include:

- Replacing rigid hierarchical organizational structures with more dynamic networks of small organizational units
- Replacing authoritarian, top down decision and policy making practices with a more consensual, bottom-up approach which facilitates the participation of as many stakeholders as possible, especially ordinary citizens
- Adopting a more 'customer'-oriented attitude to public services
- Applying market principles to enhance efficiency and productivity

There are many similarities between the above features and e-government projects in relation to what they want to achieve. This can also be reflected for instance in the strategy of the European Union, which is explicit about developing one-stop portals of electronic services to citizens and other users, under a citizen-centric perspective of e-government (European Commission, 2007). Within this strategy it is also suggested to develop portals with a view of the cycle of events of citizens (for instance birth, childhood, adulthood) so that citizens feel that the government is catering for their needs at these different events with appropriate e-services. In our view, both e-government and NPM aim to increase transparency and efficiency but this leaves us with a number of questions unanswered. What is the exact meaning of both transparency and efficiency? Does e-government help the individual goals of government, private sector and citizens at the same time? How should e-government be shaped to serve all players equally well?

To address these questions, it is important to review two key assumptions privileged by both NPM and e-government. The first one is the assumption of Individualism. Ostrom & Ostrom (1971, p. 205) note that "Work in public choice begins with methodological individualism where the perspective of representative individual is used for analytical purposes. Since the individual is the basic unit of analysis, the assumptions made about individual behaviour become critical in building a coherent theory." New Public Managements has been used to inform a broad variety of e-government projects based on individualism. The trouble with this assumption (we find), is that even a madman's actions are in general also rational and based on individualism. A similar claim can be levelled in relation to e-government and its conception of the citizen as a 'customer'. To address this shortcoming, some e-government evaluation models based on the notion of users' acceptance have included the 'public good' as an intangible that individuals care for when it comes to assess the usefulness of e-government services (Grimsley & Meehan, 2007). However this public aspect is difficult to measure or to encourage if citizens are still considered consumers of e-services.

Following from this issue of *measuring*, the second assumption to be reviewed is that any technological investment (e.g. on e-government) can be primarily evaluated by measuring costs and benefits. New public management rhetoric seems to claim that a new model of service delivery can be evaluated based on costs and benefits with a view to reduce of administrative burdens and service costs as well as make service faster. Similarly, e-government evaluation could potentially adopt a dominant view for technology investment evaluation which would privilege financial aspects (Irani, Love, Elliman, Jones, & Themistocleus, 2005; Serafeimidis & Smithson, 2003; Smithson & Tsiavos, 2004). With aspects like 'the public interest' as an intangible one, in e-government it becomes difficult to measure aspects of *relations* between citizens and governments, and how they still leave room for the exercise of action by both sides different from the one which is prescribed and which is needed for the appropriate regulation of societies (Foucault, 1978). Assessing or evaluating e-government initiatives, in our view, should also be about exploring how individuals can do 'otherwise' than expected in relation to their use of information systems and technologies (Cordoba, 2009), and this requires understanding their perceptions about what they want to do with and about government, as well as how and why they are led to evaluate e-services in particular ways.

E-Government Traditional Evaluation: Costs

Users of e-government services (not normally included in traditional cost/ benefit analyses) face two major problems with costs: 1) Reducing costs as part of their overall day-to-day activities, and 2) Tracking costs.

1. There is a risk that the cost of accessing services on-line is expensive or time-consuming because the service has been badly designed people will not want to use it. Government institutions assume that e-government users already have adequate computing facilities (web browsers and printers among others) to perform parts of the e-service delivery process. Once e-government services are being set up (with a view to reduce administrative costs by government institutions), businesses and citizens have fewer choices to influence e-government systems; acceptance then relies on how users feedback is included in the evaluation of e-government systems (Irani et al., 2005).
2. With the integration of e-government systems it has become possible to cost transactions across government institutions, and also assess if these costs compare with costs of outsourcing service delivery to ICT service providers (Dunleavy et al., 2006). However this assumes that there is co-ordination

Table 1. E-government cost methods, description and use

Method	Description	Use
Transaction costs	Uses segmentation methods to calculate use and benefits to different user groups	Quick and easy way to estimate potential cost savings from the introduction of e-government
Net present value	A straightforward method that examines monetary values and measures tangible benefits	Relatively straightforward; use when cash flows are private and benefits are tangible
Cost-benefit analysis	A flexible method that measures tangible and intangible benefits and assesses these against net total cost	Good consideration of all benefits, but can be expensive and time consuming
Cost effectiveness analysis	Focuses on achieving specific goals in relation to marginal costs	Good for considering incremental benefits against specific goals
Portfolio analysis	A complex method that quantifies aggregate risks relative to expected returns for a portfolio of initiatives	Good for consideration of risk, must use a consistent approach across a portfolio
Value assessment	A complex method that captures and measures benefits unaccounted for in traditional ROI calculations	Used by several governments to consider performance against all policy goals

between different actors involved, that cost information can be transparently obtained and shared. This assumption generates further challenges when it comes to develop more sophisticated (and integrated) e-government service portals (Layne & Lee, 2001).

These problems suggest that it is necessary to enhance the scope of e-government evaluation in relation to costs. A number of e-government evaluation techniques based on cost are given in the table below by Edwin Lau (2006). For now it suffices to say that they should be used with caution and compared in different situations, something that we will detail towards the end of the chapter.

From the table above, it can be said that an appropriate method to assign costs of e-services would be Transactional costs, because in e-government service provision, either for information, communication or transaction, there is an action performed by a user of the e-government system, therefore, cost can be based on that. This can apply to the setting up of an e-service, although in practice a government can outsource the system start-up cost to an external provider and not deal with it directly. As mentioned before, costs assignation needs to go beyond government institutions, but also consider other several aspects related to changes in procedures, and the organisation of ICT management units or organisations to administer the information management function (Dunleavy et al., 2006). For instance, significant technological and managerial problems may rise that would delay e-government implementation, increase the costs of projects, or result in their failure. Again, measuring this could present challenges.

In developing countries like those of South-East Europe and more specifically Albania, advances in networking, interfacing and online management systems have allowed governments to develop websites and platforms that contain a variety of online materials and services. In this context, based on the previous table, e-government system in Albania are replacing traditional means of access and interaction based on personal visits, phone calls, long waiting lists, hard copy materials and mail delivery with new modern ways, reducing costs in terms of time and money and increasing benefits to all players. The overall cost of e-government projects though does not only lie on their implementation, but also in the cost of bad or uncompleted projects, which undermine the system in various ways. Some of the key factors we have found that can influence e-government implementation and acceptance, and which therefore should be considered in terms of the cost

Table 2. E-government stakeholders and benefits

	Business	Citizens	Government
Cost-benefit analysis	A flexible method that measures tangible and intangible benefits and assesses these against net total cost. It is comprehensive, but can be time consuming. More specific techniques of benefits are mentioned below		
Direct Financial Benefits	**Reducing Burden:** administrative simplification	**Reducing Burden:** administrative simplification	**Reducing Efficiency Savings:** freeing resources for public and private innovation
Direct Non-Financial Benefits	**Meeting Public Expectations:** improving customer satisfaction and equity, meeting security and privacy concerns; transparency and choice		
Programe Benefits (Direct & Indirect)	**Improving Policy Effectiveness:** Achieving overall policy and programme outcomes		
Good Governance Benefits (Indirect for society)	**Supporting Growth and Legitimacy:** Good governance contributes to a sound business environment and democratic legitimacy, promotion of the information economy, supporting public sector reform; creating business opportunities		

that they could generate are (Jae-Moon, 2002; Thompson, Rust, & Rhoda, 2005): The level of regulation, costs, infrastructure, the size of the public and private institutions being involved, the rule of law, paperwork and bureaucracy, and level of corruption and black markets. E-government evaluation efforts should consider costs associated with these and other aspects.

E-Government Benefits' Analysis

We now consider a benefits-based view of e-government evaluation. The main stakeholders regarding benefits from e-government can be businesses, citizens and the government itself as explained in the table below by Edwin Lau (July 2005). We cite some benefits that could be obtained by implementing e-government systems. We leave aside costs efficiencies as mentioned in the previous section and move to describe some high level benefits that could impact stakeholders as a result of the availability of new electronic services.

As said before, aspects which could be seen as beneficial (e.g. public good) are in its majority 'intangible' and will depend on the perceptions and expectations of achievement by a number of users (business, citizens, government). This suggests that similarly to cost-based assignation and e-government evaluation, we need to *broaden* our consideration of different aspects that come into the **evaluation** of e-government systems, bearing in mind perceptions about acceptance (and not only from the perspective of individuals or user groups but the 'public' in general). Ideally, an evaluation of the system should consider the different aspects that are to be included in evaluation, as well as relationships between them. For instance, developing a new infrastructure to enable a one-stop shop e-government portal requires also preparing the human resource to manage it, and addressing the compatibility and legal aspects involved in facilitating delivery of services through such a portal

A possible alternative to e-government evaluation that considers different (tangible, intangible) aspects is the use of a technology acceptance model (TAM). In e-government projects, there could be many tensions between people, regulations and technological arrangements that need to be addressed. In this context, the Technology Acceptance Model (TAM) developed by Davis et al (1989) can be used at great efficiency to analyze the situation through perceived usefulness (PU), the degree to which a user believes that using a particular technology system would benefit him/her, and perceived ease-of-use (PEOU) the degree to which a person believes that using a particular technology system would be easy and free from any negative effect. However the Technology Acceptance Model is based on only two subjective dimensions. Perceptions can be influenced by ways in which people see how they interact with each other, in other words by the social context, something that interpretive information systems evaluation has proposed to look at as a more humane measure of information systems success (Walsham, 1993, 1999). Studying how and why people attribute usefulness to systems (e-government) could give us a broader view of different aspects that need to be addressed to make implementation of e-government initiatives more accepted by different users. This could complement and inform approaches to evaluation (e.g. cost/benefit based) and gives us insights as to how we can improve future e-government projects.

Furthermore and by considering the lack of experience with e-government systems from citizens and businesses in developing countries like Albania, we think that the government itself has the difficult task to decide on its vision of what constitutes Perceived Usefulness and Perceived Ease-of-Use of e-services. Learning from other best practices is difficult, considering that almost all developing countries in the world are at their initial steps regarding e-government reforms. On the other hand, according to Malhotra et al (2007), what works best in one country, is not supposed to work the same way in another, due to cultural and social differences. Being given such unilateral decision making power but still lacking the necessary knowledge and experience, the governments in developing countries like Albania tend to limit their risk of mistakes by limiting their own expenditures in e-government projects and allow more experienced organizations like USAID, UN or EU to finance them.

We now review the development of e-government in Albania and begin by exploring the general use of information and communications technologies.

INFORMATION AND COMMUNICATIONS TECHNOLOGIES (ICTS) IN ALBANIA

The ICT sector in Albania as in almost all developing countries has had considerable growth in this last decade. For instance, and according to the Internet World Stats web page, internet usage penetration has increased tremendously. In 2000 the percentage of population using the Internet in Albania was only 0.1 while in 2007 about 15.3% of the total population of about 3.5 millions consists of Internet users.

Almost all ICT related technology in use in Albania is imported from abroad, but with a strong tendency to adapt ICT products and services to local needs. There is a variety of choices with regard to equipment and programs developed by Albanian and foreign companies for bookkeeping, business management for bars and restaurants, hotels and warehouses, accessible and affordable for the majority of small and medium enterprises as well as for many individuals. Figure 1 is based on the Chamber of Commerce of Tirana published statistics and gives a better view of the revenue of ICT Sectors by Segments:

According to Electronic Assessment – Report on the status of information technology and communications in Albania 2002-2005, the average annual investment in ICT is about 2.5-3 million USD (or 2.2 million Euro) and all ICT revenues total about 20 million USD (or 15 million Euro). In spite of achievements that the Albanian ICT Sector has made in recent years due to private investment and companies, technological needs and ICT infrastructure remain at a considerable level. Many remote locations in the country often experience even periodic electric power interruptions for more hours a day, not to speak about internet access. Countrywide Internet coverage still remains a long term achievement which can take place only after a fast and reliable network infrastructure and a dedicated ICT policy by the government.

Figure 1. ICT sector in Albania

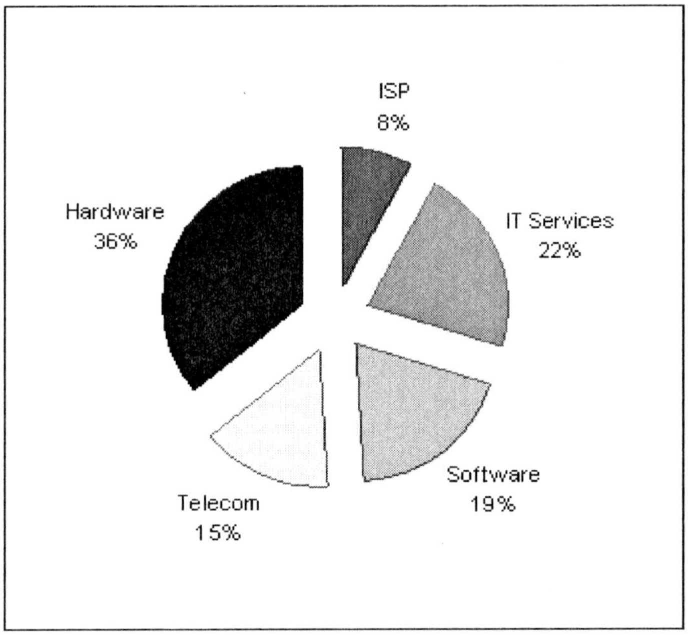

Current Policies and Projects Related to E-Government

There are many projects already underway in Albania. The following is a description of the most relevant. We leave out of our descriptions projects related to the implementation of an ID cards system, a national data bank and statistics, a health system, a land and property electronic registry, an e-accounting system, an e-schools system, a human resource database and many others. The main projects / programmes / policies / strategies are as follows.

National Information and Communication Technology (ICT) Strategy

In 2003, the Albanian Government, with UNDP assistance, drafted and adopted an ICT Strategy with the main objective to identify mechanisms and legal and fiscal frameworks needed to implement e-government (legal, economic and administrative initiatives), e-education, e-health, social services, and e-commerce, and to stimulate the development of ICT infrastructure. The Strategy, revised at the end of 2006, was developed within the context of the overall National Strategy for Social and Economic Development, the European Stabilization and Association Agreement, and sub-regional activities under the EU Stability Pact (UNDP Albania, 2008, Technical Assistance and Capacity Building Support to the National Agency for Information Society Project)

National Agency for Information Society (NAIS): June 2007

This was decreed from the Council of Ministers as an institution with dedicated staff and an allocated budget from the Office of the Prime Minister. (UNDP Albania, 2008, Technical Assistance and Capacity Building Support to the National Agency for Information Society Project). Its mission is to promote the development of an Information Society in Albania by guiding the implementation of the National ICT Strategy and coordinating the development and administration of the state information systems, so that the Government can best serve citizens by implementing modern ICT capabilities, in order to reflect on the existing European best practices.

E-Procurement

Albania's public procurement system used to be one of the most problematic ones, showing the lowest level of transparency and the highest level of corruption almost in all state structures. Now, from March 2008 e-procurement (www.app.gov.al) has become the general government practice. On the other hand, this new system is also easier and faster. According to USAID, this MCC program has generated 70 competitive awards worth more than $29 million, 6% of total procurement value and 2% of the total number of awards in the last year.

The bidding process is planned to be automated and through the internet over secure web page connection. On the other hand the Procurement Advocate is an independent procurement oversight agency supported responsible in overseeing the procurement process and handling complaints. However, the system has been subject to some criticism on many newspaper articles lately. The problem is not with the system itself, the secrecy of offers and data stored therein, but more with requests for bids that are sometimes claimed to be designed in a favorable way for one bidder only: the "lucky" one. On the other hand, the system might be discriminatory for those bidders who do not necessarily have internet access, but that would have given a better offer.

E-Business: National Registration Centre (QKR)

The E-Business registration system is operational from 2008 in Albania. According to USAID, two years ago, the process of registering a new business in Albania was costly and time consuming, requiring almost 42 days and a court decision. Although some formalities from the old system still remain, that system has been replaced with a simplified one-day procedure in the National Registration Center in Tirana or any of its 10 branch offices located throughout Albania, serving as a one-stop-shop. Businesses can check their status online at any time at www.qkr.gov.al and at the same place third parties can search for information on any registered business in Albania. The National Registration Center serves as an online business registry and is connected to all state structures on business activities. This new system has resulted very effective in reducing paperwork, bureaucracy and bribery, increasing transparency and efficiency at the same time. Also according to USAID, after one year of operation, this system has registered over 18,000 new businesses, all of which are simultaneously tax-registered, reducing informality while expanding the tax base.

Regardless of it many benefits, the system is a one-stop-shop for business registration, but not for business functioning. During business as usual activities, this is one more office, except the Local Authorities, Tax Office, Social and Health Insurance Institute, Work Inspectorate etc, added to the long list of offices a business has to go. Therefore, adding one more institution, doesn't really serve always to the "doing business climate", as long as "starting up a business" is almost nothing compared to "doing business" as usual.

E-Taxes

The E-taxes declaration and procedures for all businesses including Small Businesses is operational since July 2008 in Albania. This is another MCC Program that supports e-government New Public Management reform noted by the World Bank in the Doing Business Report, that includes e-filing of corporate taxes, first for large businesses, and now, even for medium and small enterprises (www.tatime.gov.al). The biggest benefit of this system is the noted reduction of personal contact needed between citizens on the business side and tax inspectors on the other, reducing the exploitative relationship between officials and taxpayers. According to USAID, on average, more than 61,000 tax forms are downloaded per month as of October 2008, and over 2,100 monthly declarations are e-filed. These numbers keep rising. This modern tax administration reform is oriented to greater efficiency and better service. This system is also related to the former one, the National Registration Center.

However, the system is not perfect yet. Businesses can only file online their taxes, but complains and other requests still have to be handled manually. At the same time the system does not include any e-tax payment but this has to be done most of the times in person in any commercial bank, which on the other hand, will charge you a fee for this service.

E-Accounting

The objective of this program financed by UNDP Albania (2008) is to review the Chart of Accounts (CoA) in compliance with International Public Sector Accounting Standards (IPSASs)-IMF&GFS, to draft a new accounting law and to design a pilot software project. Some of the main outcomes of this project are expected to be the following:

- Implementation of a web based IT application is planned to be done for Inventory management for the Albanian Ministry of Finance - Accounting Department
- Support of the Accounting Department in Conceptualizing the accounting principles and drafting of the National Public Accounting Law, waiting for approval by the Ministry of Finance
- Support the Accounting Department in the public sector to adopt IPSASs

In this context, e-government systems in Albania are being applied with great efficiency to facilitate information, communication and transactions. However there is still a lot more to be done to bring these systems to desired levels. It is clear that the government is working more closely with the private sector player and citizens to meet their needs, in the light of New Public Management and reforms. For the first time these other players do not perceive the government as the tax collector only, but as a closer partner trying to serve also to their business or private needs at least by providing a more convenient information and communication infrastructure. E-government in this case is aiming to bring the government, private sector and citizens even closer to each other.

CHALLENGES TO E-GOVERNMENT

As a result of the above review of new public management inspired projects / programmes / strategies/ policies, we now present a number of challenges for e-government in Albania and its evaluation. We divide them into internal, external and system-related ones.

Internal Challenges

- **Budget:** How much will a new e-government system cost and who will pay for it? Part of the e-government projects in Albania are being paid by the government and most of them by foreign donors like USAID (Millennium Challenges Program) for QKR, Tax Directorate, UNDP for the National Agency for Information Society etc. This might be an easy solution in the short run, but in the long run, the PU is not at the maximum level if they are perceived as financially dependent from abroad and subject to change.
- **Legal barriers:** Updating the legal system and laws according to international law and

Aquis Communitaire is a challenge for all South East Europe developing Countries, including Albania. In this context, the PEOU of e-government systems operating in a changing legal environment is not at the highest level.

- **Infrastructure:** What is the cost and knowledge needed build the infrastructure for new technologies that goes also with the price of replacing existing ones? There is still a lot of progress to be made in this field. GovNet a project of UNDP in relation to NAIS aims to connect via optical fibres all ministries in an intranet that will serve as a platform for many other e-government applications in Albania, however, the systems are difficult to use and to access.
- **Compatibility:** With what computer and information technology platforms should e-government systems be compatible with? Microsoft programs and platforms are the most widely used in Albania and compatibility with e-government systems should not be a problem in general and usually this is not a challenge.
- **Human resources management:** How to deal with the many public servants who would have to adapt to the new e-government systems or find alternative jobs? For the moment there are training for public servants related to the online tax system, public procurement, and registration centre, however, there is still no evidence on the real cost and problems with the reallocation of the human capital from the implementation of these new systems. These undergoing reforms lower the PEOU and efficiency of the systems.

An ideal e-government evaluation process should enable the definition of goals in these areas, and continuous monitoring of achievements in each of them. It should also allow for the inclusion of people's perspectives (experts and non experts alike). Finally, it should also enable people to continuously re-design existing plans with a view that costs and benefits can also relate to the external environment that is to be influenced (Ackoff, 1981), and which will include the public in general. Challenges in the external environment (partly[1] composed by government organisations, citizens and businesses) are considered now as follows.

External Challenges

The external challenges depend a lot on e-government readiness. Albania ranks 86th in 2008, compared to 102nd in 2005 worldwide (UN E-government survey 2008 p. 59). This means a lot of improvement but still a lot to be done.

- **Accessibility:** How to make e-government systems not only user friendly, but also accessible by all people? Albania has in 2008 about 15 internet users, 1.73 PCs, about 49 mobile phone subscribers, 11.3 main phone lines, and only 0.01 broadband connections per 100 users. (UN E-government survey 2008 p. 215). This is still very little to allow a broad use of e-government systems, showing a low level of accessibility and PEOU, especially in rural areas.
- **E-literacy:** What is the cost of creating an e-literate public and how should this be done, at central education levels, with campaigns or on a need basis? Although Albania has a 98.7 adult literacy in 2008 (UN E-government survey 2008 p. 222), this is not so much concentrated on information technology. E-government systems have a low PU at the beginning, which tend to change as long as users understand how to use and benefit from them.
- **Trustworthiness:** What is the biggest challenge, to provide secure systems or to convince the public that they are such, dealing

meanwhile with the possible mistakes from human factor also? There is still little trust on e-systems in Albania and therefore a low PU, mainly due to low level of e-literacy, internet use and accessibility.

- **Transparency**: Should e-government systems be only managed and controlled by the state ore should other stakeholders have access to ensure their transparency? E-government systems are increasingly being used to reduce corruption and increase transparency. There is a high Perceived Usefulness in this direction since these are two of the biggest problems in developing countries like Albania. However, the level of accessibility is still crucial in this context for them to be effective.
- **Benchmarking vs. users' opinion:** Should the regulation of e-government systems be made based on best practices from other governments or systems, or come from the needs of the users? Considering that there is little possibility to get any feedback in this initial phase of e-government services in Albania, most of the implemented systems are based on other countries best practices[2] or special needs.

System Challenges

- **Information availability and preservation:** What is the optimal level of information preservation and availability of e-government systems to make them better than existing systems but still not "everything controlling"? This is a difficult question even for developed countries, where e-government systems have much broader use. However, the main challenge in this context is the management of change and transition from the old systems to the new modern ones.
- **Competition and cooperation: private vs. public e-systems:** To what extent should the state limit or encourage private competition vs. its e-government system in information, communication and transaction systems? Most governments around the world, including that of Albania started their e-government initiatives with a focus on providing information and services to the citizen while service delivery platforms remained separate and parallel across various government agencies or often private bodies. In this model, service delivery was built around individual agency functions, structures, information, systems and capabilities. Now citizens and businesses around the world are increasingly demanding that their governments follow suit.

MOVING FORWARD E-GOVERNMENT EVALUATION IN ALBANIA

In Albania, it is clear that information and communication technologies in e-government services, despite their associated costs and difficulties, are having a good impact. Based on New Public Management (market like) principles like segmentation, market study, product, price, promotion and place for e-government solutions and looking back at the policies and projects being formulated, it is possible to say that a lot has been done in this field. However, as Douglas Holmes (2002) states in his article: "Governments don't print t-shirts. They don't bang drums, beat chests, shout from rooftops. Marketing isn't their thing. In fact, they're terrible at it. (Name one memorable government marketing campaign.) But it's a skill they need to learn, and quickly, if they ever hope to interact electronically with most of the population." What can e-government evaluation do in Albania to improve the implementation and acceptance of e-government systems?

We have two possible answers to the above question. The first one is to take what we see as useful principles and ideas of New Public Management reforms and make them work better, so that the pace of economic development is also benefited through e-government services. This can help people consider that e-government systems are not anymore forced by the state at all cost, but necessary and useful services. The Albanian government has realized that they first need to be made known, liked and in a free market might always face competition and have substitutes provided by the private sector as shown for example in tourism services or information networks. Therefore the government too is being considered and working as a business player based on New Public Management principles.

The second answer is to review traditional approaches to evaluation based on cost / benefit analyses. As said before, we can widen our understanding of different cost and benefits related aspects in order to embrace more fully the complexity of e-government systems implementation and acceptance; but that also needs to include people's perceptions about what they expect from e-government systems so that we can build more humanly oriented criteria for their acceptance.

In other words, we are suggesting rethinking e-government services as 'systems', composed of parts (technical and non-technical) that interact, and whose emergent properties (outputs and consequences) lead us to work jointly with experts and non-experts in several areas of activity (infrastructure, compatibility, training, education and participation among others). Evaluation approaches can be designed according to the complexities of e-government projects situations and we can employ a number of systems approaches to help us address different types of complexities (Gregory & Jackson, 1992).

These complexities, and the challenges encountered in Albania also makes us think that evaluation can help us to become more critical of existing reforms, and work within spaces in which see that our action is possible so that we can address the ethical issues resulting from e-government systems implementation at a number of levels. The door is open to develop critical and interpretive evaluation of e-government services in Albania, and to facilitate participation in their implementation and further acceptance.

CONCLUSION

This chapter has developed a number of insights in relation to e-government systems conception, implementation, use and evaluation. Based on these, governments show a pattern (business like) of development of e-government systems which has greatly altered the nature and degree of interaction of different players (government organizations, citizens and businesses). With ideas of new public management (NPM) being shared with e-government, the latter seems to adopt a view based on cost and efficiencies under the discourse of citizen-centric government. This makes e-government services (e-services) evaluation problematic if a perspective based on costs/benefits is adopted. We have started exploring alternative ways to evaluate e-services, one of them being a technology acceptance model (TAM) which gives a good prospective based on Perceived Usefulness and Perceived Ease-of-Use for e-government implementations. However, there is room to include perceptions of people which could lead us to understand why and how they accept (or reject) e-services as well as how they decide to do so.

A great number of e-government systems (for instance e-procurement, e-registration of businesses and other services) are at hands of citizens and private companies, and they seem to have provided benefits in terms of overcoming the bureaucratic state administration structures, although often not thanks to the experience, knowledge and long term vision of the governments in the developing countries, but instead that of international organizations giving help and support. We would like to understand how they have built success in e-government. Moreover, we would like to enable developing countries' governments and citizens to create their own capacities, vision and knowledge in terms of e-government, and we see this as an aim to pursue in further research studies.

REFERENCES

Ackoff, R. (1981). *Creating the Corporate Future: Plan or to be Planned for.* New York: John Wiley and Sons.

Albania, U. N. D. P. (2008). *GovNet Project.* Retrieved January 19, 2009, from http://www.govnet.undp.org.al/

Albania, U. N. D. P. (2008). *E-accounting Project.* Retrieved January 19, 2009, from http://eaccounting.undp.org.al/

Albania, U. N. D. P. (2008). *Technical Assistance and Capacity Building Support to the National Agency for Information Society Project.* Retrieved January 19, 2009, from http://www.undp.org.al/index.php?page=projects/project&id=176

Albanian Council of Ministers. (2003, April). *National Information and Communication Technologies Strategy of the Republic of Albania.* Albanian government publication.

Albanian Government. (n.d.). *Programme 2006-2009 by the Ministry of Foreign Affairs.* Retrieved: February 11, 2009, from http://mfa.gov.al/web/pub/programien2100.pdf

Bannister, F. (2007). The curse of the benchmark: an assessment of the validity and value of e-government comparisons. *International Review of Administrative Sciences, 73*(2), 171–188. doi:10.1177/0020852307077959

Barzelay, M. (2001). *The New Public Management - Improving Research and Policy Dialogue.* Berkely, CA: University of California Press.

Carter, L., & Bélanger, F. (2005). The utilization of e-government services: citizen trust, innovation and acceptance factors. *Information Systems Journal, 15*, 5–25. doi:10.1111/j.1365-2575.2005.00183.x

Chamber of Commerce of Tirana. (n.d.). *Albania Statistics.* Retrieved October 16, 2008, from http://www.cci.gov.al

Davis, F. (1989). Perceived usefulness, perceived ease of use, and user acceptance of information technology. *MIS Quarterly, 13*(3), 319–340. doi:10.2307/249008

Dunleavy, P., Margetts, H., Bastow, S., & Tinkler, J. (2006). *Digital Era Governance: IT Corporations, the State, and E-Government.* Oxford, UK: Oxford University Press.

Dunleavy, P., Margetts, H., Bastow, S., & Tinkler, J. (2008, March). Australian e-Government in comparative perspective. *Australian Journal of Political Science, 43*.

European Commission. (2007). *Information Society and Government: Linking European Policies.* Brussels, Belgium: European Commission. eSEEurope. (n.d.). *Agenda for the Development of the Information Society, Stability Pact – A cooperative effort to implement the Information Society in South Eastern Europe.* Retrieved March 21, 2009, from http://www.eseeuropeconference.org/agenda.pdf

Gordon, T. F. (2002, January). E-Government – Introduction. *ERCIM News No.48.* Retrieved November 7, 2008, from http://www.ercim.org/publication/Ercim_News/enw48/intro.html

Gregory, A., & Jackson, M. C. (1992). Evaluation methodologies: A system for use. *The Journal of the Operational Research Society, 43*, 19–28.

Grimsley, M., & Meehan, A. (2007). e-Government information systems: evaluation led design for public value and client trust. *European Journal of Information Systems*, *16*(2), 134–148. doi:10.1057/palgrave.ejis.3000674

Heichlinger, A. (2004). *eGovernment in Europe's Regions: Approaches and Progress in IST Strategy, Organisation and Services, and the Role of Regional Actors*. Maastricht, The Netherlands: European Institute of Public Administration.

Holmes, D. (2002, May 1). Marketing e-Government to a skeptical public. *CIO Governments' Review.* Retrieved December 12, 2008, from http://www.itworldcanada.com/a/CIO-Gov.-Review/d118b39c-29b1-49b3-a98c-656626ce4946.html

Insigh, B. (n.d.). *IMF slashes Albania growth for2009.* Retrieved April 13, 2009, from http://balkaninsight.com/en/main/news/17998/

Irani, Z., Love, P. E., Elliman, T., Jones, S., & Themistocleus, M. (2005). Evaluating E-government: Learning from the experiences of two UK local authorities. *Information Systems Journal*, *15*(1), 61–82. doi:10.1111/j.1365-2575.2005.00186.x

Jae-Moon, M. (2002). The evolution of e-government among municipalities: rhetoric or reality? *Public Administration Review*, *62*(4), 424–433. doi:10.1111/0033-3352.00196

Lau, E. (2005, July). *eGovernment Economics Project - eGovernment Measurement Agenda.* Brussels, Belgium: OECD eGovernment Project.

Lau, E. (2006, February 6). *OECD E-government Project: Proposal for work on Inventory of E-government Business Case Indicators* (Doc. Nr. GOV/PGC/EGOV). Public Governance and Territorial Development Directorate - Public Governance Committee.

Layne, K., & Lee, J. (2001). Developing fully functional E-government: A four stage model. *Government Information Quarterly*, *18*, 122–136. doi:10.1016/S0740-624X(01)00066-1

Malhotra, Y., & Galletta, D. F. (2007). *Extending the Technology Acceptance Model to Account for Social Influence: Theoretical Bases and Empirical Validation.* BRINT Research Institute, University of Pittsburgh.

Ostrom, V., & Ostrom, E. (1971). Public choice: A different approach to the study of public administration. *Public Administration Review*, *31*(2), 203–216. doi:10.2307/974676

Republic of Albania. (2006). *Electronic Assessment – Report on the status of information technology and communications in Albania 2002-2005.* Albanian Government publication. SCORE Project EU. (n.d.). *The ICT Research environment in ALBANIA.* Retrieved February 21, 2009, from http://consultations.score-project.eu/attach/ictcr_al_en.pdf

Serafeimidis, V., & Smithson, S. (2003). Information systems evaluation as an organizational institution - experience from a case study. *Information Systems Journal*, *13*, 251–274. doi:10.1046/j.1365-2575.2003.00142.x

Smithson, S., & Tsiavos, P. (2004). Re-constructing information systems evaluation. In C. Avgerou, C. Ciborra, & F. Land (Eds.), *The Social Study of Information and Communication Technology: Innovation, Actors and Contexts* (pp. 207-230). Oxford, UK: Oxford University Press.

Stats, I. W. (n.d.). *Internet Statistics on Albania.* Retrieved March 25, 2009, from http://www.internetworldstats.com/euro/al.htm

Thompson, D., Rust, R., & Rhoda, J. (2005). The business value of e-government for small firms. *International Journal of Service Industry Management*, *16*(4), 385–407. doi:10.1108/09564230510614022

United Nations E-government survey. (2008). *From E-government to Connected Governance* (Doc. Nr. ST/ESA/PAD/SER.E/112,). New York: Department of Economic and Social Affairs Division for Public Administration and Development Management.

USAID Press Release Tirana. (n.d.). *A Modern Home for E-Governance: New IT Agency will host e-government services*. Retrieved May 19, 2008, from http://albania.usaid.gov/shfaqart/255/92/Press_Release.htm

West, D. (2004). E-government and the transformation of service delivery and attitudes. *Public Administration Review*, *64*(1), 15–27. doi:10.1111/j.1540-6210.2004.00343.x

World Bank. (2003). *Doing Business in 2004: Understanding Regulation.* A World Bank Publication.

World Bank. (2009). *Doing Business 2009 Report: Albania.* A World Bank Publication.

ENDNOTES

1. The boundary of who constitutes the main users of e-governments can also be redefined in e-government evaluation.
2. The most significant projects have been in cooperation with the Italian and the US entities.

Chapter 18
Internet and Political Parties in Chile

Eduardo Araya Moreno
University of Chile, Chile

Diego Barría
Leiden University, The Netherlands

Gustavo Campos
University of Chile, Chile

ABSTRACT

Due to the importance that the Internet has gained as a means of communication, literature on political communication has incorporated it as one of its preferred topics of focus. Literature stems almost entirely from Europe and the United States. Very little is known about the political use of new information and communication technologies (NICTs) in other parts of the world. The present chapter aims to provide evidence in that line, starting from the study of the incorporation of the Chilean political parties to the Internet. In specific, the following questions are answered: In what extent do factors such as the organizational characteristics of the political parties explain their greater or lesser adoption of NICTs? What do parties use NICTs for? Furthermore, although briefly, the authors will try to answer the question whether the parties have experienced change in their interaction with the citizenry and their bases because of the usage of NICTs.

INTRODUCTION

In *The Internet Galaxy*, Manuel Castells showed how Internet entered the daily lives of people. It is especially remarkable that, as Castells declares, based on a series of studies, people use the web to broaden their non virtual social relationships. The complementarity between the Internet and 'real life' does not only occur at an interpersonal level. Economy, for instance, has witnessed the emergence of a virtual space of economic exchange, which has integrated with the traditional market.

Politics do not escape this phenomenon. For some decades, the growing importance that media has played in different political systems has been perceptible (Blumler & Kavanagh, 1999). With the Internet, traditional communication forms coexist with new forms of political action. Castells (1997)

DOI: 10.4018/978-1-60566-860-4.ch018

has shown that the web has been used as a vehicle for political action by social actors so different, as are the pro-gun associations in the United States and the Zapatista Army of National Liberation, in Mexico.

Due to the importance that the Internet has gained as a means of communication, literature on political communication has incorporated it as one of its preferred topics of focus (Dahlgren, 2005). In this fashion, studies have emerged on how governments, parliaments, parties and politicians use the Internet. In the case of political parties, literature has sought to understand how political parties are beginning to use the Internet, both to connect with the citizens, to perform their traditional duties and to resolve internal management problems.

The active interest that researchers - from the different disciplines - have put on the political use of NICTs is caused by the fact that, as some authors suggest, studying how technologies are used entails understanding how these affect the functioning of democracy (Hansen, Pedersen & Wahl Jorgensen, 2005). This proves interesting primarily because of a strong feeling of displeasure that currently exists, generalized throughout the Western world, regarding the performance of the political systems (Dahlgren, 2005).

Literature on these topics stems almost entirely from Europe and the United States. Very little is known about the political use of NICTs in other parts of the world. The present paper aims to provide evidence in that line, starting from the study of the incorporation of the Chilean political parties to the Internet.

Despite of the changes they have experienced and the fact that they have had to face the competition of NGOs, study centres and social movements, which sometimes intend to replace them, political parties continue fulfilling the role of intermediaries between the political system and society. Hence, studying how NICTs are used involves knowing how they are embedded in the functioning of the political system.

The Chilean case is interesting to analyze because of two reasons. The first one is the degree of institutionalization of a party system such as the Chilean, which has endured over time and, due to its deep roots in society, makes parties the main players in the political process (Mainwaring & Scully, 1995: 64).

The second reason is the Internet penetration level in the country. By September 2008, 32.3% of households had internet access in their homes (SUBTEL, 2009). Thanks to the reference made by traditional media (television, radio, newspaper) about situations that occur on the web - or about information that travels through it – the presence of the Internet in daily life is greater than what the usage statistics suggest. This statement is consistent with the fact that, by 2006, 49% of the population claimed to feel "being a part of the world of new technologies" (UNDP, 2006: 81).

In particular, this chapter is intended to answer the following questions: ¿In what extent do factors such as the organizational characteristics of the political parties explain their greater or lesser adoption of new information and communication technologies (NICTs)? ¿What do parties use NICTs for? Furthermore, although briefly, we will try to answer the question whether the parties have experienced change in their interaction with the citizenry and their bases because of the usage of NICTs.

This chapter will present an analysis on the websites of seven political parties: Communist Party of Chile (PC), Socialist Party of Chile (PS), Party for Democracy (PPD), Radical Social Democrat Party (PRSD), Christian Democrat Party (DC), National Renovation Party (RN) and Independent Democrat Union (UDI). The selection meets the criteria of identification of relevant parties, propounded by Sartori (1976). All of these parties, except for the Communist, have parliamentary representation. Nevertheless, the PC has been included because - due to the distribution of the votes among the parties - it has a certain blackmail potential. This results of the fact that the governing *Concertación de Partidos por la Democracia* (Coalition of Parties for Democracy)—which gathers PS, PPD, PRSD and DC and is in power since 1990—has obtained the presidency of the Republic thanks to the communist votes.

The paper is based on a review of websites of the different parties, as well as interviews with political leaders, specifically the chairmen of the party organization that exists in the Metropolitan Region of Santiago, or with the chairmen of the local structures that exist in the communes of La Reina, Ñuñoa and Providencia. Regional chairmen were chosen because they fulfil the function of connecting the national chairmanships with the local units, while the chairmen of those communes were selected because they represent districts with high socioeconomic level, though not the highest of the country (MIDEPLAN-UNDP, 2006).

The structure of the chapter is as follows. Next is an analytical framework to understand the political parties and the use they give NICTs. In the third section, a review is made on the characteristics of the Chilean political party system, as it is since 1990, both ideologically and organizationally. Section four analyzes the use Chilean political parties give NICTs. Section five emphasizes, in a broad perspective, the issue of citizen participation via the Internet in Chile. In the last part, some conclusions are presented.

POLITICAL PARTIES AND NICTS

Until recently, some politologists were disenchanted by the absence of a theory of general application to study the political parties and they continued discussing the approach strategies from which to address the issue (Montero & Gunther, 2003). This has occurred in moments when new topics have entered the research agenda, such as the usage of NICTs. In this topic of interest, the ground is even less fertile, since it was built based on a series of hypotheses, more programmatic than academic, which suggested that technologies would democratize the public activity. Many of these proposals have been refuted as a consequence of critical analysis in light of political theory, mainly from the perspective of Rational Choice (see Cardenal & Batlle, 2006). Similarly, empirical studies have demonstrated the limited democratizing impact of NICTs in different countries (Castells, 2001).

Nonetheless, far from creating further problems, the abandonment of optimistic views has proven beneficial, since it has allowed to focus attention on more restricted topics, such as the reasons for the use of NICTs and how technologies allow political parties to perform their traditional functions.

Regarding the incorporation to the Internet, several explanations have been ventured. Montero & Gunther (2003) suggest that the massification of the Internet is one of the challenges that parties must face, especially due to the increasing complexity of communications with the citizens. This adds to the decrease in party militancy, the presence of phenomena such as the individualization of politics, deinstitutionalization of parties and the emergence of new participation forms. All of the aforementioned occur simultaneously with the massification of the Internet (Margetts, 2001). In this context, although the militants still constitute the founding myth of the parties, in practice they lose importance (Margetts, 2001). Another incentive for incorporating into the web is the possibility to project an image associated with being 'modern'.

Regarding the ways in which NICTs allow political parties to perform their functions, existing studies start from what party theory has established. It has generally been stated that political parties perform two different types of functions: those related with representation and those related with government. This distinction starts from the premise that the fundamental role of political parties—at least during much of the twentieth century—has been to connect society and particular groups of interest with the political system. This, primarily through the development of functions such as the aggregation and filter of social demands, social integration and mobilization, as well as the structuring of the vote. The connection with the system is performed through the government functions, mainly through the formulation of policies contained in the programmes suggested to the citizenry, as well as the recruitment of the political staff that will perform the government functions (Malamud, 1996).

Margetts (2001), poses that 'Cyber parties' (i.e. parties that have entered the Internet), can continue fulfilling their traditional functions of recruiting leaders, aggregating and articulating interests, taking positions regarding discussion topics and giving guidelines to governments. The contribution of technology to these functions is given because they can improve the ability to aggregate and articulate interests since the interest groups and social movements are more active online. In the same way, web-pages can allow to disseminate the positions of the parties on different topics. This proves interesting because a logic appears in which the websites end up being a sort of reflection of what the parties are.

Römele (2003) does also attempt to provide an analytical framework on the performance of political parties through the web. In the same line as Margetts, Römele explores how the NICTs help parties to fulfil basic functions such as forming opinions, mediating interests (thanks to the possibility of bi-directional communication) and improving the internal organization, as technologies allow to improve the quality and time of internal processes, as well as the possibility to coordinate actions[1].

Regarding the party-citizens interaction, literature has stated that NICTs could erase the boundaries between militants and voters fade as the websites focus on a broad audience (Hansen, Pedersen & Wahl Jorgensen, 2005). Other authors, on the other hand, have claimed that with technologies the bases can obtain greater protagonism, due to the facts that they can connect directly with the leaderships (Ward, Lusoli & Gibson, 2002) and open spaces for internal discussion (Löfgren, 2003; Römele, 2003). Nevertheless, at least in the cases of Sweden and Denmark, empirical evidence calls this proposal into question and even leads to think that NICTs could strengthen already existing structures (Löfgren, 2003).

The accumulation of studies developed in the field in the past few years has resulted in the abandonment of the '*technological determinism*', which claimed that NICTs, by themselves, would introduce change to the political systems. With this accrued evidence, the research agenda has turned to analyze technologies in relation to the contexts in which they are used. This entails understanding both the socioeconomic characteristics of a country (Hernson, Kai Stokes-Brown & Hindman, 2007), and the ones of their political system (Anstead & Chadwick, 2008). For the specific case of political parties, there are variables to consider, such as the competition logics among parties, the political culture of the political system, as well as the distinctive features of the parties.

Regarding the particular characteristics of the parties, Conway & Corner (2004) point out that it is necessary to pay attention to aspects such as the size of the parties and their longevity. On the other hand, Batlle, Borge, Cardenal & Padró-Solanet (2006) incorporate other elements, such as the type of party, according to the typologies of mass or catch-all party, their ideology and their position within the political system. Shortly, they state that the usage parties give NICTs – in this case, of websites - depends both on the party ideology, on its organizational characteristics as well as the characteristics of the electoral market. These three variables influence the conformation of a party communication strategy, which will in turn have an effect on the characteristics of the websites.

Although some studies have shown particular cases where organizational characteristics of parties do not have an impact on the use of NICTs

Table 1. Types of parties, characteristics and uses of NICTs. Source: Own development, based on Blyth & Katz (2005), Katz & Mair (1995) and Malamud (1996).

	Mass Party	Catch-All Party	Cartel Party
Characteristics	- Party seeks to represent specific social groups, which are part of the structure of the party (intermediary function). - Predefined ideology - Internal discipline - Funding comes from the militants	- Abandonment of representation of specific groups. - More moderate policies (representation of multiple groups) - Relative financial independence from militants	- Parties closer to the State than their traditional intermediary role - Policies defined with cartel logic -Emphasis on managerial skills - Financial independence from militants (State funding)
Party communication channels	Party has its own communication channels to connect its structure	Party competes to access non-party communication channels.	Party obtains access to communication channels regulated by the State
Expected uses of NICTs	- Websites focused on militants, with information primarily dedicated to proposed policies for the traditionally represented sectors. - Available communication spaces within the internal structure or strengtheners of militant identity. -Intranet or other meeting space	- Websites focused on 'citizens', with information regarding specific problems and proposals related to them. - If communication spaces exist, they should not distinguish between citizens and militants (hence, Intranet should not exist) -Possible availability of Intranet.	- Websites with information regarding political cadres of the party, and their leadership and management skills. - If policy documents exist, they should be based on pragmatic analyses - Absence of internal party coordination spaces.

(Batlle, Borge, Cardenal & Padró-Solanet, 2006) - these variables should be attended for the reason that they could affect not only the websites, but also the use of other applications, such as e-mail, the availability of discussion forums outside the portals and applications of the own site that are not always visible, such as the Intranets. Both the particular administrative structures, such as traditions, and organizational cultures of the parties can determine the logics of the uses of these applications.

Research on the usage political parties give NICTs demands its own analytical models. So far, the problem has been the absence of consensus within the academic community regarding study methodologies. This differs from the case of e-Government literature, which has developed in the establishment of typologies on stages of progress in the incorporation of NICTs to governmental management. Here, the evolutional model of Lane & Lee (2001) of four stages (presence, transaction, vertical integration and horizontal integration) has been widely spread. Each one of these stages represents a higher level of NICT use by the public agencies to render services to the community. In the most advanced levels, the isolated logic of each public agency is surpassed, promoting their integration. There, the territorial and sectorial boundaries of each organization are erased and the main focus lies on systems conceived in accordance with the requirements of the citizens.

Although the ideas of evolution and level of complexity of the rendered services can be useful to assess the level of development of the political party websites, the different functions performed, by public agencies (to render services) on the one hand, and political parties (to represent and govern), on the other, makes it recommendable to seek for other analysis approaches.

In the present paper, websites have been analyzed based on the methodology developed by Casero (n.d.), designed specifically to study party websites. This option is based on the fact that it provides advantages in the analysis of parties, since it reports on the existing relationship between the functions performed by the parties and the communicational strategies (*top-down* and *bottom-up*).

To the extent that they perform government functions, parties should present government programmes to the citizenry, while they intend to form an opinion and establish positions regarding different discussion topics in the political arena (Malamud, 1996; Margetts, 2001; Römele, 2003). This demands that the parties communicate with the citizenry, through *top-down* logic. Thus, the website is conformed as one more element in the communicational strategy of parties (Casero, 2007) and its use will depend on the strategy objectives.

On the other hands, representation functions demand *bottom-up* communication logics. To be able to fulfil its intermediary function between the political system and society, as well as being able to aggregate positions, it is necessary, firstly, to hold information collecting mechanisms. In this sense, the web may assist the contact with the citizens, e.g. to search for policy proposals, as well as to perform the coordination functions with the militants themselves (Casero, 2007).

Based on these definitions, Casero considers 39 indicators. In the present study we will only measure the presence of these indicators on websites, refraining from assessing the characteristics of their development. A third dimension proposed

Table 2. Indicators of the dimension Information Instruments. Source: Casero (n.d.)

Dimension	Sub-dimension	Total Number of Indicators	Indicator
Information Instruments	Information Supply	13	1. History of the Party
			2. Biography and profile of candidate/head leader
			3. Greeting/presentation of candidate/head leader
			4. Biography and profile of other candidates/main leaders (who is who)
			5. Organizational structure of the party
			6. Values or ideology of the party (principles, bases...)
			7. Programme of the party (electoral)
			8. Comparison between the programme of the party with the one of rival parties
			9. Speeches and declarations from head leader/leaders (programme related)
			10. Political documents of the party (resolutions, declarations, manifestos...)
			11. Data and statistics regarding their public policies
			12. Frequently asked questions (FAQ) on the party
			13. Addresses of the party's headquarters
	Circulation of Current Information	11	1. Publication of news (press releases)
			2. News archive (historic)
			3. Activity schedule (updated)
			4. Press summary (press dossier)
			5. Newsletter
			6. Corporative magazines and publications
			7. Downloadable photographs (current)
			8. Downloadable audio files (current declarations)
			9. Downloadable video files (current)
			10. Podcast
			11. Streaming audio-video (live: IPTV)

by this methodology –regarding characteristics of the websites - has also been excluded, since it is considered to answer web design matters and not those of party characteristics[2].

In the following section, we will approach the main characteristics of the party system that exists in Chile, from 1990 onwards (when democracy became operational again), putting special emphasis on the basic consensus of the parties and their ideological and organizational features. This way, we will later be able to analyze the type of usage that political parties give NICTs.

THE CHILEAN PARTY SYSTEM

Characteristics of the System

Since the 1980's, the countries in South America experienced processes of transition towards democracy, after years of military dictatorships. Challenges such as democratic consolidation and responding to a series of problems of socio-economic nature had a preponderant place in the agenda. Shortly after, however, both the political and academic debate began to focus on the problems of the democracy to consolidate, and to improve its ability to meet the social requirements, i.e. in terms of quality and capacity to consolidate itself[3].

The Chilean case is characterized by an early formation of political parties and a party system (A. Valenzuela & J.S. Valenzuela, 1983, J.S. Valenzuela, 1995, Scully, 1992, Luna, 2008). Although they were persecuted by the military dictatorship of Augusto Pinochet (1973-1990), at the return of democracy they once again took their fundamental place as political actors (Siavellis, 1999). Nevertheless, a few changes had occurred. During the 20th century, the main segmentation in the system had lied in a socioeconomic conflict (Luna, 2008) and the active participation of parties, as representatives of specific social groups, which could have held important militant

Table 3. Indicators of the dimension Participation Instruments. Source: Casero (n.d.).

Dimension	Sub-dimension	Total Number of Indicators	Indicator
Participation Instruments	Online Inter-activity	8	1. Execution of online polls and surveys
			2. Discussion forums or free chats
			3. Discussion forums or chats with candidate/leaders
			4. Sending of questions or suggestions via e-mail
			5. Blogs of head leader/ leaders
			6. Possibility to add comments to blog
			7. Publication of short messages (SMS)
			8. Wiki (collective writing)
	Online Mobi-lization	7	1. Intranet (restricted area)
			2. Online registration to party
			3. Contribution request (financial or human resources)
			4. Possibility to obtain party e-mail
			5. Possibility to create an own blog, linked to the one of the party
			6. Online organization of offline events (calls, announcements, notifications…)
			7. Party merchandising (sale of party goods)
	Total	15	

cadres. However, post-Pinochet Chile presents phenomena of depolitization and privatization of social relationships. Instead of acting politically, citizens seem to understand that their personal wellbeing lies in their performance in the market (Silva, 2004).

These changes can be seen in a series of alterations in the relationship parties establish with the electorate. Although high-class levels and those who have a greater degree of political socialization vote according to their party identity, lower levels and those who have not been politically socialized "...connect with their representatives based on personal ties, strongly localized and centred on the provision of private goods which are traded for contingent electoral support" (Luna, 2008: 111). This adds to the deideologization of the electorate and the increase of those who define themselves as independents (Siavellis, 1999).

In this context, the current party system moves around some basic consensus between the governing *Concertación* and opposing *Alianza por Chile* (Alliance for Chile), such as the acceptance of the market and democracy (Drake & Jäksic, 1999). However, it presents some fissures regarding the position of the parties concerning the democratic break-up of 1973 (Allende v/s Pinochet), the extent of the social policies of the State, the position of the parties regarding the violation of human rights performed during the dictatorship, as well as the moral values (Fuentes, 1999; Siavellis, 1999).

Together with the construction of agreements, a moderation process has occurred within the programme platforms of the parties[4] (Fuentes, 1999) and in the competition logics (J.S. Valenzuela, 1995). The electoral system gives incentives with the intention that two greater political-electoral alliances are formed, and to exclude any party that acts outside these coalitions. These great alliances have structured the new party system starting from the democracy/authoritarianism segmentation (Fuentes, 1999; Siavellis, 1999; Luna, 2008).

Organizational Features of the Political Parties

From the formal point of view, Chile holds a legal precept (Constitutional Organization Law 18.603, on Political Parties) which within its articles defines some internal bodies that political parties must possess. Despite of a predefined structure, the law provides a space of freedom for the parties to define their own organization charts. The PDC, for instance, shows internal structures that "... are not directly conditioned by the law and prove the existence of particular institutions and a previous organizational model" (Müller et. al., 2008: 419). The organizational characteristics of the parties are summed up in the following chart.

Table 4. Parties and voting results in the 2001-2005 parliamentary elections. Source: www.elecciones.gob.cl, accessed July 2009.

Party	Coalition	% of Deputy Voting 2001	% of Deputy Voting 2005
PC	Juntos Podemos	4,56	4,71
PS	Concertación	8,73	9,21
PPD	Concertación	11,12	14,12
PRSD	Concertación	3,54	3,24
PDC	Concertación	16,52	19,02
RN	Alianza por Chile	12,02	12,94
UDI	Alianza por Chile	22,00	20,48

Table 5. Organizational features of Chile's political parties. Source: Müller et. al., (2008).

Party	PS	PPD	PRSD	PDC	RN	UDI
Indirect election bodies	-	General Council	General Council	National Board	National Executive Committee District and regional councils	General Council National Executive Committee
Authorities elected by the militants	Central Committee National, regional and communal Executive Committees	National, regional and communal Executive Committees	Regional Assembly Regional Council	National, regional and communal Executive Committees	Regional and communal Executive Committees National Executive Committee	-

Unfortunately, we do not possess information on the PC. Apparently, it is not of interest to those who have analyzed the organizational characteristics of Chilean parties (See Müller et. al., 2008; Alcántara, 2003), given that they do not hold parliamentary representation since the 1973 elections.

The structures of the parties, especially PPD, PS and PDC, show a greater level of participation among their militants in the election of the district and regional executive committees than the rest of the parties. This adds to the fact that the PDC and PS open greater spaces for participation for the militants in the choice of candidates. This is a consequence of a more complex internal life within the parties, due to an internal democracy tradition and the existence of regular groups that compete for the control of the party. Moreover, these parties have, within their structures, mass-party features, such as the existence of representative units of specific groups, for instance, workers. This obeys to a long tradition of these parties with this social sector, which is revealed by the fact that they have participated in Chile's main workers union for over 50 years (Barría, 1971, Angell, 1974). These bonds appear clear by the structure of the PDC National Board, where the Workers' Front holds representation by its own right (Müller et. al., 2008).

The parties in the *Alianza por Chile* stand out for having greater indirect base participation mechanisms, while direct participation mechanisms are almost inexistent in the UDI. For this reason, the national decision centres hold autonomy regarding their bases. In the case of RN, the participation culture occurs on the lower levels of the organization chart, while on the top, direct voting by the militants only happens when conflicts between internal groups are to be solved (Müller et. al., 2008).

WEBSITES OF CHILEAN POLITICAL PARTIES

In contrast with other countries, the Chilean political scientist community has not entered the debate on the relationship between political activity and the appearance of NICTs. The only available piece of work on party websites is the Chilean Human Development Report by UNDP (United Nation Development Programme) (2006). It uses a methodology internationally implemented by the United Nations, which analyzes the information supply, access to public information, interaction degrees and service rendering. The UNDP findings indicate that, towards 2006, 1) the UDI website was the most developed one; 2) there were no greater differences in terms of information, except for the PPD; 3) the UDI and RN - the latter sharing positions with the Humanist Party (PH) and PDC, held the greatest degrees of interactivity. This measurement also reveals that the middle left wing parties (PS, PPD and PRSD), as well as

Table 6. Content index of websites of political parties in the Parliament. Source: UNDP (2006: 179).

Party	Information	Access to public information	Interaction	Services	Total
UDI	0,90	0,00	0,83	0,57	2,30
PH	0,70	0,25	0,67	0,43	2,05
PRSD	0,70	0,25	0,33	0,57	1,95
RN	0,80	0,00	0,67	0,29	1,76
PDC	0,60	0,00	0,67	0,29	1,56
PC	0,80	0,00	0,27	0,43	1,40
PPD	0,40	0,50	0,17	0,29	1,36
PS	0,80	0,00	0,00	0,43	1,23

left wing (PC) held lower levels of interactivity than the others.

Table 7 shows the results obtained in our own assessment, regarding year 2008, based on the use of Casero's (n.d.) methodology, in accordance with the adjustments propounded in the Introduction.

From the comparison between our results and the ones of the UNDP (2006), it clearly appears that in reference to interactivity, the parties in the *Alianza por Chile* hold the most complex websites among the parties, except for the PS. These differences are not so noticeable when it comes to the development of information instruments as, according to our assessment, RN shares the first position with the PDC and PPD, and the remaining parties come shortly after. Notwithstanding, it is notorious in the development of what Casero (n.d.) identifies as participation instruments. In this matter, both RN and UDI hold the first place, followed by the PS. The remaining parties show a scarce development of these instruments.

One possible interpretation of these results could point out that, given that parties such as the PC, PPD, PRSD and PDC hold a greater militant tradition – which leads to the existence of a more complex internal party life – the development of these instruments becomes less relevant. It ought to be noticed that, except for the PPD, all of these parties were created several decades ago[5] and even managed to survive the persecutions of the military dictatorship, between 1973 and 1990. In the case of the PPD, despite its youth, it was formed by militants of other parties, such as the PS or the MAPU (a 1970's party). The PS is the only party with these characteristics which has made progress in the development of participation instruments.

Table 7. Websites of Chilean political parties. Source: Own development, based on the scrutiny of the websites of each party between July 04 and 21, 2008.

Dimension	Sub-dimension	PC	PS	PPD	PRSD	PDC	RN	UDI
Information Instruments	Information Supply	8/13	6/13	9/13	8/13	10/13	8/13	5/13
	Circulation of Current Information	6/11	7/11	5/11	5/11	4/11	6/11	5/11
	Sub-dimension Total	12/24	13/24	14/24	13/24	14/24	14/24	10/24
Participation Instruments	Online interactivity	1/8	6/8	1/8	1/8	3/8	6/8	3/8
	Online Mobilization	0/7	2/7	1/7	2/7	1/7	5/7	7/7
	Sub-dimension Total	1/15	8/15	2/15	3/15	4/15	10/15	10/15
	Total	13/39	21/39	16/39	16/39	18/39	24/39	20/39

It is difficult to hypothesize about why it holds this development, unless we resort to the thesis posed in the second section, i.e. that, to a great extent, the development of web-pages responds to uncertain factors (Batlle, Borge, Cardenal & Padró-Solanet, 2006).

The above mentioned results make it necessary to retake another important point of the second section. Ideology is a factor that shows uneven results concerning the characteristics of the websites. Previously, we emphasized that certain authors state that left wing parties should use NICTs with greater intensity, especially when it comes to citizen participation (Cardoso, et. al., 2004, Chappelet et. al. 2004). As to the Chilean case, both the UNDP (2006) results and the ones here provided, show that it is possible to state the opposite relation. The PC shows the same use of information instruments as the rest of the parties. In this regard, a section concerning the political formation of new militants stands out. However, only 1 out of 15 applications allows participation. This is probably related to the organizational and cultural characteristics of the PC, to give predominance to its internal structure and to its militants - typical of mass parties – instead of chasing votes among "citizens" in a broader sense.

RN holds an important advancement in the interactivity sub-dimension, thanks to the development of free discussion forums with leaders, sending of e-mail, existence of blogs and development of instant messaging (SMS). The UDI, on the other hand, holds all of the online mobilization applications described by Casero (n.d.).

How can it be explained that these parties have a greater development than the rest in participation instruments? The interpretation of these results must take a series of characteristics into account. On the first hand, both the UDI and RN are semi-structured parties (Alcántara, 2003), with no party life tradition - in contrast with the PC and the *Concertación* parties – and possess structures in which the decision centres are hosted in specific units at a central level, unconnected from their bases (See Müller et. al. 2008). On the second hand, these parties display a distinctive preference for market economy (Fuentes, 1999). It is reasonable to believe that this ideological vision could bring these parties to have a more managerial concept of political activity. In this logic, the website management has a professional nature, which becomes apparent in the case of RN.

Considering aforementioned elements together, it is understandable that these parties hold more complex websites than the rest, thanks to the fact that they work with professional webmasters. RN, for instance, has a webmaster that encourages leaders to enter the different now existing technological applications, such as Twitter. The party has also made investments on web services, and currently has its server hosted by Google - with enough capacity to create 100,000 e-mail accounts - and offers pages for their militants to have their own blogs with the address name.rn.cl. The development of participation instruments is explained by the previous and by the fact that, in the absence of militant tradition, the websites are conceived as a means to connect with a broader audience.

At a local level, some party structures have made progress in the construction of websites. These initiatives do not obey communication policies stemmed from the national leaderships - as a matter of fact, some base leaders criticize them for not using the NICTs with greater intensity (Lepe, 2008, personal interview) - but the will of some militant that succeeds to convince the local committees. The fact that the web-pages are conceived and administrated by amateurs has brought a series of problems. In the RN structure of District 24 (communes of La Reina and Peñalolén), for instance, the website is not used because the central leadership is unsatisfied with its quality. Another problem that arises from this way of managing the sites is that the rules regarding the use of the sites occasionally become unclear. At the PS in La Reina, for instance, an internal conflict was generated because the newly elected committee considered that the webmaster was using the local structure for the benefit of some militants and not for serving the collective interests.

Outside the sphere of the own sites, during 2008 parties have made progress in their incorporation to Facebook. Most of the interviewees have spontaneously expressed that this has been the year of this programme and that their party had had experiences in its use, especially strong among younger militants. RN has created a group for the political commission of the party and the UDI has started to gain ground thanks to José Antonio Kast, one of the candidates for the 2008 election for the presidency of the party, who made his campaign through this medium. Furthermore, in the context of municipal elections, a great number of candidates have created their own profiles. According to one candidate, this has been helpful since, thanks to Facebook, he has been able to rely on non-militant activists which he could never otherwise have contacted (Soto, 2008, personal interview).

E-mail has become an important tool for the coordination of actions. Nevertheless, the age of the militant base has arisen as a factor that inhibits a more intensive use of the NICTs. Leaders of RN, PDC and PRSD claimed that they could not perform all the communication via e-mail or websites, given that they hold a significant number of older militants, which do not use the Internet. As a result, although leaders have attempted to make progress in the use of NICTs, they have had to complement it with traditional communication media, such as telephony. Otherwise, they claim, a summoning to a meeting would have no effect (Lepe, 2008, personal interview).

E-mail is mostly used to download information from the national committees to the district levels of the parties. On the whole, our interviewees have expressed that information transmission speed is higher with the NICTs, which allows the militants to be better informed. Equally, technologies allow the leaders to receive information from the bases; nevertheless, some key informants expressed that the bottom-up information flow is smaller and it doesn't affect the decision making processes.

In this way, instead of redefining organizational designs and power distribution within the parties, technology-based information flows tend to reproduce pre-existing structures. For example, RN began to use Facebook as a discussion space, but in a close group in which only members of the Political Commission of the Party participate.

Once again, we need to focus our attention on the characteristics of the parties. In those who have an internal life tradition, with militants that discuss and group together into different sectors within the parties, debate nets have appeared. Although these nets open new spaces for militants to debate on the position of the party on different matters, they do not manage to insert themselves into the decision making processes of existing structures.

The PS Central Committee created a virtual discussion net, named *Socialist Debates*. Nonetheless, the Committee does not participate in it and it has been dominated by militants who are critic towards the party line. Apparently, this space has tended to reproduce the division of the internal politics of the PS and to group the different sectors, according to their particular identities. This issue is not problem-free in a party with institutionalized internal currents which compete for supremacy within it. The PDC, on the other hand, has created an e-mail-based debate net, in which militants from different cities participate, as well as certain members of parliament, although the latter correspond to specific cases. Through this space, militants from different territorial structures can connect and inform themselves on the news from other districts.

Although isolated, one experience in which NICTs were used to support participation and internal party discussion processes was the one of the Fifth Ideological Congress of the DC, held in 2007 and prepared since 2006. In this encounter, the main positions of the DC for the following years, regarding a series of matters, were discussed. A website was activated for the event and through it, as well as e-mail-based communication, documents were prepared for each topic committee. As stated by participants in the encounter, during the previous year at least 200 documents were distributed, and were being modified according to the debate carried out on the net. During the days of the Congress, members of each committee gathered to agree on three or four points that had previously been agreed upon during the entire year, through virtual debate (Silva, 2008, personal interview).

The use of technologies outside the sphere of the particular websites is not open for non-militants. It is understandable in the case of the PDC and PS, because internal issues are discussed. This could also explain, for instance, that when the PS in the commune of La Reina created its own website, it was conceived as a reunion space among militants and not to connect with the citizenry. This adds to another factor, exogenous to parties: the attitude towards politicians by those who participate in debates. As said by a RN leader, the party formerly used to hold open discussion forums, but external people entered them to insult politicians in general, not only those of the party. This situation led to the desertion of these open spaces and generated distrust towards them within the party (Desbordes, 2008, personal interview).

INTERNET AND POLITICAL PARTICIPATION IN CHILE. OPEN SPACES FOR POLITICAL DISCUSSION?

Literature on NICTs in Politics has paid special attention to the possibilities that technologies provide for the citizens to participate. As it was suggested in section two, recent analyses approach this issue understanding that technologies act in specific contexts. In the Chilean case, consensus has been reached regarding the deficit of spaces for citizen participation, both because the State restricts them (Delamaza, 2009; Espinoza, 2009), and due to the presence of features of an increasing depolitization of the citizenry (Silva, 2004), which becomes apparent among the youths (Riquelme, 1997; González et. al., 2005).

In these circumstances, the low intensity with which political parties use instruments to promote interaction with the citizens via Internet, should not be understood as an exclusive issue of these organizations, but as a feature of the Internet incorporation of all of the actors in the political system, as a whole.

For instance, most of the Ministries do not open spaces for the citizens to give their opinion on bills or public policies through websites, except for limited exceptions, and they conceptualize citizen participation as the right to obtain information on ministry administration and, preferably, on non virtual spaces (Araya & Barría, in press).

The Senate is an exception due to the fact that, in 2003, it created an application (*Senador Virtual – Virtual Senate*) which allows citizens to express their opinion regarding the bills that the different parliamentary committees will discuss. In addition, it has the virtue that the citizen proposals are delivered to the parliamentarians. Nevertheless, the lack of regulations regarding the way in which this information should be considered during the legislative process restricts the results of citizen participation. Furthermore, this experience has had little impact, since the participation levels are low (Araya & Barría, 2009).

One issue not yet analyzed in Chile is the fact that, despite the alleged depolitization of the citizenry, the Internet has proven to be a prosperous space for political mobilization. The clearest example is the community www.farkas2010.com, created by a group of people who wanted to convince Leonardo Farkas, an eccentric Chilean entrepreneur and philanthropist, to become a presidential candidate. This movement, originated on the web, gained such strength that in December 2008, the Survey of the Chilean Centre for Public Studies (CEP, the most important referent in the country) showed that 5% of the population wanted this entrepreneur to become the next President of the Republic (CEP, 2008). Eventually, Farkas declined, despite of the good omens of the surveys and having over 313,000 followers on Facebook.

This community was created under an apolitical speech, where philanthropy was the main instrument for social politics, in detriment of State actions. The degree of support that Farkas gained within a few months allows hypothesizing on the potential of the Internet in Chile for the incorporation of the political debate to sectors that have abstained from acting in politics through traditional spaces.

The above stated leads to notice the fact that the Internet proves to be a suitable space to promote citizen participation - 63.6% of Internet users in Chile claim to constantly visit State websites (UC-CCS, 2006:16) – which could be used to recruit citizens on programmatic discussions, or for them to make proposals to the government and parliament in public policy matters. Although it also seems that the disenchanted citizens are willing to participate, as long as there is a level of innovation in the political system.

CONCLUDING REMARKS

As has been stated in this paper, in Chile, the Internet is present in the daily life of people, through references made by traditional media. The actors of the political system have also entered the Internet, mainly to deliver information.

Notwithstanding, citizen participation is an aspect that is not yet on their websites. In the case of political parties, literature has stated that NICTs erase the boundaries between militants and citizens, given that the parties focus on wider audiences, which adds to the suggestion that the intermediate structures of parties should lose importance, since technologies allow the contact between base militants and the local leaderships (through the websites). Although this does not hold true in the case of political parties with an internal life tradition (on the contrary, NICTs have opened new spaces for the militants), in the parties of the *Alianza*, boundaries are in fact erased between militants and citizens, since their websites appeal to a broad audience.

Literature has also speculated about whether the organizational variable is able to explain the way in which political parties incorporate to the Internet. In the case of the Chilean parties, it seems clear that those parties that hold a long militant tradition and have different tendencies in their internal lives, such as the PS and PDC, use the NICTs to develop internal debates among their militants and present a low development of interactive applications on their websites. On the opposite end, RN and UDI, who hold structures that give more autonomy to the leaderships from their bases and have a scarce presence of regular internal divisions, do not use NICTs for internal debates and, in contrast, they present interactive applications on their websites. The PS is an exception, since it is a party with internal life tradition and, simultaneously, it presents a website almost as complex as those of the *Alianza*.

In the analysis offered on Table 7, it can be seen that there are no greater differences among parties when producing or distributing information. Nevertheless, the websites of the *Alianza por Chile* are more complex; adding to the fact that, for instance, RN has made an important investment in technological equipment and holds a webmaster that apparently presents higher professionalization degrees than remaining parties. However, it is not clear whether this is a result of higher financial resources, a communication policy or a specific work schedule of the person in charge of the website.

Finally, we would like to highlight a topic which has political implications for the Chilean political system but, at the same time, may originate a new discussion on the role of NICTs in politics. Internet may, potentially, be a source for populist leaders to compete from outside the system. Although Leonardo Farkas declined to become a candidate, the two months of speculations on his entrance to the political competition were sufficient to see the political effectiveness of the Internet in Chile. Moreover, as the political parties have not themselves used the Internet, it continues being a sort of 'no man's land'.

This issue leads us to notice that NICTs could be related not only with democratic expansion, but also with the weakening of the political system and the promotion of populist leaderships. This becomes clearer in Latin America, where this form of making politics has a long tradition. Therefore, the debate should begin with the acceptance of the fact that NICTs are not only functional for democracy and that options are varied: in an extreme vision, they could even lead to plebiscitary authoritarianisms.

REFERENCES

Alcántara, M. (2003). La Ideología de los Partidos Políticos Chilenos, 1994 – 2002: Rasgos, Constantes y Peculiaridades. *Revista de Ciencia Política, 23*(2), 68–87.

Altman, D. (2006). (Algunas) Reformas Institucionales para el Mejoramiento de la Calidad de la Democracia en Chile del Siglo XXI. In C. Fuentes & A. Villar (Eds.), *Desafíos Democráticos*. Santiago, Chile: LOM.

Angell, A. (1974). *Partidos Políticos y Movimiento Obrero en Chile*. México: Ediciones Era.

Anstead, N., & Chadwick, A. (2007). *Parties, Election Campaigning and the Internet: Toward A Comparative Institutional Approach* (Political and International Relations Working Paper 5). Royal Holloway University of London.

Araya, E., & Barría, D. (2009). E-Participación en el Senado Chileno. ¿Aplicaciones Deliberativas? *Convergencia. Revista de Ciencias Sociales, 52*, 239–268.

Araya, E., & Barría, D. (in press). E-Government and Citizen Participation in Chile. The Case of Ministries Websites. In C. Reddick (Ed.), *Politics, Democracy, and E-Government: Participation and Service Deliveryt.* Hershey, PA: IGI Global.

Barría, J. (1971). *Historia de la CUT.* Santiago, Chile: Prensa Latinoamericana.

Batlle, A., Borge, R., Cardenal, A. S., & Padró-Solanet, A. (2006). *Reconsidering the Analysis of the Uses of ICTs by Political Parties: an Application to the Catalan Case*. Paper presented at the 4th ECPR General Conference, 6-8 September, Pisa, Italia.

Blumler, J., & Kavanagh, D. (1999). The Third Age of Political Communication: Influences and Features. *Political Communication, 16*, 209–230. doi:10.1080/105846099198596

Blyth, M., & Katz, R. (2005). From Catch-all Politics to Cartelisation: The Political Economy of the Cartel Party. *West European Politics, 28*(1), 33–60. doi:10.1080/0140238042000297080

Blyth, M., & Katz, R. (2005). From Catch-all Politics to Cartelisation: The Political Economy of the Cartel Party. *West European Politics, 28*(1), 33–60. doi:10.1080/0140238042000297080

Cameron, M. (2007). Citizenship Deficits in Latin America Democracies. *Convergencia. Revista de Ciencias Sociales, 14*(45), 11–30.

Cardenal, A. S., & Batlle, A. (2006). La Utopía Virtual. Una Crítica al Cyberoptimismo desde la Teoría de Elección Racional. *IDP. Revista de Internet . Derecho y Política, 3*, 1–12.

Cardoso, G. (2004). Ministers of Parliament and Information and Communication Technologies as a Means of Horizontal and Vertical Communication in Western Europe. *Information Polity*, *9*(1-2), 29–40.

Casero, A. (2007). *E-campaña 2007: Internet en las Elecciones Autonómicas Valencianas*. Paper presented at the VIII Congreso AECPA 2007, Valencia, España.

Casero, A. (n.d.). Propuesta de Metodología para el Análisis de las Webs de los Partidos Políticos. Unpublished working paper.

Castells, M. (1999). *La Era de la Información. Economía, Sociedad y Cultura. Volumen 2: El Poder de la Identidad*. Madrid, Spain: Siglo XXI.

Castells, M. (2001). *La Galaxia Internet*. Madrid, Spain: Plaza y Janés.

Chappelet, J. L. (2004). The Appropriation of E-Mail and the Internet by Members of the Swiss Parliament. *Information Polity*, *9*(1-2), 89–102.

Conway, M., & Corner, D. (2004). An Evaluation of New Zealand Political Party Websites. *Information Research, 9*(4).

Criado, J. I., & Ramiló, M. C. (2003). E-Government in Practice. An Analysis of Web Site Orientation to the Citizen in Spanish Municipalities. *International Journal of Public Sector Management*, *16*(3), 191–218. doi:10.1108/09513550310472320

Dahlgren, P. (2005). The Internet, Public Spheres, and Political Communication: Dispersion and Deliberation. *Political Communication*, *22*, 147–162. doi:10.1080/10584600590933160

Delamaza, G. (2009). Participation and *Mestizaje* of State-Civil Society in Chile. In P. Silva & H. Cleuren (Eds.), *Widening Democracy. Citizens and Participatory Schemes in Brazil and Chile* (pp. 249-271). Leiden, The Netherlands: Brill.

Drake, P., & Jäksic, I. (1999). Introducción. In P. Drake & I. Jäksic (Eds.), *El Modelo Chileno. Democracia y Desarrollo en los Noventa* (pp. 7-38). Santiago, Chile: LOM.

Espinoza, V. (2009). Citizens' Involvement and Social Policies in Chile: Patronage or Participation? In P. Silva & H. Cleuren (Eds.), *Widening Democracy. Citizens and Participatory Schemes in Brazil and Chile* (pp. 273-294). Leiden, The Netherlands: Brill.

Fuentes, C. (1999). Partidos y Coaliciones en el Chile de los '90. Entre Pactos y Proyectos. In P. Drake & I. Jäksic (Eds.), *El Modelo Chileno. Democracia y Desarrollo en los Noventa* (pp. 191-220). Santiago, Chile: LOM.

González, R. (2005). Identidad y Actitudes Políticas en Jóvenes Universitarios: El Desencanto de los que no se Identifican Políticamente. *Revista de Ciencia Política*, *25*(2), 65–90.

Hansen, K., Pedersen, K., & Wahl-Jorgensen, D. (2005). *Cyber-campaigning. The Characters, Development and Consequences of parties´ Electioneering in Cyber-space*. Paper presented at the Nordic Political Science Association's conference, Reykjavik, Iceland.

Herson, P., Kai Stokes-Brown, A., & Hindan, M. (2007). Campaign Politics and the Digital Divide. Constituency Characteristics, Strategic Considerations and Candidate Internet Use in StateLegislative Elections. *Political Research Quarterly*, *60*(1), 31–42. doi:10.1177/1065912906298527

Katz, R., & Mair, P. (1995). Changing Models of Party Organization and Party Democracy. The Emergence of Cartel Party. *Party Politics*, *1*(1), 5–28. doi:10.1177/1354068895001001001

Layne, K., & Lee, J. (2001). Developing Fully Functional E-Government: A Four Stage Model. *Government Information Quarterly*, *18*, 122–136. doi:10.1016/S0740-624X(01)00066-1

Löfgren, K. (2003). *Intraparty Use of New ICTs – Bringing Party Memberships Back In?* Paper presented at the ECPR 2003 joint sessions, Edinburgh, Scotland.

Luna, J. P. (2008). Partidos Políticos y Sociedad en Chile. Trayectoria Histórica y Mutaciones Recientes. In A. Fontaine, et al (Eds.), *Reforma a los Partidos Políticos en Chile* (pp. 75-124). Santiago, Chile: UNDP, CEP, Libertad y Desarrollo, Proyectamerica y CIEPLAN.

Lusoli, W., & Ward, S. (2003). *Digital Rank-and-File: Party Activists' Perceptions and use of the Internet.* Paper presented at the Annual Meeting of the American Political Science Association, Philadelphia, United States.

Mainwaring, S., & Scully, T. R. (1995). La Institucionalización de los Sistemas de Partidos en América Latina. *Revista de Ciencia Política, 17*(1-2), 63–102.

Malamud, A. (2003). Los Partidos Políticos. In J. Pinto (Ed.), *Introducción a la Ciencia Política* (pp. 321-350). Buenos Aires, Argentina: EUDEBA.

Margetts, H. (2001). *The Cyber Party.* Paper presented to workhop 'The Causes and Consequences of Organisational Innovation in European Political Parties,' ECPR Joint Sessiones of Workshops, Grenoble, France.

MIDEPLAN (Ministry of Planification of Chile) & UNDP. (2006). *Las Trayectorias del Desarrollo Humano en las Comunas de Chile (1994-2003).* Santiago, Chile: MIDEPLAN-UNDP.

Montero, J. R., & Gunther, R. (2003). *Los Estudios sobre los Partidos Políticos: una Revisión Crítica* (Estudio/Working Paper 12). Colección en la Red de Cuadernos de Trabajo. Universidad Autónoma de Madrid.

Morlino, L. (2007). Explicar la Calidad Democrática: ¿Qué tan Importante son las Tradiciones Autoritarias? *Revista de Ciencia Política, 27*(2), 3–22.

Müller, G., et al. (2008). Democracia Interna y Selección de Autoridades. In A. Fontaine, et al. (Eds.), *Reforma a los Partidos Políticos en Chile* (pp. 413-428). Santiago, Chile: UNDP, CEP, Libertad y Desarrollo, Proyectamerica y CIEPLAN.

O'Donell, G. (1992). *Delagative Democracy?* (Working Paper 172). Kellog Institute for International Studies, University of Notre Dame.

Riquelme, A. (1999). ¿Quiénes y Por Qué 'No Están Ni Ahí'? Marginación y/o Automarginación en la Democracia Transicional. Chile, 1988-1997. In P. Drake & I. Jáksic (Eds.), *El Modelo Chileno. Democracia y Desarrollo en los Noventa* (pp. 261-280). Santiago, Chile: LOM.

Römmele, A. (2003). Political Parties, Party Communication and New Information and Communication Technologies. *Party Politics, 9*(1), 7–20.

Sartori, G. (1976). *Parties and Party Systems: A framework for analysis.* New York: Cambridge University Press.

Scully, T. (1992). *Los Partidos de Centro y la Evolución Política Chilena.* Santiago, Chile: Notre Dame University-CIEPLAN.

Siavellis, P. M. (1999). Continuidad y Transformación del Sistema de Partidos en una 'Transición Modelo'. In P. Drake & I. Jäksic (Eds.), *El Modelo chileno. Democracia y desarrollo en los noventa* (pp. 223-259). Santiago, Chile: LOM.

Silva, P. (2004). Doing Politics in a Depoliticised Society: Social Change and Political Deactivation in Chile. *Bulletin of Latin American Research, 23*(1), 63–78. doi:10.1111/j.1470-9856.2004.00096.x

Subsecretaría de Transportes (SUBTEL). (2009). *Estadísticas e Indicadores por Tipo de Servicio. Internet, Series Conexiones Internet.* Retrieved March 26, 2009, from http://www.subtel.cl/prontus_subtel/site/artic/20070212/pags/20070212182348.html#T1

UNDP. (2006). *Desarrollo Humano en Chile. Las Nuevas Tecnologías: ¿Un Salto al Futuro?* Santiago, Chile: UNDP.

Universidad Católica (UC) y Cámara Chilena de la Construcción. (CCS). (2006). *Monitoreando el futuro digital: resultados encuesta WIP-Chile, 2006.* Retrieved March 16, 2007, from http://www.wipchile.cl/estudios/WIP_Chile_2006___informe.pdf

Valenzuela, A., & Valenzuela, J. S. (1983). Los Orígenes de la Democracia. Reflexiones Teóricas sobre el Caso de Chile. *Estudios Públicos, 12,* 7–39.

Valenzuela, J. S. (1995). Orígenes y Transformaciones del Sistema de Partidos en Chile. *Estudios Públicos, 58,* 5–80.

Ward, S., & Gibson, R. (2003). On-line and on Message? Candidate Websites in the 2001 General Election. *British Journal of Politics and International Relations,* *5*(2), 188–205. doi:10.1111/1467-856X.00103

Ward, S., Lusoli, W., & Gibson, R. (2002). Virtually Participating: A Survey of Online Party Members. *Information Polity, 7,* 199–215.

ENDNOTES

1 Literature has also focused on other topics. Political cyber campaigns have been one of the issues that have received most attention. See Ward & Gibson (2003) and Hansen, Pedersen & Wahl Jorgensen (2005). A series of studies have also focused their attention on the possible changes that NICTs can produce in the organizational aspect of parties, as well as in their internal structure. See Batlle, Borge, Cardenal & Padró-Solanet (2006), Hansen, Pedersen & Wahl Jorgensen (2005), Löfgren (2003), Lusoli & Ward (2003), Römele (2003) and Ward, Lusoli & Gibson (2002).

2 In other studies, although with a different approach than Casero (n.d.), the design of the websites has been related to the level of participation that citizens can reach (Criado & Ramiló, 2003).

3 Regarding the debate related to different aspects of modern democracy in Latin America, see Altman (2006), Cameron (2007), Morlino (2007) or O'Donell (1992).

4 Although consensus has been generated among the programme platforms, we could not speak about a cartelization of the system, according with the characteristics described in the second section. Especially, considering the existence of a party such as the PC which acts outside a possible collusion among remaining actors.

5 The PR (the former structure of the current PRSD) was created in 1863; the PC, in 1921; the PS, in 1933; the PDC, in 1957; the PPD, in 1987; RN, in 1987; the PPD, in 1988; and the UDI, in 1989.

Chapter 19
Information and Communication Technologies (ICTs) and Government: A Challenge to the Concept of Citizenship?

Alejandro Elias Ochoa-Arias
Universidad de Los Andes, Mérida, Venezuela

Mariángela Petrizzo
ENDOLOGICA, Open Informatics and Professional Services Cooperative, Venezuela
SPIRALIA C.A. Networking Learning Solutions, Venezuela

ABSTRACT

This chapter re-interprets the development of electronic participation and electronic government in the context of an alternative ideology. Bringing back the critique of previous chapters about government as a technology, the chapter shows how it is possible to generate new and socially oriented spaces for democratic processes in which technologies are tools for transformation. This requires expanding the concept of citizenship, of the state and of society.

INTRODUCTION

This paper unfolds two contrasting perspectives regarding the way in which the incorporation of ICTs in governmental sectors influences and affects the notion of citizenship, and also how such incorporation implies and nurtures new trends in current society.

The paper begins its argument on showing that e-government and e-democracy have been regarded as independent processes of incorporation of technology in the functioning of society which is a misleading conception that assumes independence upon processes that has been historically closely related. Consequently, we attempt to bridge the gap between these two concepts: e-government and e-democracy by showing that government practices in society are in themselves a very powerful technology. Therefore, the process of implementing Information and Communication Technologies (ICTs) in governmental practices should be regarded as a very highly complex and social-historical bounded process that requires a careful comprehension of governmental practices,

DOI: 10.4018/978-1-60566-860-4.ch019

beyond the usually classical approach associated with the improvement of efficiency and efficacy in common operations regarding the management of public resources.

For this purpose, we start showing the governmental technology as a way of managing human conduct at collective and individual level that has provided a process of economization of the government as a goal in itself. This is what we called governmentalization, following Foucault's use of the term, to refer to the historical process through which the State has been conducting human affairs until the present.

A second part of this paper concentrates in showing how the incorporation of ICTs is playing an increasingly relevant role in the definition of boundaries between citizens and the State and their relationships. It is argued that those boundaries are not a result of a political debate regarding the role of the State and the citizens, but instead, they are based upon a taken for granted assumption that this State should be efficient in the performance of current tasks. The process of building citizenship is becoming influenced and controlled by the limits of ICTs, when ICTs could be conceived as tools for the transformation upon the way in which public affairs are managed by allowing a richer process of debate due to an increase of information, flows and instantaneous access to knowledge and decision making processes.

The final part of this paper is concerned with developing a proposal to bring together the processes of e-government and e-democracy by sustaining a need of a re-thinking the concept of citizen as being a more capable and better informed member of society to take part and become responsible of the process of government in local, regional and national affairs. The concept of citizenship is proposed to be expanded instead of being minimized to that role of customer which is the dominant conception at the moment of introducing ICTs into the organizational structure of the State and its relationships with society.

ICTS AND GOVERNMENT: A COMPLEX MIX

The arrival of internet and other information and communication technologies (ICTs) to large sectors of population and almost every corner of the world, open a wide range of possibilities to the states, governments and citizens alike. Such possibilities range from an improvement of administrative procedures as to the introduction of changes in the relationships between citizens and the political milieu, basically through a more formal and deeper intercourse between those spheres of social action. They are what has been called e-government and e-democracy (Araya, 2005).

A large part of the studies devoted to the implementation of ICTs in the relationships between the State and citizens, are based on the assumption that administrative and political actions are mutually independent, and in some cases, they are linked but not in a very substantial way. Therefore, there is a dichotomy between administration and democracy which ignores the process in which the citizenship is built in a democratic environment and how it affects the ways in which the citizens are engaged in the activities of e-government and e-democracy (Petrizzo, 2005). In the constitution of citizens and the structures and practices of government, the whole set of elements will be affected notwithstanding their administrative or political nature.

In the case of Latin-american societies, such a separation is not only analytical, but also formal in some cases. In these situations, a divorce between public administration and political action has implied the existence of government as an alien element which can be used and controlled by those holding political power and their sectional interests. The absence of a will towards a more ambitious policy orientated to enhance a continuous improvement of the system of activities and procedures of the public administration was inevitable.

In the following, a reflection upon some of the issues related to the social milieu in which the incorporation of ICTs is generating the so called, "society of information" will be presented, in order to show a need for a social articulation and participation in the generation and social appropriation of technology in order to configure the so called, "society of knowledge". It is done by considering some characteristics regarding social institutions, technology and citizenship in order to grasp the way in which such conceptions conditions the links between citizens, institutions and political sphere. The purpose is to unveil how an instrumental bias in considering social impact of ICTs could condition the arrangement and social legitimation of the State. On the other hand, to unveil that an acknowledgment of this instrumental perspective as a partial and not conclusive account of technology could mark a transformation of political and administrative practices, and the constitution of a notion of citizenship engaged with the building and articulation of the State for a reconstitution of the public good.

TECHNOLOGY AND THE RELATIONSHIPS BETWEEN THE STATE AND CITIZENS

From a commonly accepted perspective, technology is understood as an instrument useful for the effective realization of tasks that have been designed and established previously in terms of realizing a given goal, taken for granted. From this perspective, technology is considered as a neutral element regarding the social milieu in which it happens to occur. This account of technology is not enough to understand the role played by technology itself when it is used in internal social processes through which society considers its choices and paths for development. Basically, it happens because the neutral condition of technology is unveiled as a fiction. It is a fiction because every instruments or technological processes are necessarily bounded by the society in which they are created. The non neutrality of technology becomes a critical factor to be taken into account. In particular, in those societies in which there is not generation of technology but only consumption and implementation of foreign technologies, that issue becomes even more relevant.

From a distinct account of a fragmented and pragmatic account for the application of ICTs for government and assuming technology as a social product, it is relevant to ask about the society in which ICTs are used and the forms and conditions of the institutions (public institutions at the forefront) that are compatible with such implementation. It implies a reflection on the "society of knowledge". Even more, to ask ourselves if it is directing us towards a technological society. What are the institutional conditions that allow a functioning of the State in a society of knowledge? How do relationships develop between the State and citizens?

It implies at least, two tasks. First, it is required to appraise the incorporation of ICTs in the State in order to develop a critique of the dominant account upon which the incorporation of ICTs relies. That is, to consider the experts orientation and self-limited application of ICTs in the operational structures of the government without taking into account other aspects such as attributes and roles of social actors in relation to the State and in relation to the modes in which these social actors interact among them. Secondly, this fragmented and specialized account on the use of technology regarding the functions of the State, determined by a perspective which assumes the incorporation of technology as independent field from the social context and its management, implies a necessary debate about the notion of technological society.

This debate should consider at least three conceptions that seems to explain technology in relationship to society as a whole, based on the concepts of: the State, citizen and the mechanisms of articulation between both elements:

- **First view:** An Instrumentalist account of State, Citizen and their relationships

The core assumption of a theory called instrumental lies on assuming technology as a neutral instrument that can be used without any concern about the social context in which it is applied. Its use, impact and appropriation will exclusively depends on the economic costs associated with its implementation. Therefore, technology is universal in terms of its principles, products and criteria for its justification.

The assumption that technology is neutral implied that the whole process of incorporating technology in managing the State will be ruled by the efficiency and cost-benefits relationship. In this sense, it should become paramount to the process of incorporating technology (in particular, ICTs) in the State to ensure that it would improves the whole performance of the State notwithstanding social contexts upon which the State is running its own affairs.

According to this view, a separation of political issues from technical demands is understandable. In this sense, the horizon of expectations regarding the technological drive of the State is constituted in the way in which the State becomes instrumental for the flourishment of society. The State will use technology if it reduces uncertainty and improves their capability to respond efficiently to its environment and users. In this sense, it becomes clear that technology driven from an universal context of application will imply a basic set of tasks orientated to ensure that information flows work properly between State and citizens. A clear definition of roles and responsibilities hold by every actor of society will be a cornerstone for such reduction of uncertainty. In this sense, it is required that technological incorporation should be concentrated in those procedures of the State that has been clearly consolidated and acknowledged as objective among the different actors. The administrative feature of the State will play a key role as it constitutes a very technical and widely accepted role of the State. Procedures such as taxation, payrolls and the whole administrative set of tasks will be an area where ICT is welcome and fits properly. Briefly, technical demands to rationalize everyday procedures in running public affairs will become the rate upon which incorporation of technology should develop.

The conception of the State from such standpoint will be based on an ever-increasing rationalization of collective procedures to satisfy material demands established through the specific and consolidated channels of communication between the State and its citizens. Mainly, administrative features based on monitor and control of expenditures, resources and demands will be of a central concern for the use of technology. The tendency of technology to orchestrate activities of the State will grow from a technical-oriented will, to make rational and transparent those procedures through which the State provides its services. That means that levels of establishing relationships with the citizens will be based on capabilities of providing services through homogeneous and anonymous procedures.

The citizens in this conceptual framework will be defined in terms of their capability of behaving anonymously and following a technical rationalization of their behavior and demands toward the State. The anonymous behavior is essential on the basis that it will be evaluated on the grounds of impartiality, homogeneity, self-restraint, and repetitive patterns. In this sense, citizenship (e-citizenship) becomes a regulated channel of demands and duties clearly defined and orchestrated in order to preserve the stability and operational continuity of the organizational structure and procedures, in particular, those based on the use of ICTs.

It is worthy to mention that mechanisms of relation between the State and citizens are regulated from the dynamics of technological advances upon the administrative structure of the State, and the development of patterns of behavior related to ways of access to the citizens and from the citizens to the State. The universality of the context of application of technological innovation implies that those areas considered as unstable in relation to the performance of the State and citizens should be kept away from technological implementation.

Finally, it seems that the process of incorporation of ICTs in the running of public affairs seems to be self-contained to what is called the system, (see Haberma's account on Systems and Lifeworld (Habermas, 1983) in the sense that the rationality behind the system patterns of behavior is highly objective and usually related to an end-means rationality. According to Feenberg (2002), such a conception of technology seems to be unaware of the social construction of technology. However, its relevance in this debate lies in their ability to provide a framework to understand one of the trends related to the implementation of ICTs in governmental milieu. The dominance of an administrative orientated perspective and technical self-contention of ICTs incorporation could be considered to reproduce and even to increase a lack of pertinence and focus on the State by attending some issues that are not critical for the success of the State in the constitution of an actor capable of enhancing the capability of citizens to develop a social project for the whole society. However, such a criticism seems to lies upon a different conception of technology from that implied by an instrumentalist notion of technology.

- **Second view:** The State and citizens as technological tools

The basis of a so called substantive theory about technology and society (Feenberg, 1992) is to regard technology as a historical and cultural process that proposes homogeneity and standardization as privileged social form in order to conceive whatever is the case according to its utility. In this sense, technology becomes a way of living and conceiving the whole world as devices. This puts the human being at the verge of living an illusion of freedom as there is a guaranteed access to use the world, without realizing that human being in individual and collective terms, becomes also a device. Technology becomes a "false" determinism in the way of conceiving the world as an instrument.

The conception of technology as being a cultural and historical process upon which human being is unveiled as a tool and, therefore, every social intercourse should be regarded in terms of its utility, implies that the State and citizens are valued as purposeful agents. It could be argued that it is a radicalization of the previous conception of technology. Indeed, a radical twist lies on making the whole set of actors involved in social intercourse (State, citizens and technology) as being driven by a will to disclose themselves as instruments. In this regard, there is not need of separating political and technical issues as all of them: social relations, the definition of collective answers and creation of institutions and organizations responds to a technical rationalization. In this regard, a key factor is to acknowledge technical rationality as the ground upon which the claims of good and true are based. By technical rationality it is understood as a a pragmatic rationality grounded on functionality and utility of different components involved in the execution of a function. They become devices for control and security in order to ensure that the right dispositions of things (including the State and citizens), would help the process of orchestrate the means for the realization of the goals of some actors who are not clearly identifiable in this framework. It could be argued that the current order based on freedom as the main source for legitimacy and the building of institutions allowed the whole set of institutions to promote its availability to use. Let us put forward how we consider will become the State and the citizen devices according to this framework.

In order to grasp the idea of the State as a technological device, it should become clear that as a technological device, the State plays a double role. On the one hand, it becomes accessible and available for the use of anyone who requires it for his/her own purposes. On the other hand, it uses and manipulates other devices in order to reach its own condition of device. Therefore, it seems that this double loop in which an instrument is used by others and it uses other devices, requires a set of control and monitor defined as a control of second order. In this sense, there is a hidden power upon which the control of the State lies.

Although it is not our purpose to unveil the nature of such power, it is relevant to recall that it is what Foucault called as governmentalization in order to grasp the rationality upon which the "ensemble formed by the institutions, procedures, analysis and reflections, the calculations and tactics that allow the exercise of this very specific albeit complex form of power" (Foucault, 1991 p. 102). A form of power which is exerted over the population. The State is upon the control and surveillance deploying a dispositive of security by addressing the care and monitor of people from their most common basic premises: the biological conditions in the frame of society. Natality, education, health and death became the major sources for the building of a system of regulation or government. The State governs and orders the individuals qua population and disposes of them asdevice for the production and reproduction of mechanisms through which the exercise of government and the disposal of them in economic terms are possible. The economic dimension is the historical dimension upon which utility has been based. In this regard, the State driven by a substantive notion of technology would become a dispositive of government and control over the population (see Ochoa, 2009 for a detailed account of this).

The other element of society to be conceived as a technological device is citizenship. It is clearly constituted by the population. This implies to conceive citizens as anonymous units considered from a biological perspective upon which they demand and require to use the State as a device for the realization of their specific goals. In this sense, a way of using the State in order to fulfill the need of citizens is defined and contained to be exclusively related, or at least justifiable, from those demands related to preserve basic needs. An important element among those considered to be basic needs refers to the condition of security and control of uncertainty. In this regard, it seems almost natural that a priority for the incorporation of ICTs in running public affairs refers to the issue of control of the State and citizens by each one of their counter-parties. Therefore, the emergence of very sophisticated and highly appreciated devices to control public expenditures, mobilization of human beings, transmission of diseases, birth rate and other variables related to the government of population and to restrain the State of becoming very intrusive, plays a key part in the process of requesting ICTs in the public domain.

- **Third view:** The State and citizenship based on a critical account of technology

This account is called critical because it proposes to overcome the limits of a progressive conception of technology (instrumental) or pessimistic (substantive) of the technology, by assuming that in choosing technology we become what we are as society, which will be also influential on our future choices as a collectivity. This act of choosing is profoundly conditioned and cannot be understood as a "free" use as it is pretended by the instrumental theory, but also it cannot be comprehended as a inevitable destiny. On the contrary, it is a process in which multiple perspectives are at stake. From a critical perspective, technology is not a destiny but a conflict arena, a space of social struggle in which different alternatives for the civilization are subject of debate and decision.

This conception of technology conceived as a socio-historical outcome of humankind upon which a rich and complex set of possibilities appear to be possible, offers possible paths for the disclosing of human being in connection to new accounts of the world. It can be considered to be a third possibility to an assumed linearity of the historical development of human being as controller of the world, or a most fatalistic account of human being as a device of the most powerful device: technology itself. In this regard, it is our contention that a third path of reflection upon the role of technology in the present could start by asking once again about the boundaries of the so called system of activities upon which technology could be considered to be a helpful tool or rather, a deceitful device upon which an illusion of freedom is the facade of an "iron cage". It has been called by Feenberg (1992) as subversive rationalization. It implies to approach the incorporation of technology as a constructivist and undetermined way of providing a political arena to carry on the dispute upon the ideological frameworks informing human actions at individual and collective level.

Here, technology provides tools for an opening of a space of possibilities for social interaction and towards the building of society as a whole. In a way, technology instead of being considered from the standpoint of a rationality of ends-means is asked to be considered from a dimension that somehow resembles a communicative rationality upon which Haberma's account of the Lifeworld lies (Habermas, 1983). Let us briefly explain this claim. Subversive rationalization refers to the attempt to overcome and render as inapplicable or at least, injustifiable, a dominance of the rationality ends-means in the social intercourse. This subversion requires to render the technical-instrumental framework as relative and it is only possible if there is a practical rationality propelling such subversion. By practical rationality we understand a rationality engaged in comprehend the current state of affairs from a perspective that could render comprehensible different frameworks in which a given set of actions could be meaningful. In this regard, subversive rationalization starts by considering as relative the framework upon which a given account of the current situation is considered to be truthful and good on the grounds of its utility. Therefore, the process of assuming a practical framework of inquiry will become critical if it is capable of rendering the dominance of rationality of modern world as one among others. In this case, the issue at stake, technology, could be rendered as a social construction if the separation of the political and the technical is overcome in favor of a notion of politics tied to the classical concept of it, instead of a modern account of politics that refers inevitably to the strategic calculus in order to preserve power.

The notion of the State based on such an account of technology will be based on assuming the State as a social construction in itself which appears to orchestrate procedures and patterns of behavior in order to preserve a social and political order that has been assumed as good. However, the condition of a "good order" is not to be considered as a dogmatic true. On the contrary, it is required to evaluate and criticized- such order from a standpoint in which the pursuit of social well being and the appropriateness of such order to all citizens is possible. In this sense, the definition of technology as a purposeful device is too narrow to include a constructivist approach upon which technology is only an advanced stage upon which society lies to build new possibilities as a whole. In a sense, it implies a critique of the boundaries of technology as a redefined concept related to the end-means rationality. In the account of the State as a purposeful agent, boundaries critique would imply the opening of a space in which the State is built as a device of control and governance. Indeed, the State will become a privileged space for the evolution of innovative forms of using technologies in order to make the world (not only the physical but also the human and symbolic worlds) a dwelling place.

Citizenship engaged in such a social construction of the space in which society builds up its future, as a space of possibilities implies that each citizen becomes a practical reasoner engaged in evaluating and improving the way in which society as a whole is a human construction. Following this line of reasoning, citizenship is not exerted from a technical standpoint in which boundaries are defined in terms of duties and rights. Indeed, exercise of citizenship is an agonic pursuit of a betterment on the conditions upon which society is capable to create and recreate itself in material terms, but also in relation to the development of signs and discourses upon which the exercise of the dominion of a exclusive rationality is continuously criticized.

From above, it seems clear that the way in which we attempt to carry on a reflection upon the process of incorporating ICTs in government will be mainly conducted based on this third possibility. However, it is our contention that the incorporation of ICTs in the management of government implies a crossroad situation. On one path, it seems that ICTs as technology provides with a whole set of characteristics capable of enframing its use in terms of efficiency and efficacy by providing a sort of freedom based on information and knowledge in technical terms. On the other path of development, there is a possibility to understand the use of ICTs as a condition to allow processes of emancipation, even of those forms of technical control and prediction, by enhancing the presence of concurrent information and different accounts of the "same" reality from different points of view.

The implementation of ICTs in the government of society considered from a critical conception of technology poses a challenge to the critical understanding and systemic comprehension of the technological issue beyond the use of technology and it will require a critical appraisal of citizenship and the governmental organizations. This step implies a challenge to both, citizens and government.

ICTS AND KNOWLEDGE SOCIETY

It has been argued before that understanding of ICTs and their incorporation in governmental practice and in relation to citizens requires a previous understanding of the historical evolution of the social impact for an ever growing activity related to the information management and their technologies.

It is commonly accepted that current society is driven by information. Therefore, the label "information society" becomes a common term and also tied up to other terms such as "knowledge society" and so on. Such a label has been somehow a resemblance of what happened with "industrial society" or "industrial revolution". That is, a way of characterize a whole society as being ordered and structured in terms of a central activity that is considered as a critical factor for the current economic success and a better future of society as a whole.

But, What does convert current society into a information society instead of knowledge society? Apparently, the reason stems from considering a criticality of the activity in terms of economic growth or improvement in the material performance of society. Indeed, information society is a "new" label of what was called in the past as "new economy", in order to emphasize that the impact of ICTs goes beyond the economic sphere and seems to configure a new social relationship. On the contrary, if we pursue to label it as a "knowledge society", we will be compelled to incorporate a broader cultural context and social activities in which creation and dissemination of knowledge will be more complex. Therefore, it is our claim that information society is more appropriate term to deal with the current transition process towards an authentic knowledge society. For the sake of the argument, it could be argued that knowledge has been considered as a key factor for economic growth. However, the next step is still to be taken for society as a whole.

Again, there is a need to inquiry about the differences between a social relationship as being defined in terms of a dominance of economic rationality in society and information society, and even more, in terms of the knowledge society. By considering some definitions of this last term, we can disclose a will of going beyond the economic nature of knowledge as a driven force of society. However, it is our main concern to show some of the limitations that the use of technology could impose upon the State. Let us consider some definitions.

(...) the concept of 'knowledge societies' includes a dimension of social, cultural, economical, political and institutional transformation, and a more pluralistic and developmental perspective. In my view, the concept of 'knowledge societies' is preferable to that of the 'information society' because it better captures the complexity and dynamism of the changes taking place. (...) the knowledge in question is important not only for economic growth but also for empowering and developing all sectors of society (Plathe y Mastrangelo, 2003)

"It is about a code to talk about a socio-technological transformation, as every society has been based on knowledge". Indeed, every society of what we have got knowledge, have been conditioned on information and knowledge. Whether in power, wealth, social organization.. In that sense, it seems confusing to talk nowadays about "society of knowledge". Are we coming from social realities without knowledge?. It would mean that we reached the top of knowledge. Therefore, we should take the concept "society of knowledge" from a perspective less attached to the terminology, as something more general upon which we conceive the constitution of our reality (Castells, 2002) (translated by the authors)

"[introduces] a concept of knowledge-value, based on subjective perceptions, a concept that refer to a "price of knowledge" as the "value created from that knowledge". In other words, is the price or value that a society relates to that product generated by a creative use of knowledge, a sort of a subjective "use-value". (Boisier, 2001) (translated by the authors)

In opposition to the idea of a Society of knowledge, the society of information assigns a role to the economic factor and wealth generation derived from information management (findings, generation, distribution, acceptance and recycling) in the process of constituting relationships between citizens and the State. In this case, there is a dominance of the efficiency and efficacy values as world wide accepted standards to judge an activity as being performed properly.

The apparent consensus about distinctive features between information and knowledge is based on the involvement of a living organism in using the former to generate knowledge, which is then vulnerable and changeable which is something in common with humankind. In the case of knowledge, it could be argued that its major risks are related to inaccessibility and fragmentation of knowledge. By the way, that is how knowledge became institutionalized in current society and which somehow diminishes the social mechanism through which knowledge develops: learning networks. However, the institutional framework regarding access to knowledge refers to a system which does not provide education from a critical and open perspective. Knowledge and its diffusion through social mechanisms already institutionalized reflects more a concern for preservation rather than innovative use, creation and alternative knowledge. Therefore, a framework dominated by an instrumentalist account of technology, government and society as a whole becomes dominant.

In any case, it is evident that human societies had been based and are already grounded upon knowledge. What is the difference regarding the present situation? For some, it is about a "democratization of knowledge", as an euphemism to refer that ICTs guarantees access to stored data and information. It reveals a ground dominated by an instrumental account of technology, government and knowledge. Knowledge becomes a commodity with an universal face value. Although it could be valid for technical accounts of reality, it is not in those situations in which knowledge is involved in making judgments, considering values and ordering norms.

Therefore, an instrumental dimension of knowledge acquires a major presence when knowledge is intended to become a commodity and, consequently, a key factor in the ordering of society. In this scenario, access to technology alone is not enough for the building of a society of knowledge. Even more, to provide access with restrictions to data and information reinforces a technological dependence. This constitutes a situation with high levels of vulnerability and risks for every society and becomes even more harmful for the development of relationships among citizens, government, and the State.

Will be enough ICTs to enhance that freedom required in the society of knowledge?, Is ICT enough to allow technology to become a space of confrontation and debate about the alternatives of civilization?. Considering that technology is a historical outcome of the social dynamics of society, any answer attempted requires a more comprehensive category of analysis: citizenship, and the modes and spaces in which it relates to the State and its institutions. This is developed in the following section.

E-GOVERNMENT AND E-DEMOCRACY: CITIZENSHIP AS A KEY LINK

Due to the intrinsic value of electronic data transmission in terms of enhancing links among peers and bringing together people and organizations, ICTs reveal themselves as technological tools for the development of complex societies. Development and expansion of ICTs made possible a ubiquity of services and functions of the State, and simultaneously an expansion of a set of considerations required in the framework of rights and duties. For example, searching for standards of accessibility to digital information to guarantee the inclusion of every citizen without any constraint regarding handicapped people, or the recognition of digital rights and duties and their fulfillment by private corporations. In this context, it is easy to understand the positive impact using ICTs in administration of the State, as it provides a clear and direct between government and the citizens (tax payments, official documents request and so on). These are some of the positive results of digitalization of procedures connecting citizens and the State.

A massive use of ICTs in public offices and functions of government is known as "e-government". It began in late 1980s, as a toolbox for the enhancement of administrative field of national public offices, defining "(...) the ways in which governments use the new technologies in order to provide a better access to information and governmental services, improve public services and open up mechanisms to participate in democratic procedures and institutions" (Lopez and Leal, 2002) (translated by the authors). E-government refers to the provision of services and information from the government to the public through electronic media. In this context, "public" comprises every actor outside of public administration (citizens, NGOs, private organizations) in relation to such administration.

This basic notion of e-government implies that strategies used to articulate it has got as main customers the following sectors:

- **Citizens:** The privileged beneficiaries of strategies of e-government regarding the provision of information and transactions
- **Private sector:** In any of its manifestations, either individual or collective
- **State:** Through its own organizational entities (in differente levels, national, regional or local) and its public servants

Therefore, strategies of e-government (including plans and policies) will be orientated to attend any of these sectors in terms of an incremental and progressive process taking into consideration the peculiarities of each sector. Meanwhile, strategies of e-government for its internal implementation

Table 1. Levels of performance and application of ICTs in public offices

Stage	Components
Basic Information	Use of electronic devices. Digital information. An exclusive channel of access without facilities for exchange of information.
Interaction	Mobile-gov, electronic debates, issue networks and communities, orientation on services and provision of specific information to different sectors of society. Use of Web 2.0 for control of governmental activities as Twitter or Facebook. Bi-directional communication (citizens, public institutions, private sector)
Transactions	Taxes, payments, request of official documents. A more "effective" exchange between government and user allows electronic transactions.
Transformation	Improvement and re-engineering of internal processes

constitute in itself a complex network of activities, functions and tools. It requires an adaptive capability regarding its environment, training of citizens and public servants, and a permanent revision on its levels of performance and appropriateness.

From the perspective of an incremental application of strategies of e-government, there are clearly differentiated stages:

- **Information Visibility:** In which different public institutions and agencies offer information about themselves in the internet to every public
- **Bi-directional Communication:** It offers exchange of information between users (their own information) and facilitating exchange of relevant information
- **Effective Exchange:** It includes the realization of administrative transactions involving information exchange (I.e, payment of taxes and electronic certification of documents)
- **Service Integration:** It allows a common site for access and functional performance covering every issue in which interactions with government and different public services are managed

This provides an schematic account for the implementation of ICTs in the activities and functions of the State. However, introduction of ICTs has not been able to overcome, at least in the political discourse, a differentiation between administrative work (bureaucratic) in relation to transactions regarding external actors to the government; and those actions defined as "democratic" and related to the exercise and constitution of democracy through processes of deliberation, participation, collective decision making, public control and, electoral procedures. This differentiation results from a sort of an implicit duality in which the citizen is conceived as an user of public administration and the citizen as a co-builder of his/her own destiny through common political practices.

This duality has been reflected in an expertise and analysis regarding a re-engineering of the bureaucratic structure through the so called, electronic democracy. It embodies strategies for the implementation of ICTs in the management and execution of procedures related to promote and enhance a democratic will. In conclusion, those categories used to the analysis of ICTs, government and the State, have not been helpful in the process of understanding the role of citizens in this complex phenomenon. Therefore, it is far from being an integral and coherent account of the citizens. Instead, it seems that a fragmented and incoherent account of the citizens is becoming dominant and, in some cases, reduced to be users.

This has been illustrated by Araya (2005) by referring to a reductionism in approaching the relationship between electronic democracy and electronic government. "The first one is understood as informatics related to management and, regarding the second one, it is considered as an utopia regarding political relationships between State and citizenship" (translated by the authors). It suggests that a new approach searching for complementarity between both concepts is required in order to enhance a responsible building of society in which citizens play a critical role beyond the dimension of user of electronic devices and services, which incidentally, enhances an instrumental conception of citizen and State, already explained.

It should become evident that there are circumstances in which administrative and exercise of a democratic will are required to work together. Such requirements configure specific characteristics for the citizenship. In another paper (Ochoa and Petrizzo, 2006), we argued that there is double-limitation imposed on normative accounts of citizenship (the modern liberal account being dominant). First, the liberal account reduces citizenship to the transactions in order to formalize the attachment of an individual to a political community, if the citizen did not earn it as a birth right. Such an affiliation implies a legal bonding to duties and access to rights guaranteed by the State. Secondly, there is a reduction of responsibilities of each individual regarding the preservation and enhancement of public spaces and the search of the "public good" (understood as the ground upon which good is conceived for the well being of all, see Fuenmayor, 2001)

The overcoming of this artificial division between administrative and democratic activities from a viewpoint of the citizens, requires an evaluation of the notion of citizenship. It should go beyond the formal discourse to include everyday activities in which citizens are constituted. It implies "(..) to incorporate the sphere of communal bonds as an active part of the process of regaining social spaces and building of new political instances in order to articulate collective actions. Citizens in their active exercise of duties and responsibilities are called to constitute a new civil society. It implies a complex network of relationships among citizens and their communities, and the political institutional framework by orchestrating political actions of social organization and community organizations as well" (Ochoa and Petrizzo, 2006) (translated by the authors).

The constitution of a new society should be aware of the historical condition of modern State based on spaces of individual autonomy and the relevance of issues such as exchange of goods and their value, that is, an account of society based on a system regulated by administrative instances and money as the main vehicle of objective interaction (Habermas, 1983). Instead, this new society is assumed to be built upon the constitution of spaces of collective action beyond the sphere of the State, but without excluding it. On the contrary, public debate should focus a relevant part of its efforts to determine the definition of the State as a whole. Therefore, citizens should be considered as co-builders of communal bonds, tasks and goals even in those aspects considered to be exclusively related to the administration of the State. As a consequence, "the exercise of citizenship requires a balanced action among the duties developed by a historical becoming of society and the virtues and characteristics of this citizenship built in everyday activities" (Ochoa and Petrizzo, 2006) (translated by the authors). It implies a mutual recognition between citizens and the State, not only regarding an institutional framework, but also by considering social practices and local collective capabilities from a critical account of their own internal processes. Among these processes, it is a key issue to consider the role of technology in enhancing such capabilities, virtues and functionalities. In such a case, exercise of citizenship and democracy are constituted in a double-loop learning process. This should overcome the artificial separation of the activities of the State in "administrative tasks" (so called, government) and "democratic procedures" (so called, citizenship)

Table 2. Levels of application of ICTs in the relationships between the State and citizens

Stage	Components
Basic Information	Digital information and access to ICTs. Communication channels in order to provide information about citizens, social actors, enterprises and the government.
Interaction	Electronic forum, virtual communities, electronic transactions, transactional ties and transformation of citizens, social agents, social movements, private sectors and government.
Deliberation and Political Engagement	Public debate of laws, Online introduction of proposals and political initiatives. Improvement of political links between the State and citizens, and among social movements for the definition of specific actions of policies for the constitution and re-definition of the public good.

From this account, the concept of e-government and an incremental process described above can be conceived not only in a progressive movement in the dimension of administrative transactions by incorporating new actors and expanding the space of action. Indeed, it should be considered also as a process of evolving in the political dimension as well. This is a process of transformation based on the application of ICTs in sectors that are beyond technical or administrative transactions of Public Administration. Because of a need of a closer link between citizens and the State, it is proposed to conceive a modified scheme for the strategies of e-government and e-democracy:

- **Stage of basic information:** In which the different governmental agencies offer information regarding their own purposes and procedures using the web. It is intended also to incide on the formation of the citizens under the premises already commented. There is not just access to date and information, it also includes, learning. This learning constitutes a citizen with basic knowledge about the modes of relationships between the State and institutions. In this stage, there is also a role of formation of political consciousness through the incorporation of other organizations of civil society and social movements with presence in the web.
- **Stage of interaction:** It is orientated to surpass the transactional component between the State and society. It is about a conversion of the citizen into an agent of social change. A social change conceived as collective processes of participation in the definition of public policies, norms and public decision making procedures.
- **Stage of deliberation and political engagement:** In this stage, citizens become active actors by engaging actively and politically into the constitution of society through the betterment of social ties and institutional relationships orientated to the constitution and reinforcing of the "public good". In it, the double loop learning reaches its major activity in terms of reconstituting citizenship, the State and relationships between them.

In this framework, the purpose of using ICTs is aware of the pressumption of a non neutrality of technology and simultaneously, the building of a citizen whose actions promotes the development of the relationships among the State and the democratic exercise beyond the electoral dimension. In this sense, the use of ICTs can be conceived as a mechanism to enhance the joint exercise of both activities (administration and deliberation) as a primordial part for the constitution of a critical citizenship.

From the non-neutral condition of technological devices, ICTs are not independent of the social reality in which they emerged, in which they are applied, and those social conditions that can be enhanced by using technology. Therefore, in the case of a social framework in which it is intended to have an integral account of the citizenship and the public good, it is required for any technological device used in order to enhance both activities (the formation of the citizens and its real exercise), to review and evaluate from a critical perspective the meaning of such a technological device for the realization of a given project of society.

Finally, it is relevant to inquiry about how the knowledge generated by and through the use of ICTs can be observed as relevant in the constitution of the meaning of public. Besides, what does make possible to tie up administrative and deliberative practices as one practice?. How does knowledge management is possible? How can it be performed in the public domain? These questions reveal that the role of knowledge in the application of ICTs in public domain requires a critical approach about technology and beyond an instrumental dimension.

CONCLUSION: THE CHALLENGE TO CITIZENSHIP

The central challenge implied by the incorporation of ICTs in the public domain refers to a remaking and reformulation of the central role played by the citizens in the constitution of administration of the State and the building of democratic institutions. In this case, it seems that a propellant of the use of ICTs in the administrative field holds an account of citizens in which their role is regarded as users interested in an instrumental use and relationship with the State. It is clear, that such an account is insufficient and even counterproductive to the constitution of a new social order.

Regarding the analysis of government, current literature of social sciences, as far as our knowledge allowed us to judge, there is not an explicit analytic account to justify a separation between management and the managed, that is, between administration of the State and the public. This separation is harmful to a process of an active engagement of citizens in building the public good and in building their own exercise of citizenship.

In this context, to consider political deliberation as a foundational factor of democracy is critical to overcome a modern notion of democracy self-contained to the election of political representatives and speakers. This overcoming will enhance the generation of a set of questions and proposals about the conditions upon which political community is constituted and enriched as a whole. This deliberative process would be enriched if an appropriate use of ICTs could be reached by assuming that ICTs constitutes also a part of the social milieu in which citizens are critical participants and users of ICTs.

The political deliberation goes beyond the electoral stage and even it could allowed the evaluation of public decisions which hardly can be accessed by citizens. This deliberation becomes a substantive element in the constitution of citizenship in which not only access to information, but also, a capability to disseminate and share information and knowledge in a synergistic network with fellow citizens constitutes a critical factor in building society as a complex and egalitarian network. In other words, political deliberation potentiates each citizen through her exercise as co-builder of society through the use, generation and dissemination of information flows related to the constitution and formation of citizenship. It is our contention that a rebuilding of citizenship requires a holistic and comprehensive perspective. In this regard, the starting point is to recreate the political practice, by assuming it to pursue the definition of a project of society capable to distinguish and to make problematic current society from an open debate about what a good society is. In this case, the notion of public good is central. Therefore, it has to go back beyond the economic dominance for the definition of the good. In this case, the public is not the good as it affects to everyone on a similar grounds, but rather, that issues that allowed citizens to recognize themselves and the others from an open dialog even of those differences that could be central for the building of social and individual identity (Arendt, 1997).

From this building of the public and public good as a whole, a positive step towards the solution of basic organizational issues arising in contemporary society is reached by offering a virtual space in which governmental and non governmental institutions could link meaningfully with the citizens. These are the citizens which Arendt (1958) acknowledges as responsible of the building and enrichment of public space by being also the promoters of a dialog process involving institutions, social actors and citizens themselves in a framework of a citizenship conceived as active, compromised and civic. This creates the conditions for the promotion of a citizen conceived as an active actor engaged in building and preserving the public from her own local space, without being self-contained only to local issues. That is, a social actor compromised with the local and pursuing an expansion of the public good to reach society as a whole (Ochoa and Petrizzo, 2006).

The use of ICTs in this deliberative process and in the building of citizenship becomes critical in the framework of evaluating the impact of information in the exercise of citizenship in a democratic milieu. It will constitute what we could call "informed democracy", beyond the technical implications of the term. That is, a democracy that empower and enhances a better citizens through the enrichment of the deliberative processes and the media used to reach it: a democracy formed from inside and towards the internal reflection of itself as a whole. A critical account of technology is central to such an account of democracy. The challenge posed to citizens and ICTs alike is then to find an appropriate place and mode in which ICTs could become a central propellant to such projects of society and citizenship.

REFERENCES

Araya, R. (2005). Internet, política y ciudadanía. In *Proceedings of the Nueva Sociedad N? 195 Enero-Febrero 2005*, Caracas, Venezuela (pp. 56-71).

Arendt, H. (1958). *The Human Condition*. Chicago: University of Chicago Press.

Arendt, H. (1997). *Qué es la política?* Barcelona, Spain: Paidós.

Boisier, S. (2001). Sociedad del conocimiento, conocimiento social y gestión territorial. *Interacoes . Revista Internacional de Desenvolvimento Local, 2*(3), 9–28.

Castells, M. (2002). *La dimensión cultural de internet" Speech available at Debates Culturales de la UOC*. Retrieved January 20, 2009, from http://www.uoc.edu/culturaxxi/esp/articles/castells0502/castells0502.html

Coleman, S., & Gotze, J. (2005). *Bowling together: Online public engagement in policy deliberation*. UK: Hansard Society.

Feenberg, A. (1992). Subversive Rationalization: Technology, power and democracy. *Inquiry, 35*(3/4).

Foucault, M. (1991). Governmentality. In G. Burchell, C. Gordon, & P. Miller (Eds.), *The Foucault Effect* (pp. 87-104). London: Harvester and Wheatsheaf. London.

Fuenmayor, R. (2001). Venezuela: Su enfermedad y Crisis Actual. In F. Parra (Ed.), *Estudios de Derecho Público: Libro Homenaje a Humberto J. La Roche Rincón* (pp. 335-358).

Habermas, J. (1983). *The Theory of Communicative Action*. Boston, MA: Polity Press.

López, J., & Leal, I. (2002). *e-Gobierno. Gobernar en la sociedad del Conocimiento*. Vitoria, España: Instituto Vasco de la Administración Pública.

Ochoa, A., & Petrizzo, M. (2006). Ciudadanía y Desarrollo Endógeno. In A. Ochoa (Ed.), *Aprendiendo en torno al desarrollo endógeno* (pp. 53-74). CDCHT, FUNDACITE Mérida, CSI-ULA.

Petrizzo, M. (2005). Participación ciudadana y tecnologías de la información y comunicación: hacia la administración pública relacional. In *Proceedings of the Nueva Sociedad N 195,* Caracas, Venezuela (pp. 88-101).

Pilonieta, C., & Ochoa, A. (2006). El desarrollo endógeno sustentable. Una aproximación conceptual. In A. Ochoa (Ed.), Aprendiendo en torno al desarrollo endógeno (pp. 21-36). CDCHT, FUNDACITE Mérida, CSI-ULA.

Plathe, A., & Mastrangelo, J. (2003). Interviu with Abdul Waheed Khan, Towards Knowledge Societies. *A world of science . RE:view, 1*(4).

Vial, A. (2005). Nosotros los ciudadanos a la intemperie. In *Proceedings of the Nueva Sociedad N 195,* Caracas, Venezuela (pp. 42-53).

Compilation of References

Abranson, M., & Means, G. E. (2001). *e-Government 2001 – IBM Endowment for The Business of Government*. Oxford, UK: Rowman & Littlefield Publishers, Inc.

Acevedo, M., Callicot, B., Monticino, M., Lyons, D., Palomino, J., & Rosales, J. (2007). Models of Natural and Human Dynamics in Forest Landscapes: cross-site and cross-cultural synthesis. *Geoforum*. doi:.doi:10.1016/j.geoforum.2006.10.008

Acevedo, M., Rosales, J., Delgado, L., Ablan, M., Dávila, J., Callicot, J. B., & Monticino, M. (2007, September). Modelos de Interacción Humano-Ambiental: El Enfoque de la Biocomplejidad. *Ecosistemas, 16*(3). Retrieved April 2009, from http://www.revistaecosistemas.net/articulo.asp?Id=507

Ackoff, R. (1981). *Creating the Corporate Future: Plan or to be Planned for*. New York: John Wiley and Sons.

Ackoff, R. (1981). The Art and Science of Mess Management. *Interfaces, 11*(1), 20–26. doi:10.1287/inte.11.1.20

Administration on the net. (2004 June). An ABC guide to e government in Austria Federal staff unit for ICT strategies, Federal Chancellery Austria, Alexander, I., & Robertson, S. (2004). Understanding Project Sociology by Modelling Stakeholders. *IEEE Software, 21*(1), *23–27*.

Afele, J. S. (2003). *Digital Bridges: Developing Countries in the Knowledge Economy*. Hershey PA: Idea Group Inc.

Agune, R., & Carlos, J. (2005). Governo eletrônico e novos processos de trabalho. In E. Levy & P. Drago (Eds.), *Gestão Pública no Brasil Contemporâneo*. São Paulo, Brazil: Fundap.

Ahumada, J. (1969). Notas para una teoría general de planificación. In *Cuadernos de la Sociedad Venezolana de Planificación Números 4 y 5*. Caracas: SVP.

Aja, E. (2001). Spain: Nation, Nationalities, and Regions. In J. Loughlin (Ed.), *Subnational Democracy in the European Union* (pp. 229-253). Oxford, UK: Oxford University Press.

Albania, U. N. D. P. (2008). *E-accounting Project*. Retrieved January 19, 2009, from http://eaccounting.undp.org.al/

Albania, U. N. D. P. (2008). *GovNet Project*. Retrieved January 19, 2009, from http://www.govnet.undp.org.al/

Albania, U. N. D. P. (2008). *Technical Assistance and Capacity Building Support to the National Agency for Information Society Project*. Retrieved January 19, 2009, from http://www.undp.org.al/index.php?page=projects/project&id=176

Albanian Council of Ministers. (2003, April). *National Information and Communication Technologies Strategy of the Republic of Albania*. Albanian government publication.

Albanian Government. (n.d.). *Programme 2006-2009 by the Ministry of Foreign Affairs*. Retrieved: February 11, 2009, from http://mfa.gov.al/web/pub/programien2100.pdf

Alcántara, M. (2003). La Ideología de los Partidos Políticos Chilenos, 1994 – 2002: Rasgos, Constantes y Peculiaridades. *Revista de Ciencia Política, 23*(2), 68–87.

Allan, B. (2007). *Blended Learning: tools for teaching and learning.* London: Facet Publishing.

Al-Maimani, A. S. (2000). *Information Impact on socio-economic decision-making (national economy of Sultanate of Oman).* MBA dissertation, the University of Hull.

Alterman, R. (Ed.). (2001). *National Level Planning in Democratic Countries. An International Comparison of City and Regional Policy-Making.* Liverpool, UK: Liverpool University Press

Altman, D. (2006). (Algunas) Reformas Institucionales para el Mejoramiento de la Calidad de la Democracia en Chile del Siglo XXI. In C. Fuentes & A. Villar (Eds.), *Desafíos Democráticos.* Santiago, Chile: LOM.

Alvesson, M., & Willmott, H. (2003). Introduction. In M. Alvesson & H. Willmott (Eds.), *Studying Management Critically* (pp. 1-22). London: Sage.

Amit, R., & Zott, C. (2001). Value creation in e-business. *Strategic Management Journal, 22*(6-7), 493–520. doi:10.1002/smj.187

Anderson, J. (1975). *Public Policy-Making.* New York: Praeger.

Angell, A. (1974). *Partidos Políticos y Movimiento Obrero en Chile.* México: Ediciones Era.

Ansoff, I. (1965). *Corporate Strategies.* New York: McGraw-Hill Education.

Anstead, N., & Chadwick, A. (2007). *Parties, Election Campaigning and the Internet: Toward A Comparative Institutional Approach* (Political and International Relations Working Paper 5). Royal Holloway University of London.

Araya, E., & Barría, D. (2009). E-Participación en el Senado Chileno. ¿Aplicaciones Deliberativas? *Convergencia. Revista de Ciencias Sociales, 52,* 239–268.

Araya, E., & Barría, D. (in press). E-Government and Citizen Participation in Chile. The Case of Ministries Websites. In C. Reddick (Ed.), *Politics, Democracy, and E-Government: Participation and Service Deliveryt.* Hershey, PA: IGI Global.

Araya, R. (2004). Tres perspectivas para observar el gobierno electrónico. In R. Araya & M. Porrúa (Eds.), *América Latina Puntogob. Casos y Tendencias en Gobierno Electrónico* (pp. 23-42). Chile: FLACSO Chile – OEA – AICD.

Araya, R. (2005). Internet, política y ciudadanía. In *Proceedings of the Nueva Sociedad N? 195 Enero-Febrero 2005,* Caracas, Venezuela (pp. 56-71).

Arendt, H. (1958). *The Human Condition.* Chicago: University of Chicago Press.

Arendt, H. (1997). *Qué es la política?* Barcelona, Spain: Paidós.

Arenilla, M. (2000). Introduction (Introducción). In M. Arenilla (Ed.), *Modernization Processes in Public Administrations. La Rioja Case* (El Proceso de Modernización en las Administraciones Públicas. El Caso de La Rioja) (pp. 23-34). Logroño, Spain: Gobierno de La Rioja - Consejería de Desarrollo Autonómico y Administraciones Públicas.

Arthur, B. W. (1994). *Increasing Returns and Path Dependence in the Economy.* Ann Arbor, MI: University of Michigan Press.

Ashby, W. R. (1964). *An Introduction to cybernetics.* London: Methuen.

Attiya, H., & Welch, J. (2004). *Distributed Computing: Fundamentals, Simulations, and Advanced Topics (2nd Ed.).* Hoboken, N.J.: Wiley-Interscience.

Axelsson, B., & Easton, G. (Eds.). (1992). *Industrial Networks: A New View of Reality.* London: Routledge.

Babbie, E. (1999). *Métodos de pesquisa de Survey.* Belo Horizonte, Brazil: Editora UFMG.

Bailey, K. D. (2005). Beyond system internals: Expanding the scope of living systems theory. *Systems Research and Behavioral Science, 22*(6). doi:10.1002/sres.659

Ballestero-Diaz, F., & Font-Jaume, A. (2004). El despertar del e-procurement en las administraciones públicas.

Información Comercial Española (ICE) . Revista de Economía, 813, 73–88.

Bandara, W., Gable, G. G., & Rosemann, M. (2006). Business Process Modelling Success: An Empirically Tested Measurement Model. In D. Straub & S. Klein (Eds.), *Proceedings of International Conference on Information Systems*, Milwaukee, Wisconsin, USA (pp. 1-20).

Bandara, W., Tan, H. M., Recker, J., Indulska, M., & Rosemann, M. (2007). *Bibliography of process modeling: An Emerging research field*. Retrieved from http://eprints.qut.edu.au

Bañez, T. (2003). Ciudadanía y Participación. In M. J. Benruz & R. Susín (2003), *Ciudadanía: dinámicas de pertenencia y exclusión* (pp. 97-112). Universidad de la Rioja, Servicio de Publicaciones.

Banister, F. (2004). *Deep E-government*. Paper presented at the European Group of Public Administration (EGPA) Annual Conference, Ljubljana.

Bannister, F. (2007). The curse of the benchmark: An assesment of the validity and value of e-government comparisons. *International Review of Administrative Sciences, 73*(2), 171–188. doi:10.1177/0020852307077959

Barría, J. (1971). *Historia de la CUT*. Santiago, Chile: Prensa Latinoamericana.

Barzelay, M. (2001). *The New Public Management - Improving Research and Policy Dialogue*. Berkely, CA: University of California Press.

Baskerville, R., & Myers, M. (2002). Information systems as a reference discipline. *MIS Quarterly, 26*(1), 1–14. doi:10.2307/4132338

Batlle, A., Borge, R., Cardenal, A. S., & Padró-Solanet, A. (2006). *Reconsidering the Analysis of the Uses of ICTs by Political Parties: an Application to the Catalan Case*. Paper presented at the 4th ECPR General Conference, 6-8 September, Pisa, Italia.

Baym, N. K. (1995). The Performance of humor in computer-mediated communication. *Journal of Computer-Mediated Communication, 1*(2).

Beck, E., Madon, S., & Sahay, S. (2004). On the margins of the Information Society: A comparative study of mediation. *The Information Society, 20*, 279–290. doi:10.1080/01972240490481009

Becker, J., Pfeiffer, D., & Räckers, M. (2007). *Domain Specific Process Modelling in Public Administrations – The PICTURE-Approach*. Berlin: Springer.

Beer, S. (1959). *Cybernetics and Management*. London: English Universities Press.

Beer, S. (1979). *Heart of Enterprise*. Chichester, UK: John Wiley & Sons.

Beer, S. (1981). *Brain of the Firm (2nd Ed)*. Chichester, UK: Wiley.

Beer, S. (1985). *Diagnosing the System for Organisations*. Chichester, UK: John Wiley and Sons Ltd.

Beer, S. (1994). *Beyond Dispute: The Invention of Team Syntegrity*. Chichester, UK: John Wiley and Sons.

Behn, R. D.(1998). O novo paradigma da gestão pública e a busca da accountability democrática. Revista do Serviço Público, Brasília, DF, 49*(4).*

Bekkers, V. J. J. M., & Korteland, E. (2005). *Governance, ICT and the Modernization Agenda of Public Administration: a Comparison of some European Policy Initiatives*. Paper presented at European Group of Public Administration Annual Conference, 1Bern, Zwitserland.

Bellamy, Ch., & Taylor, J. A. (1998). *Governing in the Information Age*. Buckingham, UK: Open University Press.

Bijker, E., Hughes, T., & Pinch, T. (Eds.). (1987). *The Social Construction of Technological Systems: New Directions in the Sociology and History of Technology*. Cambridge, MA: MIT Press.

Billiets, M., et al. (2006). D2.2 – Legislative, Institutional and EU Policy Related Requirements. *IST STREP PROJECT SemanticGov*. Retrieved from http://www.semantic-gov.org/

Bloomfield, B., & Coombs, R. (1992). Information technology, control and power: The centralization and decentralization debate revisited. *Journal of Management Studies*, *29*(4), 459–484. doi:10.1111/j.1467-6486.1992.tb00674.x

Blumler, J., & Kavanagh, D. (1999). The Third Age of Political Communication: Influences and Features. *Political Communication*, *16*, 209–230. doi:10.1080/105846099198596

Blyth, M., & Katz, R. (2005). From Catch-all Politics to Cartelisation: The Political Economy of the Cartel Party. *West European Politics*, *28*(1), 33–60. doi:10.1080/0140238042000297080

Boehm, B. (1988). A Spiral Model of Software Development and Enhancement. In *IEEE Computer*, 21(5), 61-72.

Boisier, S. (2001). Sociedad del conocimiento, conocimiento social y gestión territorial. *Interacoes . Revista Internacional de Desenvolvimento Local*, *2*(3), 9–28.

Bolton, D., Jones, S., Till, D., Furber, D., & Green, S. (1994). Using domain knowledge in requirements capture and formal specification construction. In M. Jirotka & J. Goguen (Eds.), *Requirements Engineering: social and technical issues* (pp. 141-162). Amsterdam: Academic Press.

Bonina, C. M., & Lepore, W. (2005). Desafíos y oportunidades del gobierno electrónico en el marco de la nueva gestión pública: el caso de Argentina. In *Proceedings of the X Congreso Internacional del CLAD sobre la Reforma del Estado y de la Administración Pública*, Santiago, Chile.

Borja, J. (2002). La ciudad y la nueva ciudadanía. *La Factoría,* N° 17. Retrieved January 15, 2009, from http://www.lafactoriaweb.com/articulos/borja17.htm

Boston, J., et al. (1991). *Reshaping the State*. Oxford, UK: Oxford University Press.

Bowling, D., & Espejo, R. (1992). An intervention with the Cybernetic Methodology in Regent Engineering. *Transactions of the Institute of Measurement and Control*, *14*(1), 17–28. doi:10.1177/014233129201400104

Boyd, D., & Ellison, N. (2007). Social network sites: Definition, history, and scholarship. *Journal of Computer-Mediated Communication, 13*(1), Article 11. Retrieved January 2009, from http://jcmc.indiana.edu/vol13/issue1/boyd.ellison.html

Brans, M., & Rossbach, S. (1997). The autopoiesis of administrative systems: Niklas Luhmann on public administration and public policy. *Public Administration*, (75): 417–439. doi:10.1111/1467-9299.00068

Brasil. (2001). *A política de governo eletrônico no Brasil* [online]. Comitê Executivo do Governo Eletrônico, Brasília. Retrieved from http://www.governoeletronico.gov.br/governoeletronico/index.html

Brasil. (2002). *2 anos de governo eletrônico: balanço de realizações e desafios futuro* [online]. Comitê Executivo do Governo Eletrônico, Brasília. Retrieved from http://www.governoeletronico.gov.br/arquivos/Balanco_2_Anos_Egov.pdf

Brasil. (2002). *Levantamento das ações de governo eletrônico* [online]. Comitê Executivo do Governo Eletrônico, Brasília. Retrieved from http://www.governoeletronico.gov.br/arquivos/Anexo_Balanco_Inventario.pdf

Bresciani, P., Perini, A., Giorgini, P., Giunchiglia, F., & Mylopoulos, J. (2004). Tropos: An Agent-Oriented Software Development Methodology. *Autonomous Agents and Multi-Agent Systems*, *8*, 203–236. doi:10.1023/B:AGNT.0000018806.20944.ef

Bresser Pereira, L. C. (2005). *Construção do Estado e Administração Pública: uma abordagem histórica.* Relatório do GV Pesquisa, Fundação Getulio Vargas de São Paulo – EAESP/FGV, 27.

Brewer, G., & Deleon, P. (1983). *The Foundations of Policy Analysis.* Monterey, CA: Brooks/Cole.

Brocklesby, J., & Cummings, S. (1996). Foucault plays Habermas: An alternative philosophical underpinning for critical systems thinking. *The Journal of the Operational Research Society*, *47*(6), 741–754.

Brown, J., & Duguid, P. (1999). *The Social Life of Information*. Cambridge, MA: Harvard University Press.

Bujanda, M. (2005). *Ciudadanía y deliberación democrática en la escuela primaria con apoyo de las tecnologías digitales. Una experiencia de investigación y desarrollo curricular*. Paper presented at La Educación y la formación de una ciudadanía democrática, Harvard Graduate School of Education and Fundación Arias para la Paz y el Progreso Humano, Costa Rica, San José.

Bula, G. (2004). Observations on the development of cybernetic ideas in Colombia: A tribute to Stafford Beer. *Kybernetes: The International Journal of Systems and Cybernetics*, *33*(3-4), 647–658. doi:10.1108/03684920410523625

Calista, D. J., & Melitski, J. (2007). E-Government and E-Governance: converging constructs of public sector information and communication technologies. *Public Administration Qualterly*, *31*(1), 87–120.

Cameron, M. (2007). Citizenship Deficits in Latin America Democracies. *Convergencia. Revista de Ciencias Sociales*, *14*(45), 11–30.

Capanegra Vallé, H. (2002). El Gobierno Electrónico: Hacia una verdadera Reforma del Estado. In *Proceedings of the XVI Concurso de Ensayos y Monografías del CLAD sobre Reforma del Estado y Modernización de la Administración Pública Gobierno Electrónico.*

Capano, G. (2003). Administrative Traditions and Policy Change: when Policy Paradigms Matter. The Case of Italian Administrative Reform during the 1990s. *Public Administration*, *81*, 781–801. doi:10.1111/j.0033-3298.2003.00371.x

Capgemini. (2006). *Architecture and the Integrated Architecture Framework*. Retrieved from http://www.capgemini.com/resources/thought_leadership/architecture_and_the_integrated_architecture_framework/

Capra, F. (1996). *The Web of Life: A New Synthesis of Mind and Matter*. London: Harper Collins.

Carayannis, E., Kwak, Y., & Anbari, F. (2003). *The Story of Managing Projects: an interdisciplinary approach.* Quorum Books.

Cardenal, A. S., & Batlle, A. (2006). La Utopía Virtual. Una Crítica al Cyberoptimismo desde la Teoría de Elección Racional. *IDP. Revista de Internet . Derecho y Política*, *3*, 1–12.

Cardoso, G. (2004). Ministers of Parliament and Information and Communication Technologies as a Means of Horizontal and Vertical Communication in Western Europe. *Information Polity*, *9*(1-2), 29–40.

Carroll, J. S. (1998). Organizational learning activities in high-hazard industries: The logics underlying self-analysis. *Journal of Management Studies*, *35*(6), 699–717. doi:10.1111/1467-6486.00116

Carter, L., & Bélanger, F. (2005). The utilization of e-government services: citizen trust, innovation and acceptance factors. *Information Systems Journal*, *15*(1), 5–25. doi:10.1111/j.1365-2575.2005.00183.x

Casero, A. (2007). *E-campaña 2007: Internet en las Elecciones Autonómicas Valencianas*. Paper presented at the VIII Congreso AECPA 2007, Valencia, España.

Casero, A. (n.d.). Propuesta de Metodología para el Análisis de las Webs de los Partidos Políticos. Unpublished working paper.

Castells, M. (1996). *The Information Age: Economy, Society and Culture*. Oxford, UK: Blackwell Publishers.

Castells, M. (1999). *La era de la información. Economía, Sociedad y Cultura*. Madrid, Spain: Alianza Editorial.

Castells, M. (1999). *La Era de la Información. Economía, Sociedad y Cultura. Volumen 2: El Poder de la Identidad*. Madrid, Spain: Siglo XXI.

Castells, M. (2001). *La Galaxia Internet*. Madrid, Spain: Plaza y Janés.

Castells, M. (2001). *The Internet Galaxy: Reflections on Internet, Business and Society*. Oxford: Oxford University Press.

Castells, M. (2002). *La dimensión cultural de internet□ Speech available at Debates Culturales de la UOC*. Retrieved January 20, 2009, from http://www.uoc.edu/culturaxxi/esp/articles/castells0502/castells0502.html

CEC. (1997). *Green paper on the convergence of the telecommunications, media and information technology sectors, and the implications for regulation: Towards an information society approach.* Brussels: Commission of the European Communities (CEC).

Cegarra, J. G. (2006). *Linking e-government with organisational memory through exploitation of knowledge.* In M. A. Sicilia (Ed.), *Competencies in Organizational E-Learning: Concepts and Tools.* Hershey, PA: Idea Group Publishing.

Cegarra, J. G., & Dewhurst, F. W. (2006). Linking shared organisational context and relational capital through unlearning: An initial empirical investigation in SMEs. *The Learning Organization, 13*(1), 49–62. doi:10.1108/09696470610639121

Cegarra, J. G., Dewhurst, F. W., & Briones-Peñalver, A. J. (2007). Factors affecting the use of e-government in the telecommunications industry in Spain. *Technovation, 27*(10), 595–604. doi:10.1016/j.technovation.2007.03.003

Census of India. (2001). New Delhi: Office of the Registrar General & Office Commissioner of India.

CEPAL. (2007) Libro Blanco de interoperabilidad de gobierno electrónico para América Latina y el Caribe. Versión 3.0, Septiembre.

Chadwick, A., & May, Ch. (2003). Interaction between States and Citizens in the Age of Internet: "e-Government" in the United States, Britain, and the European Union. *Governance: An International Journal of Policy and Administration, 16*(2), 271–300.

Chamber of Commerce of Tirana. (n.d.). *Albania Statistics.* Retrieved October 16, 2008, from http://www.cci.gov.al

Chapman, D. (1990). Planning for conjunctive goals. In J. Allen, J. Hendler, & A. Tate (Eds.), *Readings in Planning* (pp. 537-558). San Mateo, CA: Morgan Kaufmann Publishers, Inc.

Chapman, J. (2004). *System Failure: Why Governments Must Learn to Think Differently (2nd Ed.).* London: Demos.

Chappelet, J. L. (2004). The Appropriation of E-Mail and the Internet by Members of the Swiss Parliament. *Information Polity, 9*(1-2), 89–102.

Charbonneau, S. (2004). *Software Project Management: A Mapping between RUP and the PMBOK.* Rational Software. Retrieved fromhttp://www-128.ibm.com/developerworks/ rational/library/4721.html

Checci, R. M., Hsieh, J., & Straub, D. (2003). Public IT Policies in Less Development Countries: a critical assessment of the literature and a reference framework. *Journal of Global Information Technology Management, 6*, 45.

Checkland, P. (1981). *Systems Thinking, Systems Practice.* Chichester, UK: John Wiley & Sons.

Checkland, P., & Holwell, S. (1998). *Information, Systems and Information Systems: Making Sense of the Field.* Chichester, UK: John Wiley and Sons.

Checkland, P., & Poulter, J. (2006). *Learning for Action: A Short Definitive Account of Soft Systems Methodology and Its Use for Practitioners, Teachers and Students.* Chichester: John Wiley.

Checkland, P., & Scholes, J. (1990). *Soft Systems Methodology in Action.* Chichester, UK: John Wiley & Sons.

Chowdhury, R. (2005). *The Information Messiah: Issues & Debates within the NHS.* Working Papers in Social Sciences and Policy No. 15. Hull, UK: University of Hull.

Chowdhury, R., & Nobbs, A. (2008). Critical Insights into NHS Information systems Deployment. In D. Jemielniak & J. Kociatkiewicz (Eds.), *Management Practices in High Tech Environments* (pp.245-264). Hershey, PA: Idea Group Inc.

Chowdhury, R., Clarke, S., & Butler, R. (2007). Healthcare IT Project Failure: A Systems Perspective. In *Journal of Cases in Information Technology* (pp.1-15). Hershey, PA: Idea Group Inc.

Christopher, M. (2005). *Logistics and Supply Chain Management: Creating Value-Added Networks* (3rd ed.). Harlow, UK: Prentice Hall.

Churchman, C. W. (1968). *The Systems Approach*. New York: Delacorte Press.

Churchman, C. W. (1970). Operations research as a profession. *Management Science*, *17*, b37–b53.

Churchman, C. W. (1971). *The Design of Inquiring Systems, Basic Concepts of Systems and Organizations*. New York: Basic Books.

Churchman, C. W. (1979). *The Systems Approach and its Enemies*. New York: Basic Books.

Ciborra, C. (2001). *From Control to Drift: The Dynamics of Corporate Information Infrastructures*. Oxford: Oxford University Press.

Clarke, S. (2001). *Information Systems, Strategic Management: An integrated approach*. London: Routledge.

Clarke, S. (2007). *Information Systems Strategic Management: An Integrated Approach* (2nd ed.). London: Routledge.

Coakes, E., & Clarke, S. (2006). The Concept of Communities of Practice. In E. Coakes & S. Clarke (Eds.), *Encyclopedia of Communities of Practice in Information and Knowledge Management* (pp. 92-95). Hershey, PA: Idea Group Reference.

Cobb, R. W., & Elder, C. D. (1983). *Participation in American Politics: The Dynamics of Agenda-building*. Baltimore, MD: Johns Hopkins University Press.

CODESI. (2003). *Report of the Technical Committee for Development of Information Systems in the Generalitat Valenciana* (Memoria del Comité Técnico para el Desarrollo de los Sistemas de Información de la Administración de la Generalitat Valenciana. Año 2002). Valencia, Spain: Generalitat Valenciana.

Cohen, C., & Midgley, G. (1994). *The North Humberside Diversion From Custody Project for Mentally Disordered Offenders: Research Report*. Hull, England: Centre for Systems Studies.

Cohen, J., & Arato, A. (1992). *Civil Society and Political Theory*. Cambridge, MA: MIT Press.

Coleman, S., & Gotze, J. (2005). *Bowling together: Online public engagement in policy deliberation*. UK: Hansard Society.

Commission of the European Communities (CEC). (2004). *Framework to reinforce the exchange of good practices in e-government: A contribution to eEurope 2005*. Retrieved April 27, 2004, from http://europa.eu.int/information_society/programmes/egov_rd/doc/good_practices_framework_baseline.pdf

Contini, F. (2009). ICT, assemblages and institutional contexts: understanding multiple development paths. In F. Contini, & G. F. Lanzara (Eds.), *ICT and Innovation in the Public Sector* (pp. 244-271). London: Palgrave.

Conway, M., & Corner, D. (2004). An Evaluation of New Zealand Political Party Websites. *Information Research*, *9*(4).

Córdoba, J. R. (2002). *A Critical Systems Thinking Approach for the Planning of Information Technology in the Information Society*. Unpublished Doctoral dissertation, University of Hull, Hull, UK.

Córdoba, J. R. (2005). Communities and evaluation of e-government services. In S. Clarke & E. Coakes (Eds.), *Encyclopaedia of Communities of Practice in Information and Knowledge Management* (pp. 32-34). Hershey, PA: Idea Group Publishing.

Córdoba, J. R. (2006). Using Foucault to analyse ethics in the practice of problem structuring methods. *The Journal of the Operational Research Society*, *57*(9), 1027–1034. doi:10.1057/palgrave.jors.2602080

Córdoba, J. R., & Midgley, G. (2006). Broadening the boundaries: An application of critical systems thinking to IS planning in Colombia. *The Journal of the Operational Research Society*, *57*(9), 1064–1080. doi:10.1057/palgrave.jors.2602081

Córdoba, J. R., & Robson, W. D. (2003). Making the evaluation of information systems insightful: Understanding the role of power-ethics strategies. *Electronic Journal of Information Systems Evaluation*, *6*(2), 55–64.

Córdoba, J., & Midgley, G. (2008). Beyond organisational agendas: Using boundary critique to facilitate the inclusion of societal concerns in information systems planning. *European Journal of Information Systems*, *17*, 125–142. doi:10.1057/ejis.2008.4

Córdoba, J., & Robson, W. (2008). Power and Ethics in IS Evaluation. In D. Jemielniak & J. Kociatkiewicz (Eds.), *Management Practices in High Tech Environments* (Vol. forthcoming). Hershey, PA: Idea Group.

Costa, E. (2001). *Global E-commerce Strategies for Small Businesses*. Cambridge, MA: MIT Press.

Criado, J. I. (2006). *Regional Electronic Public Administration and Modernisation Agenda in Spain*. Paper presented at European Group of Public Administration Annual Conference, Milano, Italy.

Criado, J. I., & Ramiló, M. C. (2003). E-Government in Practice. An Analysis of Web Site Orientation to the Citizen in Spanish Municipalities. *International Journal of Public Sector Management*, *16*(3), 191–218. doi:10.1108/09513550310472320

Criado, J. I., & Ramilo, M. C. (2007). Local Public Administration in the Age of the Web. The Case of Spain. In P. Trevorrow, D. Griffin, & E. Halpin (Eds.), *Developments in e-Government. A Critical Analysis* (pp. 49-72). Amsterdam: IOS Press.

Criado, J. I., Hughes, O. E., & Teicher, J. (2002). *e-Government in Australia and Spain: a Study of Contrasts and Similarities*. Paper presented at VI International Research Symposium on Public Management, Edinburgh, UK.

Criado, J., & Ramilo, M. C. (2003). E-government in practice: An analysis of web site orientation to the citizens in Spanish municipalities. *International Journal of Public Sector Management*, *16*(3), 191–218. doi:10.1108/09513550310472320

Dahlgren, P. (2005). The Internet, Public Spheres, and Political Communication: Dispersion and Deliberation. *Political Communication*, *22*, 147–162. doi:10.1080/10584600590933160

Danziger, J. N., & Andersen, K. V. (2002). Impacts of Information Technology on Public Administration: an Analysis of Empirical Research from the "Golden Age" of Transformation. *International Journal of Public Administration*, *25*(5), 591–627. doi:10.1081/PAD-120003292

Danziger, J. N., Dutton, W. H., Kling, R., & Kraemer, K. L. (Eds.). (1982). *Computers and Politics. High Techology in American Local Governments*. New York: Columbia University Press.

Darier, E. (1999). Foucault and the environment: An introduction. In E. Darier (Ed.), *Discourses of the Environment* (pp. 1-33). Oxford, UK: Blackwell.

Das, S. R., & Chandrashekhar, R. (2006). *Capacity Building for E-Governance in India*. UNDP Asia-Pacific Development Information Programme. Retrieved July 21st, 2008 from, http://www.apdip.net/projects/e-government/capblg/casestudies/India-Chandrashekhar.pdf

Dasgupta, M. (2005). *Public Health in India: An Overview*. World Bank Policy Research Working Paper 3787, New Delhi. Retrieved September 13th 2006, from http://www-wds.worldbank.org/servlet/WDSContentServer/WDSP/IB/2005/12/02/000016406_20051202162921/Rendered/PDF/wps3787.pdf

Davenport, T. H., & Grover, V. (2001). General perspectives on knowledge management: Fostering a research agenda. *Journal of Management Information Systems*, *18*(1), 5–21.

David, P. (1986). Understanding the Economics of QWERTY: the Necessity of History. In W. Parker (Ed.), *Economic History and the Modern Historian* (pp. 30-49). London: Blackwell.

Davis, F. (1989). Perceived usefulness, perceived ease of use, and user acceptance of information technology. *MIS Quarterly*, *13*(3), 319–340. doi:10.2307/249008

De Bono, E. (2004). *How to Have a Beautiful Mind*. London: Vermillion.

Delamaza, G. (2009). Participation and *Mestizaje* of State-Civil Society in Chile. In P. Silva & H. Cleuren (Eds.), *Widening Democracy. Citizens and Participatory Schemes in Brazil and Chile* (pp. 249-271). Leiden, The Netherlands: Brill.

Diniz, V. (2005). *A história do uso da tecnologia da informação na gestão pública brasileira através do CONIP*. Paper presented at X Congreso Internacional del CLAD sobre la Reforma del Estado y de la Administración Pública, Santiago, Chile.

Document Object Identifier System. (n.d.). Retrieved from http://www.doi.org/

Domingo, C., Ramírez, V., Velásquez, A., & Zavarce, H. (2006). Use of Enterprises Data in Economic Planning. In *Proceedings of the Modelling, Simulation, and Optimization. MSO 2006*. Calgary, Canada: Acta Press.

Donner, J., Gandhi, R., Javid, P., Medhi, I., Ratan, A., Toyama, K., & Veeraraghavan, R. (2008). Stages of Design in Technology for Global Development. *IEEE Computer*, *41*(6), 34–41.

Dooley, K. (1997). A complex adaptive systems model of organizational change. *Nonlinear Dynamics Psychology and Life Sciences*, *1*(1), 69–97. doi:10.1023/A:1022375910940

Doolin, B. (2004). Power and resistance in the implementation of a medical management information system. *Information Systems Journal*, *14*(4), 343–362. doi:10.1111/j.1365-2575.2004.00176.x

Doyle, J. R. (1991). Problems with Strategic Information Systems frameworks. *European Journal of Information Systems*, *14*(1), 273–280. doi:10.1057/ejis.1991.48

Drake, P., & Jäksic, I. (1999). Introducción. In P. Drake & I. Jäksic (Eds.), *El Modelo Chileno. Democracia y Desarrollo en los Noventa* (pp. 7-38). Santiago, Chile: LOM.

Dunleavy, P., & Margetts, H. (2007). *Government on the Internet. Progress in Delivering Information and Services Online*. London: National Audit Office.

Dunleavy, P., Margetts, H., Bastow, S. H., & Tinkler, J. (2006). *Digital Era Governance: IT Corporations, the State and E-government*. Oxford, UK: Oxford University Press.

Dunleavy, P., Margetts, H., Bastow, S., & Tinkler, J. (2008, March). Australian e-Government in comparative perspective. *Australian Journal of Political Science, 43*.

Dunsire, A. (1996). Tipping the balance: Autopoiesis and governance. *Administration & Society*, *28*(3), 299–334. doi:10.1177/009539979602800302

Dutton, W. H., & Kraemer, K. L. (1978). Management Utilization of Computers in American Local Governments. *Communications of the ACM*, *21*(3), 206–218. doi:10.1145/359361.359364

Dutton, W., Blumler, J., Garnham, N., Mansell, R., Cornford, J., & Peltu, M. (1996). The politics of information and communication policy: The information superhighway. In W. Dutton & M. Peltu (Eds.), *Information and Communication Technologies: Visions and Realities*. (pp. 387-405). Oxford, UK: Oxford University Press.

Eastin, M., & LaRose, R. (2000). Internet self-efficacy and the psychology of the Digital Divide. *Journal of Computer-Mediated Computation*, *6*(1). Retrieved January 2009, from http://jcmc.indiana.edu/vol6/issue1/eastin.html

Edvardsson, I. (2006). Knowledge management in SMEs: the case of Icelandic firms. *Knowledge Management Research and Practice*, *4*(4), 275–282. doi:10.1057/palgrave.kmrp.8500111

Eemeren, F. H. v. (2003). Anyone who has a view: theoretical contributions to the study of argumentation. In *Argumentation library* (Vol. 8, pp. 92). Dordrecht, The Netherlands: Kluwer Academic.

Elliott, M. S., & Scacchi, W. (2003). Free software developers as an occupational community: resolving conflicts and fostering collaboration. In *Proceedings of the 2003 ACM SIGGROUP conference on supporting groupwork* (pp. 21-30). New York: ACM Press.

ENCAT. (2006). *Manual de Integração do Contribuinte: padrões técnicos de comunicação. Encontro Nacional dos Coordenadores e Administradores Tributários Estaduais* – ENCAT. Available at www.nfe.fazenda.gov.br.

Espejo, R. (2002). Systems and the Information Society: Requisite Organisation and Problem Solving. *Systems and the Information Society Network*. Retrieved November 2004 from http:\\www.itsy.co.uk

Espejo, R., & Harnden, R. (Eds.). (1989). Outside and then: An interpretative approach to Viable System Model. In *The Viable System Model: Interpretations and Applications of Stafford Beer's Viable System Model*. Chichester, UK: John Wiley & Sons.

Espejo, R., & Schwaninger, M. (Eds.). (1993). *Organisational Fitness: Corporate Effectiveness through Management Cybernetics*, Frankfurt and New York: Campus Verlag.

Espejo, R., Schumann, W., Schwaninger, M., & Billelo, U. (1996). *Organisational Transformation and Learning*. Chichester, UK: John Wiley & Sons.

Espinosa, A. (1995). *Strategic Information Systems at the Colombian President's Office; A Managerial Cybernetics' Perspective*. Doctoral Thesis, Aston Business School, Aston University, Birmingham, England, U.K.

Espinosa, A. (2007). A cybernetic re-evaluation of socio-economic development programs. *Kybernetes*, *35*(1/2), 30–44. doi:10.1108/03684920610640218

Espinosa, A., & Walker, J. (2005). Environmental Management revisited. Lessons from a cybernetic intervention in Colombia. *Cybernetics and System. International Journal (Toronto, Ont.)*, *36*(7), 573–589.

Espinosa, A., Harnden, R., & Walker, J. (2007). Beyond Hierarchy: Stafford Beer revisited. Invited paper. [Special issue]. *Kybernetes: The International Journal of Systems and Cybernetics*, *36*(3/4), 333–347. doi:10.1108/03684920710746995

Espinosa, A., Harnden, R., & Walker, J. (2008). A Complexity Approach to Sustainability: Stafford Beer revisited. *European Journal of Operational Research*, *187*, 636–651. doi:10.1016/j.ejor.2007.03.023

Espinoza, V. (2009). Citizens' Involvement and Social Policies in Chile: Patronage or Participation? In P. Silva & H. Cleuren (Eds.), *Widening Democracy. Citizens and Participatory Schemes in Brazil and Chile* (pp. 273-294). Leiden, The Netherlands: Brill.

European Commission. (2005). *Insights for the European e-Government Community*. e-Government Observatory. Retrieved from http://europa.eu.int/idabc/en/chapter/140

European Commission. (2007). *Information Society and Government: Linking European Policies*. Brussels, Belgium: European Commission. eSEEurope. (n.d.). *Agenda for the Development of the Information Society, Stability Pact – A cooperative effort to implement the Information Society in South Eastern Europe*. Retrieved March 21, 2009, from http://www.eseeuropeconference.org/agenda.pdf

Fahey, D. K., Carson, E. R., Cramp, D. G., & Muir Gray, J. A. (2004). Applying systems modelling to public health. *Systems Research and Behavioral Science*, *21*, 635–649. doi:10.1002/sres.602

Farrell, D., & Beinhocker, E. (2007). Next Big Spenders: India's Middle Class. In *McKinsey Global Institute Business Week*. Retrieved on January 5, 2008, from http://www.mckinsey.com/mgi/mginews/bigspenders.asp

Feenberg, A. (1992). Subversive Rationalization: technology, Power and Democracy. *Inquiry*, *35*(3/4).

Flood, R. L., & Jackson, M. C. (1991). *Creative Problem Solving: Total Systems Intervention*. Chichester: Wiley.

Folgueiras, P. (2005). *De la tolerancia al reconocimiento: programa de formación para una ciudadanía intercultural*. Unpublished doctoral dissertation, Universidad de Barcelona, Facultad de Pedagogía.

Force, D. O. T. (2000). *Okinawa Charter on Global Information Society, Declaration of the G8*. Retrieved May 2008, from http://www.mofa.go.jp/policy/economy/summit/2000/documents/charter.html

Foucault, M. (1977). *The History of Sexuality Volume One: The Will to Knowledge* (Vol. 1). London: Penguin.

Foucault, M. (1984a). The ethics of the concern of the self as a practice of freedom (R. e. a. Hurley, Trans.). In P. Rabinow (Ed.), *Michel Foucault: Ethics Subjectivity and Truth: Essential Works of Foucault 1954-1984* (pp. 281-301). London: Penguin.

Foucault, M. (1984b). What is enlightenment? (C. Porter, Trans.). In P. Rabinow (Ed.), *The Foucault Reader: An Introduction to Foucault's Thought* (pp. 32-50). London: Penguin.

Foucault, M. (1991). Governmentality. In G. Burchell, C. Gordon & P. Miller (Eds.), *The Foucault Effect* (pp. 87-104). London: Harvester & Wheatsheaf.

Fountain, J. (2001). *Building the Virtual State – Information Technology and Institutional Change*. Washington, DC: Brooking Institution Press.

Frey, K. (2000). Políticas públicas: um debate conceitual e reflexões referentes à prática da análise de políticas públicas no Brasil. [IPEA]. *Planejamento e Políticas Públicas*, *21*, 211–259.

Frissen, P. H. A., & Snellen, I. Th. M. (Eds.). (1990). *Informatization Strategies in Public Administration*. Amsterdam: Elsevier Science Publishers.

Fuenmayor, R. (2001). Venezuela: Su enfermedad y Crisis Actual. In F. Parra (Ed.), *Estudios de Derecho Público: Libro Homenaje a Humberto J. La Roche Rincón* (pp. 335-358).

Fuentes, C. (1999). Partidos y Coaliciones en el Chile de los '90. Entre Pactos y Proyectos. In P. Drake & I. Jäksic (Eds.), *El Modelo Chileno. Democracia y Desarrollo en los Noventa* (pp. 191-220). Santiago, Chile: LOM.

Galliers, R. (1999). Towards the integration of e-business, knowledge management and policy considerations within an information systems strategy framework. *The Journal of Strategic Information Systems*, *8*(3), 229–234. doi:10.1016/S0963-8687(00)00023-8

Galliers, R. D. (1991). Strategic Information Systems Planning: myths, reality and guidelines for successful implementation. *European Journal of Information Systems*, *1*(1), 55–64. doi:10.1057/ejis.1991.7

Galliers, R. D., & Baets, W. R. J. (1998). *Reflections on BPR, IT and Organizational Change, in Information technology and organisational transformation*. Chichester, UK: John Wiley & Sons.

Garai, A., & Shadrach, B. (2006). Processes and Appropriation of ICT in Human Development. In *Rural India: Bridging the Research and Practice Gaps*. Retrieved on August 15, 2008, from http://www.dgroups.org/groups/oneworld/OneWorldSA/docs/TICTEIV_pdf.pdf

Gartner. (2003). New Performance Framework Measures Public Value of IT. *Research Note, 8.*

Gil-García, J. R. (2004). Information Technology Policies and Standards: a Comparative Review of the States. *Journal of Government Information*, *30*(5-6), 548–560. doi:10.1016/j.jgi.2004.10.001

Gil-Garcia, J. R., & PARDO. (2005). T. A. *E-Government success factors: Mapping practical tools to theoretical foundations. Government Information Quarterly*, *22*(2), 187–216. doi:10.1016/j.giq.2005.02.001

Gil-Garcia, J., & Pardo, T. (2005). e-Government Sucess Factors: mapping practical tools to theoretical foundations. *Government Information Quarterly*, *22*, 187–216. doi:10.1016/j.giq.2005.02.001

Glassey, O. (2008). A case study on process modelling—Three questions and three techniques. *Decision Support Systems*, *44*, 842–853. doi:10.1016/j.dss.2007.10.004

Gomez, P., & Probst, G. J. B. (1989). Organizational closure in management: A complementary view to contingency approaches. *Cybernetics and Systems*, *20*, 311–320. doi:10.1080/01969728908902210

González, R. (2005). Identidad y Actitudes Políticas en Jóvenes Universitarios: El Desencanto de los que no se Identifican Políticamente. *Revista de Ciencia Política*, *25*(2), 65–90.

Gonzalo, P. (2008). *Los ataques de Mumbai, en la India, consolidan el periodismo ciudadano.* Retrieved from http://www.periodismociudadano.com/2008/11/28/los-ataques-de-mumbai-en-la-india-consolidan-la-eficacia-del-periodismo-ciudadano/

Gordon, C. (1991). Governmental Rationality: An Introduction. In G. Burchell, C. Gordon & P. Miller (Eds.), *The Foucault Effect* (pp. 1-72). London: Harvester & Wheatsheaf.

Gordon, T. F. (2002, January). E-Government – Introduction. *ERCIM News No.48*. Retrieved November 7, 2008, from http://www.ercim.org/publication/Ercim_News/enw48/intro.html

Gore, A. (1991, July 1991). A national vision. *Byte*, 188-189.

GOVERNO FEDERAL. (2007). *PAC: Programa de Aceleração do Crescimento*. Sítio Institucional. Retrieved fromhttp://www.brasil.gov.br/pac

Grabher, D., Schere, R., & Walser, M. (2000). One region, three networks: Processes of self organization as a strategy for sustainable development. In *Proceedings of the 40th European Congress of the Regional Science*, Barcelona, Spain, Aug-Sept.

Greek Administrative Process Code. (1999). *Law 2690/1999*.

Greenfield, P. M., & Subrahmanyam, K. (2003). Online discourse in a teen chatroom: New codes and new modes of coherence in a visual medium. *Applied Developmental Psychology*, *24*(6), 713–738. doi:10.1016/j.appdev.2003.09.005

Grefter, A. (2006, September 16). This is your space. *New Scientist*, 46–48. doi:10.1016/S0262 4079(06)60500-9

Gregory, A., & Jackson, M. C. (1992). Evaluation methodologies: A system for use. *The Journal of the Operational Research Society*, *43*, 19–28.

Grest, D. W., Taylor, M. J., & Lunn, J. (2005). Legal consideration in cyber-activism. *Systems Research and Behavioral Science*, *22*(6), 565–570. doi:10.1002/sres.625

Grimsley, M., & Meehan, A. (2007). e-Government information systems: evaluation led design for public value and client trust. *European Journal of Information Systems*, *16*(2), 134–148. doi:10.1057/palgrave.ejis.3000674

GVA. (1996). *Strategic Action Plan for Valencian Administration Modernization* (Ejes del Plan Estratégico de Modernización de la Administración Valenciana (PEMAV)). Valencia: Generalitat Valenciana.

GVA. (2000). *Moderniza.com. 2nd Action for Valencian Administration Modernization.* (Plan Moderniza.com. 2° Plan de Modernización de la Comunidad Valenciana. La Administración de Todos). Valencia: Generalitat Valenciana.

GVA. (2004). *Avantic 2004-2010*. Valencia: Generalitat Valenciana.

Habermas, J. (1972). *Knowledge and Human Interests*. London: Polity Press.

Habermas, J. (1983). *The Theory of Communicative Action*. Boston, MA: Polity Press.

Hamel, G. P., & Prahalad, C. K. (1994). *Competing for the Future*. Cambridge, MA: Harvard Business School Press.

Hammer, M., & Champy, J. (1995). *Re-engineering the Corporation: A Manifesto for Business Revolution*. London: Nicholas Brealey.

Handle System Architecture. (n.d.). Corporation for National Research Initiatives. Retrieved from http://www.handle.net/overviews/architecture.html

Hannan, M., & Freeman, J. (1984). Structural inertia and organizational change. *American Sociological Review*, *49*, 149–164. doi:10.2307/2095567

Hansen, K., Pedersen, K., & Wahl-Jorgensen, D. (2005). *Cyber-campaigning. The Characters, Development and Consequences of parties' Electioneering in Cyber-space*. Paper presented at the Nordic Political Science Association's conference, Reykjavik, Iceland.

Hawkins, R. (1995). Enhancing the user role in the development of technical standards for telecommunications. *Technology Analysis and Strategic Management*, *1*, 21–40. doi:10.1080/09537329508524193

Hazlett, S., & Hill, F. (2003). E-Government: the realities of using IT to transform the public sector. *Managing Service Quality*, *13*(6), 445–452. doi:10.1108/09604520310506504

Heeks, R. (1998), Information Age Reform of the Public Sector: The Potential and Problems of IT for India', In

Information Systems for Public Sector Management, working Paper Series, Paper no, 6.

Heeks, R. (1999). Reinventing Government in the Information Age. In R. Heeks (Ed.), *Reinventing Government in the Information Age. International Practice in IT-Enable Public Sector Reform* (pp. 9-21). London: Routledge.

Heeks, R. (2002). Information Systems and Developing Countries; Failure, Success, and Local Improvisations. *The Information Society, 18*, 101–112. doi:10.1080/01972240290075039

Heeks, R. (2003). *Most eGovernment-for-Development Projects Fail: How Can Risks be Reduced?* IDPM i-Government Working Paper no.14, University of Manchester, UK.

Heeks, R. (2004). *eGovernment as a Carrier of Context*. Institute for Development Policy and Management, iGovernment Series, Working paper, no. 15. Retrieved fromhttp://idpm.man.ac.uk/publications/wp/igov/index.shtml

Heeks, R. (2005). e-government as a carrier of context. *Journal of Public Policy, 25*(1), 51–74. doi:10.1017/S0143814X05000206

Heeks, R. (2006). *Implementing and Managing e-Government. An International Text*. London: Sage.

Heeks, R., & Bailur, S. (2007). Analyzing e-government research: perspectives, philosophies, theories, methods and practice. *Government Information Quarterly, 24*, 243–265. doi:10.1016/j.giq.2006.06.005

Heeks, R., & Stanforth, C. (2007). Understanding e-Government project trajectories from an actor-network perspective. *European Journal of Information Systems, 16*(2), 165–177. doi:10.1057/palgrave.ejis.3000676

Heichlinger, A. (2004). *eGovernment in Europe Regions. Approaches and Progress in IST Strategy, Organisation and Services, and the Role of Regional Actors*. Maastricht, The Netherlands: European Institute of Public Administration.

Henderson, J., & Venkatraman, N. (1999). Strategic alignment: Leveraging information technology for transforming organisations. *IBM Systems Journal, 38*(2&3), 472–484.

Her Majesty's Stationery Office. (2000). *Statutory Instrument No 896: The Local Government (best value) Performance Indicators Order*. London: Crown.

Hernes, T., & Bakken, T. (2003). Implications of self-reference: Niklas Luhmann's autopoiesis and organization theory. *Organization Studies, 24*(9), 1511–1535. doi:10.1177/0170840603249007

Herson, P., Kai Stokes-Brown, A., & Hindan, M. (2007). Campaign Politics and the Digital Divide. Constituency Characteristics, Strategic Considerations and Candidate Internet Use in StateLegislative Elections. *Political Research Quarterly, 60*(1), 31–42. doi:10.1177/1065912906298527

Hilbert, M. R. (2001). *Latin America on its path into the digital age: where are we?* Division of Production, Productivity and Management of the Economic Commission for Latin America and the Caribbean of the United Nations, ECLAC.

Ho, T.-K. (2002). Reinventing local governments and the e-government initiative. *Public Administration Review, 62*(4), 434–444. doi:10.1111/0033-3352.00197

Hodgson, D., & Cicmil, S. (2006). Making projects critical: An introduction. In D. Hodgson & S. Cicmil (Eds.), *Making Projects Critical* (pp. 1-25). London: Palgrave.

Holmes, D. (2002). *e.gov – eBusiness Strategies for Government*. London: Nicholas Brealey Publishing.

Holmes, D. (2002, May 1). Marketing e-Government to a skeptical public. *CIO Governments' Review*. Retrieved December 12, 2008, from http://www.itworldcanada.com/a/CIO-Gov.-Review/d118b39c-29b1-49b3-a98c-656626ce4946.html

Hood, Ch. (1991). A Public Management for All Seasons? *Public Administration, 69*(1), 3–19. doi:10.1111/j.1467-9299.1991.tb00779.x

Hood, Ch. (1995). Emerging Issues in Public Administration. *Public Administration, 73*(2), 165–183. doi:10.1111/j.1467-9299.1995.tb00822.x

Hughes, O. E. (2003). *Public Management and Administration* (3rd ed.). London: Palgrave Macmillan.

Human Development Report (2007/2008). *United Nations Development Program*. New York.

Information Society Commission. (1999). *Information society Ireland: Discussion document on IT access for all*. Dublin: Information Society Commission.

Insigh, B. (n.d.). *IMF slashes Albania growth for2009*. Retrieved April 13, 2009, from http://balkaninsight.com/en/main/news/17998/

Introna, L. D. (1997). *Management, Information and Power: A Narrative of the Involved Manager*. Basingstoke: Macmillan.

Introna, L. D. (2003). Disciplining information systems: Truth and its regimes. *European Journal of Information Systems, 12*, 235–240. doi:10.1057/palgrave.ejis.3000465

Irani, Z., Love, P. E., Elliman, T., Jones, S., & Themistocleus, M. (2005). Evaluating E-government: Learning from the experiences of two UK local authorities. *Information Systems Journal, 15*(1), 61–82. doi:10.1111/j.1365-2575.2005.00186.x

ISC. (1996). *Information Society Ireland: Strategy for Action*. Dublin: Information Society Commission Ireland.

ITA. (2002). *Summary report of Information Technology Authority (ITA): Towards Digital Oman*. Oman.

Jackson, M. C. (1995). Beyond the Fads: Systems Thinking for Managers. *Systems Research, 12*(1), 25–42. doi:10.1002/sres.3850120106

Jackson, M. C. (2003). *Creative Holism: Systems Thinking for Managers*. Chichester, UK: John Wiley and Sons.

Jackson, M. C. (2003). *Systems Thinking: Creative Holism for Managers*. Chichester, UK: Wiley.

Jackson, M. C. (2005). Reflections on knowledge management from a critical systems perspective. *Knowledge Management Research and Practice, 3*(4), 187–196. doi:10.1057/palgrave.kmrp.8500067

Jackson, M. C., & Keys, P. (1984). Towards a System of Systems Methodologies. *The Journal of the Operational Research Society, 35*(6), 473–486.

Jacobson, D. (1996). Contexts and Cues in Cyberspace: The Pragmatics of Naming in Text-Based Virtual Realities. *Journal of Anthropological Research, 52*(44), 461–479.

Jae-Moon, M. (2002). The evolution of e-government among municipalities: rethoric or reality? *Public Administration Review, 62*(4), 424–433. doi:10.1111/0033-3352.00196

Janis, I. (1982). *Victims of Groupthink* (2nd ed.). Boston, MA: Houghton Mifflin.

Janssen, M., & Hjort-Madsen, K. (2007). Analyzing Enterprise Architecture in National Governments: The cases of Denmark and the Netherlands. In *Proceedings of the 40th Hawaii International Conference on System Sciences*.

Japanese Government (2000). *First meeting of the digital opportunity task force (dotforce)*. Okinawa: G8.

Jeffcoate, J., Chappell, C., & Feindt, S. (2000). Attitudes towards process improvement among SMEs involved in e-commerce. *Knowledge and Process Management, 7*(3), 187–195. doi:10.1002/1099-1441(200007/09)7:3<187::AID-KPM86>3.0.CO;2-6

Jones, C. (1970). *An Introduction to the Study of Public Policy*. Belmont, CA: Wadsworth.

Jordan, G., & Schubert, K. (1992). A Preliminar/ Ordering of Policy Network Labels. *European Journal of Political Research, 21*(1-2), 9–10. doi:10.1111/j.1475-6765.1992.tb00286.x

Kaler, J. (2003). Differentiating Stakeholder Theories. *Journal of Business Ethics Dordrecht, 46*(1), 71–83. doi:10.1023/A:1024794710899

Kanungo, V. (2004). *Citizen Centric e-Governance in India: Strategies for Today, Vision for Future*. New Delhi: Society for Promotion of E-Governance. Retrieved June 22, 2008, from http://en.wikipedia.org/wiki/Languages_of_India

Katz, R., & Mair, P. (1995). Changing Models of Party Organization and Party Democracy. The Emergence of Cartel Party. *Party Politics*, *1*(1), 5–28. doi:10.1177/1354068895001001001

Keating, M. (1985). Regional Autonomy in the Changing State Order: A framework for analysis. *Regional Politics and Policy*, *2*(3), 111–132.

Keating, M., & Rhodes, R.A.W. (1979). Is there a regional level of government in England? *University of Strathclyde Studies in Public Policy* (No. 49).

Kettinger, W. J., & Grover, V. (1997). The use of Computer-mediated Communication in an Interorganizational Context. *Decision Sciences*, *28*(3), 513–555. doi:10.1111/j.1540-5915.1997.tb01321.x

Kettl, D. F. (2002). The Transformation of Governance. Baltimore, MD: John Hopkins University Press

King, S. F., & Johnson, O. A. (2006). VBP: An approach to modelling process variety and best practice. *Information and Software Technology*, *48*, 1104–1114. doi:10.1016/j.infsof.2006.02.003

Kingdon, J. W. (1995). *Agendas, Alternatives, and Public Policies* (2nd ed.). New York: Harper Collins.

Klijn, E. H. (1995). Policy Networks: An Overview. In W. J. M. Kickert & J. F. Koppenjan (Eds.), *Managing Complex Networks*. London: Sage.

Klijn, E. H. (1996). Analyzing and Managing Policy Processes in Complex Networks. *Administration & Society*, *28*, 90–119. doi:10.1177/009539979602800104

Klitgaard, R., & Light, P. C. (2005). *High-performance Government – Structure, Leadership, Incentives*. New York: RAND Corporation.

Koppenjan, J. F. M., & Klijn, E. H. (2004). *Managing Uncertainties in Networks*. London: Routledge.

Kraemer, K. L., & King, J. L. (2003). *Information Technology and Administrative Reform: Will the Time After e-Government Be Different?* Paper presented at Heinrich Reinermann Schrift Fest, Post Graduate School of Administration, Speyer.

Lasswell, H. D. (1951). The Policy Orientation. In D. Lerner & H. D. Lasswell (Eds.), *The Policy Sciences*. Stanford, CA: Stanford University Press.

Lau, E. (2005, July). *eGovernment Economics Project - eGovernment Measurement Agenda*. Brussels, Belgium: OECD eGovernment Project.

Lau, E. (2006, February 6). *OECD E-government Project: Proposal for work on Inventory of E-government Business Case Indicators* (Doc. Nr. GOV/PGC/EGOV). Public Governance and Territorial Development Directorate - Public Governance Committee.

Layne, K., & Lee, J. (2001). Developing fully functional E-government: A four stage model. *Government Information Quarterly*, *18*, 122–136. doi:10.1016/S0740-624X(01)00066-1

Lee, F. S. L., Vogel, D., & Limayem, M. (2003). Virtual community informatics: A review and research agenda. *Journal of Information Technology Theory and Application*, *5*(1), 47–61.

Leischow, S. J., & Milstein, B. (2006). Systems Thinking and Modeling for Public Health Practice. In *MPH . American Journal of Public Health*, *96*(3), 403–405. doi:10.2105/AJPH.2005.082842

Leitner, Ch. (2003). *eGovernment in Europe: the State of Affairs*. Maastricht, The Netherlands: European Institute of Public Administration.

Leitner, Ch., & Criado, J. I. (2006). eGovernment in Europe: Towars a Focus Based on Benchlearning (El Desarrollo de la e-Administración Europea. Hacia un Enfoque Basado en el Benchlearning). In L. Cotino (Ed.), *Freedom, Democracy and Electronic Government* (Libertades, Democracia y Gobierno Electrónicos) (pp. 263-299). Grenada: Comares.

Lenk, K., & Traunmüller, R. (2002). *Electronic Government: Where Are We Heading?* Paper presented at Electronic Government, I International Conference, EGOV 2002, Aix-en-Provence, France.

Leonard, A. (1999 August). A Viable System Model: Consideration of Knowledge Management. *Journal of Knowledge Management Practice*.

Levinthal, D. A., & March, J. G. (1993). The myopia of learning. *Strategic Management Journal, 14*, 95–112. doi:10.1002/smj.4250141009

Lindblom, C. E., & Cohen, D. K. (1979). Usable knowledge: Social science and social problem solving New Haven, CT: Yale University Press.

Locke, S. (2006). E-local government strategies and small businesses. *The Journal of American Academy of Business, 10*(1), 21–30.

Löfgren, K. (2003). *Intraparty Use of New ICTs – Bringing Party Memberships Back In?* Paper presented at the ECPR 2003 joint sessions, Edinburgh, Scotland.

López, A. (1998). *La Nueva Gestión Pública: Algunas Precisiones para su Abordaje Conceptual.* INAP. Serie I: Desarrollo Institucional y Reforma del Estado. Documento Nr. 68.

López, J., & Leal, I. (2002). *e-Gobierno. Gobernar en la sociedad del Conocimiento.* Vitoria, España: Instituto Vasco de la Administración Pública.

Loureiro, C. (2008). A theoretical framework for a structuration model of social issues in software development in information systems. *Systems Research and Behavioral Science, 25*(1), 99–109. doi:10.1002/sres.868

Lozada, M. (2004). El ciberciudadano: representaciones, redes y resistencias en Venezuela y América Latina. In D. Mato (Ed.), *Políticas de ciudadanía y sociedad civil en tiempos de globalización* (pp. 167-180). Caracas, FACES, UCV.

Luhmann, N. (1995). *Social Systems.* (J. Bednarz, Jr. & D. Baecker, Trans.). Stanford, CA: Stanford University Press.

Luna, J. P. (2008). Partidos Políticos y Sociedad en Chile. Trayectoria Histórica y Mutaciones Recientes. In A. Fontaine, et al (Eds.), *Reforma a los Partidos Políticos en Chile* (pp. 75-124). Santiago, Chile: UNDP, CEP, Libertad y Desarrollo, Proyectamerica y CIEPLAN.

Lusoli, W., & Ward, S. (2003). *Digital Rank-and-File: Party Activists' Perceptions and use of the Internet.* Paper presented at the Annual Meeting of the American Political Science Association, Philadelphia, United States.

Lynn, L. (1998). The New Public Management: How to Transform a Theme into a Legacy. *Public Administration Review, 58*(3), 231–237. doi:10.2307/976563

Lyytinen, K., & Hirschheim, R. (1997). Information systems failure; A survey and classification of the empirical literature. *Oxford Surveys in Information Technology, 4*, 257–309.

Lyytinen, K., & Newman, M. (2008). Explaining information systems change: a punctuated socio-technical change model. *European Journal of Information Systems, 17*(6), 589–613. doi:10.1057/ejis.2008.50

Maani, K. E., & Cavana, R. Y. (2000). *Systems Thinking and Modelling: Understanding Change and Complexity.* Auckland, New Zealand: Prentice Hall.

Mainwaring, S., & Scully, T. R. (1995). La Institucionalización de los Sistemas de Partidos en América Latina. *Revista de Ciencia Política, 17*(1-2), 63–102.

Malamud, A. (2003). Los Partidos Políticos. In J. Pinto (Ed.), *Introducción a la Ciencia Política* (pp. 321-350). Buenos Aires, Argentina: EUDEBA.

Malhotra, Y. (2000). Knowledge management for e-business performance. *Information Strategy: The Executives Journal, 16*(4), 5–16.

Malhotra, Y., & Galletta, D. F. (2007). *Extending the Technology Acceptance Model to Account for Social Influence: Theoretical Bases and Empirical Validation.* BRINT Research Institute, University of Pittsburgh.

Mansell, R. (Ed.). (2002). *Inside The Communication Revolution: Evolving Patterns of Social and Technical Interaction* (1st ed.). UK: Oxford University Press.

Mansell, R., & Silverstone, R. (1997). Introduction. In R. Mansell & R. Silverstone (Eds.), *Communication by design. The politics of information and communication technologies.* (pp. 1-14). Oxford, UK: Oxford University Press.

Mansell, R., & Steinmueller, W. (2000). *Mobilizing the Information Society: Strategies for Growth and Opportunity.* Oxford, UK: Oxford University Press.

Manuel, P. D., & AlGhamdi, J. (2003). A data-centric design for n-tier architecture. *Information Sciences, 150*, 195–206. doi:10.1016/S0020-0255(02)00377-8

March, J. G. (1991). Exploration and exploitation in organisational learning. *Organization Science, 2*(1), 71–87. doi:10.1287/orsc.2.1.71

March, J. G., & Olsen, J. P. (1989). *Rediscovering Institutions: the Organizational Basis of Politics*. New York: Free Press.

March, J. G., & Olsen, J. P. (1996). Institutional Perspectives on Political Institutions. *Governance: An International Journal of Policy and Administration, 9*, 247–264.

Marchionini, G., Samet, H., & Brandt, L. (2004). Digital government. *Communications of the ACM, 46*(1), 25–27.

Marcoccia, M. (2004). On-line polylogues: conversation structure and participation framework in internet newsgroups. *Journal of Pragmatics, 36*(1), 115–145. doi:10.1016/S0378-2166(03)00038-9

Margetts, H. (1995). The Automated State. *Public Policy and Administration, 10*(2), 88–103. doi:10.1177/095207679501000207

Margetts, H. (1999). *Information Technology in Government: Britain and America*. London: Routledge.

Margetts, H. (2001). *The Cyber Party*. Paper presented to workhop 'The Causes and Consequences of Organisational Innovation in European Political Parties,' ECPR Joint Sessiones of Workshops, Grenoble, France.

Martínez, G. (2004). Internet y ciudadanía global: procesos de producción de representaciones sociales de ciudadanía en tiempos de globalización. In D. Mato (Ed.), *Políticas de ciudadanía y sociedad civil en tiempos de globalización* (pp. 181-200). Caracas, FACES, UCV.

Mason, R., & Mitroff, I. (1981). *Challenging Strategic Planning Assumptions: Theory, Cases and Techniques*. New York: John Wiley and Sons.

Masuda, Y. (1980). *The Information Society as Post-Industrial Society*. Tokyo: Institute for the Information Society.

Mattelart, A. (2003). *The Information Society: An Introduction.* (S. Taponier, & J. Cohen, Trans.). London: Sage.

Matthews, W. (2002). E-gov falters easily. *Federal Computer Week, 13*, 43.

Maturana, H. R. (1975). The organization of the living: a theory of the living organization. *International Journal of Man-Machine Studies, 7*, 313–332. doi:10.1016/S0020-7373(75)80015-0

Maturana, H. R., & Varela, F. J. (1980). *Autopoiesis and Cognition: The Realization of the Living*. Dordrecht, Holland: Reidel Publishing Co.

May, C. (2002). *The Information Society: A Sceptical View*. Cambridge, UK: Polity.

Maya-Jariego, I. (2004). Sentido de comunidad y potenciación comunitaria. *Apuntes de psicología, 22*(2), 187-211. Retrieved January 30, 2009, from http://www.cop.es/delegaci/andocci/files/contenidos/VOL22_2_3.pdf

McCulloch, W. S. (1965). *Embodiments of Mind.* Cambridge, MA: MIT Press.

McDaniel, S. E., Olson, G. M., & Magee, J. C. (1996). Identifying and Analyzing multiple threads in computer-mediated and face-to-face conversations. In M. S. Ackerman (Ed.). *Proceedings of the 1996 ACM conference on computer supported cooperative work* (pp. 39-47). New York: ACM Press.

McLoughlin, I., & Cornford, J. (2006). Transformational change in the local state? Enacting e-government in English local authorities. *Journal of Management & Organization, 12*(3), 195–208.

McNabb, D. E. (2007). *Knowledge management in the public sector: A Blueprint for innovation in Government.* London: M.E. Sharpe.

McTaggart, J. M. E. (1964). *A commentary on Hegel's logic*. New York: Russell & Russell.

Medeiros, P. H. (2004). *E-gov no Brasil: aspectos institucionais e reflexos na governança.* Dissertação de mestrado em Administração da Universidade de Brasília (UNB), Brasília.

Medhi, I., & Kuriyan, R. R. (2007). Text-Free UI: Prospects for Social Inclusion. In *Proceedings of International Conference on Social Implications of Computers in Developing countries,* Brazil.

Medhi, I., & Toyama, K. (2007). Full-Context Videos for First-Time, Illiterate Users. In *Proceedings of ALT. CHI forum at ACM CHI '07*, San Jose, USA.

Medhi, I., & Toyama, K. (2007). Full-Context Videos for First-Time, Non-Literate PC Users. In *Proc. IEEE/ACM International Conference on Information and Communication Technologies and Development,* Bangalore, India.

Medhi, I., Pitti, B., & Toyama, K. (2005). Text-Free UI for Employment Search. In *Proceedings of Asian Applied Computing Conference*, Nepal.

Medhi, I., Sagar, A., & Toyama, K. (2006). Text-Free User Interfaces for Illiterate and Semi-Literate Users. In *Proceedings of International Conference on Information and Communication Technologies and Development*, Berkeley, PA.

Megill, K. A. (1997). *The Corporate Memory: Information Management in the Electronic Age*. London: Bowker & Saur.

Melistki, J. (2002). *The Adoption and Implementation of e-Government: the Case of e-Government in New Jersey.* Unpublished doctoral dissertation, Graduate School Newark Rutgers, The State University of New Jersey, New Jersey.

Mendoza, X. (1990). Técnicas Gerenciales y Modernización de la Administración Pública en España. In *Documentación Administrativa* (Nr. 223). Madrid.

Menzies, H. (1996). *Whose Brave New World? The Information Highway and the New Economy.* Toronto: Between the Lines.

Meyer, J. W., & Rowan, B. (1977). Institutionalized Organizations: Formal Structure as Mith and Ceremony. *American Journal of Sociology*, *83*(2), 340–363. doi:10.1086/226550

MIDEPLAN (Ministry of Planification of Chile) & UNDP. (2006). *Las Trayectorias del Desarrollo Humano en las Comunas de Chile (1994-2003*). Santiago, Chile: MIDEPLAN-UNDP.

Midgley, G. (1996). What Is This Thing Called CST? In R.L. Flood & N.R.A. Romm (Eds.), *Critical Systems Thinking: Current Research and Practice*. New York: Plenum Publishing Corporation.

Midgley, G. (1997). Mixing methods: Developing systemic intervention. In J. Mingers & A. Gill (Eds.), *Multimethodology: The Theory and Practice of Combining Management Science Methodologies.* (pp. 249-290). Chichester, UK: John Wiley and Sons.

Midgley, G. (2000). *Systemic Intervention: Philosophy, Methodology and Practice*. New York: Kluwer Academic.

Millen, D. R., & Dray, S. M. (2000). Information sharing in an online community of journalists. *Aslib Proceedings*, *52*(5), 166–173. doi:10.1108/EUM0000000007011

Milliken, F., & Lant, T. (1991). The effect of an organization's recent performance history on strategic persistence and change: The role of managerial interpretations. In J. Dutton, A. Huff, & P. Shrivastava (Eds.), *Advances in Strategic Management* (Vol. 7, pp. 125-152). Greenwich, CT: JAI Press.

Mincomunicaciones (2007). *Agenda de Conectividad - Gobierno en Línea y Plan de Acción 2007* (Presentation). Bogotá: Ministerio de Comunicaciones.

Mingers, J. (1989). An introduction to autopoiesis – implications and applications. *Systems Practice*, *2*(2), 159–180. doi:10.1007/BF01059497

Mingers, J. (1995). *Self-Producing Systems: Implications and Applications of Autopoiesis.* London: Plenum.

Ministry of National Economy (MNE). (2007). *e-Oman Bulletin*, Oman.

Ministry of National Economy (MNE). (2007). *Oman Digital Society Report*. Oman.

Mintzberg, H. (1994). *Rethinking Strategic Planning*. Long Range Planning.

Molina, A., & Kinder, T. (2001). National systems of innovations, industrial clusters and constituency-building in Scotland's electronics industry. *International Journal of Entrepreneurship and Innovation Management*, *1*(2), 241–275. doi:10.1504/IJEIM.2001.000456

Montero, J. R., & Gunther, R. (2003). *Los Estudios sobre los Partidos Políticos: una Revisión Crítica* (Estudio/Working Paper 12). Colección en la Red de Cuadernos de Trabajo. Universidad Autónoma de Madrid.

Moon, M. J. (2002). The evolution of e-government among municipalities: rhetoric or reality. *Public Administration Review*, *62*(4), 424–433. doi:10.1111/0033-3352.00196

Moore, M. (1995). *Creating Public Value: Strategic Management in Government*. Cambridge, MA: Harvard University Press.

Morlino, L. (2007). Explicar la Calidad Democrática: ¿Qué tan Importante son las Tradiciones Autoritarias? *Revista de Ciencia Política*, *27*(2), 3–22.

Müller, G., et al. (2008). Democracia Interna y Selección de Autoridades. In A. Fontaine, et al. (Eds.), *Reforma a los Partidos Políticos en Chile* (pp. 413-428). Santiago, Chile: UNDP, CEP, Libertad y Desarrollo, Proyectamerica y CIEPLAN.

Mumford, E. (1983). *Designing Human Systems for New Technology: The ETHICS Method*. Manchester: Manchester Business School.

Nakamura, R. (1987). The Textbook Policy Process and Implementation Research. *Policy Studies Review*, *7*(1), 142–154. doi:10.1111/j.1541-1338.1987.tb00034.x

National E-governance Plan (NeGP). (2006). *Framework India*. New Delhi: Ministry of Communication & Information Technology. Retrieved on June 24, 2008, from http://www.mit.gov.in/download/eu/spsingh.pdf

Negroponte, N. (1995). *Being Digital*. Boston, MA: MIT Press.

Neto, J. J., & Bocoli, F. S. (2003) *Sucesso^{SW}=CMM$_2$ + PMBOK*. PMI Journal, Seção Rio Grande do Sul, 5 (May).

Nidumolu, S., Goodman, S. E., Vogel, D., & Danowitz, A. K. (1996). Information technology for local administration support: The governorates project in Egypt. *MIS Quarterly*, *20*(2), 197. doi:10.2307/249478

Niedźwiedzka, B., Czabanowska, K., & Śmietana, R. (2008). Controlled Vocabulary in Public Health: An Overview of the Achievements to Date. *Journal of Public Health*, accepted. Retrieved November 25, 2008, from http://www.springerlink.com/content/31-5862w47215-m058/

Nielsen, J. (2006). *Participation inequality: encouraging more users to contribuite*. Retrieved February 2009, from http://www.useit.com/alertbox/participation_inequality

Nonaka, I. (1994). A dynamic theory of organizational knowledge creation. *Organization Science*, *5*(1), 14–37. doi:10.1287/orsc.5.1.14

Nonaka, I., & Konno, N. (1998). The concept of "ba": Building a foundation for knowledge creation. *California Management Review*, *40*(3), 1–15.

Nonaka, I., & Takeuchi, H. (1995). *The Knowledge Creating Company*. Oxford, UK: Oxford University Press.

Nooteboom, B. (1999). The dynamic efficiency of networks' in Interfirm networks. In A. Grandori (Ed.), *Organization and Industrial Competitiveness* (pp. 91-119). London: Routledge.

Norris, D., & Moon, M. J. (2005). Advancing e-government at the grassroots: Tortoise or hare? *Public Administration Review*, *65*(1), 64–75. doi:10.1111/j.1540-6210.2005.00431.x

Nueva votación en el Parlamento Europeo sobre patentes. (2006). Retrieved February 2, 2007, from http://www.lapastillaroja.net/archives/001122.html

O'Donell, G. (1992). *Delagative Democracy?* (Working Paper 172). Kellog Institute for International Studies, University of Notre Dame.

O'looney. J. A. (2002). *Wiring Governments – Challenges and Possibilities for Public Managers*. New York: Quorum Books.

Ochoa, A., & Petrizzo, M. (2006). Ciudadanía y Desarrollo Endógeno. In A. Ochoa (Ed.), *Aprendiendo en torno al desarrollo endógeno* (pp. 53-74). CDCHT, FUNDACITE Mérida, CSI-ULA.

Ochoa, A., & Petrizzo, M. (2009). *Information and Communication Technologies and Government: A challenge to citizenship?* In J.-R. Córdoba & A. Ochoa-Arias (Eds.), *Systems Thinking and E-Participation: ICT in the Governance of Society*. Hershey, PA: IGI Global.

Ochoa-Arias, A. (Ed.). (2006). *Aprendiendo en Torno al Desarrollo Endógeno (1st Ed.)*. Mérida, Venezuela: Universidad de Los Andes (Venezuela) and Fundacite-Mérida.

ODPM. (2003). *Local e-Government. Process Evaluation of the Implementation of Electronic Local Government in England*. London: Office of the Deputy Prime Minister (ODPM).

OECD. (2003). *The e-Government Imperative*. Paris: Organisation for Economic Co-Operation and Development.

Oldfield, P. (2002). *Domain Modelling*. Retrieved November 25, 2008, from http://www.aptprocess.com/whitepapers/DomainModelling.pdf

Olías de Lima, B. (2002). Spanish Public Administration: Assessment and Perspectives (Las Administraciones Públicas Españolas: Balance y Perspectivas). In J. Crespo & G. Pastor (Eds.), *Spanish Public Administration* (Administraciones Públicas Españolas) (pp. 367-393). Madrid, Spain: McGraw-Hill.

Oram, A. (2001). *Peer-to-Peer: Harnessing the Benefits of a Disruptive Technology*. Sebastopol, CA: O'Reilly & Associates.

Orlikowski, W. (1992). The duality of technology: Rethinking the concept of technology in organisations. *Organization Science*, *3*(3), 398–427. doi:10.1287/orsc.3.3.398

Orr, K., & Vince, R. (2009). (forthcoming). Traditions of Local Government. *Public Administration, 87*. doi:10.1111/j.1467-9299.2009.01770.x

Osborne, D. (1997). *Banishing bureaucracy: the five strategies for reinventing government*. New York: Plume.

Osborne, D., & Gaebler,T. (1992). *Reinventing Government – How the Entrepreneurial Spirit is Transforming the Public Sector*. Addison-Wesley Publ. Co.

Osler, A. (1998). Conflicts, Controversy and Caring. Young people's attitudes towards children's rights. In C. Holdem & N. Clough (Eds.), *Children as citizens. Education for participation* (pp. 113-126). London: Jessica Kingsley Publishers.

Ostrom, V., & Ostrom, E. (1971). Public choice: A different approach to the study of public administration. *Public Administration Review*, *31*(2), 203–216. doi:10.2307/974676

Owen, D. (1995). *Nietzsche, Politics and Modernity*. Thousand Oaks, CA: Sage Publications.

Palkovits, S., & Wimmer, M. (2003). *Processes in e-Government – A Holistic Framework for Modelling Electronic Public Services*. Berlin: Springer.

Pan, G., Pan, S., Newman, M., & Flynn, D. (2006). Escalation and de-escalation of commitment: A commitment transformation analysis of an e-government project. *Information Systems Journal*, *16*(1), 3–21. doi:10.1111/j.1365-2575.2006.00209.x

Parks, M. R., & Floyd, K. (1996). Making friends in cyberspace. *Journal of Computer-Mediated Communication*, *1*(4).

Pavlichev, A., & Garson, G. D. (2004). *Digital Government: Principles and Best Practices*. Hershey, PA: Idea Group Publishing.

Pérez Ríos, J. (2008). *Diseño y Diagnóstico de Organizaciones Viables. Un enfoque sistémico*. Valladolid, Spain: Iberfora.

Perez, C. (1983). Structural change and assimilation of new technologies in the economic and social systems.

Futures, 15(5), 357–375. doi:10.1016/0016-3287(83)90050-2

Peters, B. G. (1996). *The Policy Capacity of Government* (Research Paper, No. 18). Ottawa, Canada: Canadian Centre for Management Development.

Peters, B. G. (2005). *The New Institutionalism. Institutional Theory in Political Science* (2nd ed.). London: Continuum.

Peters, B. G., & Savoie, D. J. (1998). *Taking stock: Assessing Public Sector Reforms.* Montreal, Canada: Queens University Press.

Petrizzo, M. (2005). Participación ciudadana y tecnologías de la información y la comunicación: hacia una administración pública relacional. *Nueva Sociedad, 195*, 88–101.

Pfeffer, J. (1998). *The Human Equation: Building Profits by Putting People First.* Boston, MA: Harvard Business School Press.

Phang, C. W., Sutanto, J., Kankanhalli, A., Tan, C., & Hock-Hai, T. (2006). Senior citizen's acceptance of information systems: A study in the context of e-government services. *IEEE Transactions on Engineering Management, 53*(4), 555–569. doi:10.1109/TEM.2006.883710

Pierson, P. (2000). The Limits of Design: Explaining Institutional Origins and Change. *Governance: An International Journal of Policy and Administration, 13*(4), 475–499.

Pierson, P. (2004). *Politics in Time: History, Institutions, and Social Analysis.* Princeton, NJ: Princeton University Press.

Pilonieta, C., & Ochoa, A. (2006). El desarrollo endógeno sustentable. Una aproximación conceptual. In A. Ochoa (Ed.), Aprendiendo en torno al desarrollo endógeno (pp. 21-36). CDCHT, FUNDACITE Mérida, CSI-ULA.

Pinto, S., & Fernandes, C. (2005). *Institucionalização do Governo Eletrônico no Brasil.* Paper presented at X Congreso Internacional del CLAD sobre la Reforma del Estado y de la Administración Pública. Santiago, Chile.

Plathe, A., & Mastrangelo, J. (2003). Interviu with Abdul Waheed Khan, Towards Knowledge Societies. *A world of science . RE:view, 1*(4).

PMI. (2004). *A Guide to the Project Management Body of Knowledge: PMBOK Guide Third Edition.* Project Management Institute Inc., Pennsylvania, USA.

PMI. (2006). *Government Extension to the PMBOK Guide Third Edition.* Project Management Institute Inc., Pennsylvania, USA.

Pollit, Ch. (1993). *Managerialism and the Public Services: The Anglo-American Experience* (2nd ed.). Oxford, UK: Blackwell.

Pollit, Ch. (2002). Clarifying Convergence: Striking Similarities and Durable Differences in Public Management Reform. *Public Management Review, 4*(1), 471–492.

Pollit, Ch., & Bouckaert, G. (2003). Evaluating Public Management Reforms: an International Perspective. In H. Wollmann (Ed.), *Evaluation in Public Sector Reform. Concepts and Practice in International Perspective* (pp. 12-35). Cheltenham, UK: Edward Elgar.

Pollit, Ch., & Bouckaert, G. (2004). *Public Management Reform. A Comparative Analysis* (2nd ed.). Oxford, UK: Oxford University Press.

Pollit. C. (2003). *Joined-up Government: a survey. Political Studies Review, 1*, 34–49. doi:10.1111/1478-9299.00004

Porter, M., & Millar, V. (1985). How information gives you competitive advantage. *Harvard Business Review, 63*(4), 149–160.

Pouloudi, A., & Whitley, E. A. (1997). Stakeholder Identification in Inter-Organisational Systems: Gaining Insights for Drug Use Management Systems. *European Journal of Information Systems, 6*(1), 1–14. doi:10.1057/palgrave.ejis.3000252

Powell, W. W. (1991). Expanding the Scope of Institutional Analysis. In W. W. Powell & P. J. DiMaggio (Eds.), *The New Institutionalism in Organizational Analysis* (pp. 183-203). Chicago: University of Chicago Press.

Powell, W. W., & DiMaggio, P. J. (Eds.). (1991). *The New Institutionalism in Organizational Analysis*. Chicago: University of Chicago Press.

Prats, J. (2005). *De la Burocracia al Management, del Management a la Gobernanza*. Madrid: INAP.

Pressman, J., & Wildavsky, A. (1973). *Implementation*. Berkeley, CA: University of California Press.

Prigogine, I., & Stengers, I. (1984). *Order Out of Chaos: Man's New Dialogue with Nature*. New York: Bantam Books

Pryme. (2006). *Gobierno Electrónico en Chile 2000-2005: Estado del Arte II*. Ministerio Secretaría General de la Presidencia. Proyecto de Reforma y Modernización del Estado.

Puyosa, I. (2007). *Cómo cosechar información sobre los mercados y la opinión pública en el campo de la Web 2.0*. Retrieved February 3, 2009, from http://www.scribd.com/doc/5579367/Como-cosechar-informacion-sobre-los-mercados-y-la-opinion-publica-en-el-campo-de-la-web-20-

Quintero, R., Barros, R., Dávila, J., Moreno, N., Tonella, G., & Ablan, M. (2004). A Model of the Biocomplexity of Deforestation in Tropical Forest: Caparo Case Study. In C. Pahl, S. Schmidt, & T. Jakeman (Eds.), *Proceedings of the iEMSs 2004 International Congress: "Complexity and Integrated Resources Management,"* International Environmental Modelling and Software Society, Osnabrueck, Germany. Retrieved from http://www.iemss.org/iemss2004/proceedings

Rambøll Management. (2004). *User Satisfaction and Usage Survey of eGovernment Services*. EU: Top of the Web Survey.

Ramió, C., & Salvador, M. (2002). The Configuration Public Administrations in Comunidades Autónomas (La Configuración de las Administraciones de las Comunidades Autónomas). In J. Subirats & R. Gallego (Eds.), *Twenty Years of Autonomies in Spain* (Veinte Años de Autonomías en España) (pp. 99-133). Madrid, Spain: Centro de Investigaciones Sociológicas, Scott, W. R. (1995). *Institutions and Organizations*. London: Sage.

Reilly, K. (2004). Estrategias de e-gobierno en ocho países de América Latina. In R. Araya & M. Porrúa (Eds.), *América Latina Puntogob. Casos y Tendencias en Gobierno Electrónico* (pp. 71-110). Chile: FLACSO Chile – OEA – AICD.

Reinhard, N., & Dias, I. (2005) Categorization of e-Gov initiatives: a comparison of three perspectives. *X Congreso Internacional del CLAD sobre la Reforma del Estado y de la Administración Pública*, Santiago, Chile, 18 - 21 Oct.

Reinhard, N., & Dias, I. (2005). *Categorization of e-Gov initiatives: a comparison of three perspectives*. Paper presented at X Congreso Internacional del CLAD sobre la Reforma del Estado y de la Administración Pública, Santiago, Chile.

Rempel, M. (1996). Systems theory and power/knowledge: A Foucauldian reconstruction of Niklas Luhmann's systems theory. *The International Journal of Sociology and Social Policy*, *16*(4), 58–90. doi:10.1108/eb013250

Reponen, T. (1998). The role of learning in information systems planning and implementation. In R. D. Galliers & W. R. J. Baets (Eds.), *Information Technology and Organisational Transformation* (pp. 133-149). Chichester: John Wiley and Sons.

Republic of Albania. (2006). *Electronic Assessment – Report on the status of information technology and communications in Albania 2002-2005.* Albanian Government publication. SCORE Project EU. (n.d.). *The ICT Research environment in ALBANIA*. Retrieved February 21, 2009, from http://consultations.score-project.eu/attach/ictcr_al_en.pdf

Resource Renewal Institute. (2009). *Green Plans in Action: Netherlands*. Retrieved July 20, 2009 from http://www.rri.org/greenplans_netherlands.html

Rheingold, H. (1994). *The virtual community: finding connection in a computerized world.* London: Secker & Warburg.

Rhodes, R. A. W. (1997). *Understanding Governance: policy Networks, Governance Reflexivity and Accountability*. Buckingham, UK: Open University Press.

Ribes, X. (2007). El valor de los metadatos y la inteligencia colectiva. In *Telos. Cuadernos de Comunicación e Innovación. de la Fundación Telefónica*. Retrieved from http://www.campusred.net/TELOS/articuloperspectiva.asp?idarticulo=2&rev=73

Riley, T. B. (2003). *E-Government vs E-Governance: Examining the differences in a changing public sector climate. International tracking survey report '03. Number four*. Commonwealth Centre for E-Governance. Retrieved May 15, 2009, from http://www.electronicgov.net

Riquelme, A. (1999). ¿Quiénes y Por Qué 'No Están Ni Ahí'? Marginación y/o Automarginación en la Democracia Transicional. Chile, 1988-1997. In P. Drake & I. Jáksic (Eds.), *El Modelo Chileno. Democracia y Desarrollo en los Noventa* (pp. 261-280). Santiago, Chile: LOM.

Riquelme, H. (2002). Commercial internet adopting in China: Comparing the experience of small medium and large businesses. *Internet Research: Electronic Networking Applications and Policy, 12*(3), 276–286. doi:10.1108/10662240210430946

Rivera, E. (2003). *Nueva Economía, Gobierno Electrónico y Reforma del Estado*. FLACSO, Editorial Universitaria.

Robb, F. F. (1989). The limits to human organisation: The emergence of autopoietic systems. In M.C. Jackson, P. Keys & S. Cropper (Eds.), *Operational Research and the Social Sciences*. New York: Plenum Press.

Robb, F. F. (1989). The application of autopoiesis to social organization – A comment on John Mingers' 'An introduction to autopoiesis – implications and applications.' . *Systems Practice, 2*(3), 343–348. doi:10.1007/BF01059979

Robertson, R. (2003). Glocalización: tiempo-espacio y homogeneidad-heterogeneidad. In *Cansancio del Leviatán: problemas políticos de la mundialización*. Madrid: Trotta.

Robey, D., & Boudreau, M. (1999). Accounting for the contradictory organizational consequences of information technology: Theoretical directions and methodological implications. *Information Systems Research, 10*(2), 167–185. doi:10.1287/isre.10.2.167

Rogers, E. (1995). *Diffusion of innovations* (4th ed.). New York: Free Press.

Römmele, A. (2003). Political Parties, Party Communication and New Information and Communication Technologies. *Party Politics, 9*(1), 7–20.

Ronaghan, S. (2002). *Benchmarking e-Government: a global perspective. Assessing the progress of the UN member states*. United Nations Division for Public Economics and Public Administration & American Society for Public Administration.

Rose, R. (2005). The internet and governance in a global context. *Journal of Public Policy, 25*(1), 1–3. doi:10.1017/S0143814X05000267

Ruiloba, J. M. (2007). Un marco para el análisis de las políticas públicas: características y elementos definitorios del modelo de las corrientes múltiples de J. W. Kingdon. *Revista Ciencias Jurídicas y Sociales, 3*.

S/A. (2009, February 12). Facebook es la red social más utilizada. In *Google Noticias*. Retrieved February 2009, from http://www.google.com/hostednews/afp/article/ALeqM5jOlNcddVSFdkJiJs_HArLf1D5JiQ

Sabherwal, R., Hirschheim, R., & Goles, T. (2001). The dynamics of alignment: Insights from a punctuated equilibrium model. *Organization Science, 12*(2), 179–197. doi:10.1287/orsc.12.2.179.10113

Sabucedo, L. A., & Rifon, L. A. (2006). *Semantic Service Oriented Architectures for eGovernment Platforms*. American Association for Artificial Intelligence. Retrieved from http://www.aaai.org

SAP. (2000). *R/3 System SAP Solution Map*. [cited Dec. 2000]; Available from: www.sap.com.

Sappenfield, M., & Chopra, A. (2007). India's super-rich get even richer. *The Christian Science Monitor*. Retrieved October 11, 2008, from http://www.csmonitor.com/2007/1218/p01s04-wosc.html

Sartori, G. (1976). *Parties and Party Systems: A framework for analysis*. New York: Cambridge University Press.

Savvas, I., & Bassiliades, N. (2008). A Process-Oriented Ontology-Based Knowledge Management System for Facilitating Operational Procedures in Public Administration, accepted for publication in *Expert Systems with Applications*.

Savvas, I., Pimenidis, E., & Sideridis, A. (2007). Proposing a high-level requirements mapping framework, for testing implementation compatibility in e-government projects. In *Proceedings of ECEG. The 7th European Conference of e-Government.*, Den Haag, The Netherlands (pp. 459-468).

Savvas, I., Pimenidis, E., & Sideridis, A. (2007). A Review of E-Governance Models in the EU. In *Proceedings of Advances in Computing and Technology (AC&T) Conference*, University of East London, England (pp. 110-119).

Savvas, I., Pimenidis, E., & Sideridis, A. (2008). Using egov systems to remedy public service failings: in search of a transformation algorithm. In *Proceedings of The e-Government Workshop '08(eGOV08)*, Brunel University, UK.

Saxena, K. B. C. (2005). Towards excellence in e-governance. *International Journal of Public Sector Management*, *18*(6), 498–513. doi:10.1108/09513550510616733

Schedler, K., & Scharf, M. C. (2001). *Exploring the Interrelations Between Electronic Government And the New Public Management. A Managerial Framework for Electronic Government*. Switzerland: Institute for Public Services and Tourism at the University of St. Gallen.

Schein, E. (1993). How can organisations learn faster? The challenge of entering the green room. *Sloan Management Review*, *34*(2), 85–92.

Schrock, A. (2009). Examining social media usage: technology clusters and social network sites membership. In *First Monday*, 14(1). Retrieved February 3, 2009, from http://firstmonday.org/htbin/cgiwrap/bin/ojs/index.php/fm/article/view/2242/2066

Schumann, W. (1990). Strategy for information systems in the film division of Hoechst A.G. *Systems Practice*, *3*(3).

Schware, R., & Deane, A. (2003). Deploying e-government programs: the strategic importance of "I" before "E". *info*, *5*(4), 10-19.

Scott Morton, M. S. (1994). The state of the art of research. In F. W. McFarlane (Eds.), *The Information Systems Research Challenge Proceedings* (pp. 13-45). Cambridge, MA: Harvard Business School Press.

Scully, T. (1992). *Los Partidos de Centro y la Evolución Política Chilena*. Santiago, Chile: Notre Dame University-CIEPLAN.

SEFAZ/SP. (2006). *Prêmio Mário Covas: Nota Fiscal Eletrônica, a modernização do fisco à serviço da sociedade*. Secretaria da Fazenda do Estado de São Paulo. Internal Document.

SEFAZ/SP. (2007). *Projeto Nacional da Nota Fiscal Eletrônica, NF-e: resultados obtidos em 2006*. Secretaria da Fazenda do Estado de São Paulo, 2007. Internal Document.

Selnes, F., & Sallis, J. (2003). Promoting relationship learning. *Journal of Marketing*, *67*(July), 80–95. doi:10.1509/jmkg.67.3.80.18656

Senge, P. (1999). *The Fifth Discipline: The Art and Practice of the Learning Organisation*. London: Random House.

Senge, P. M., et al. (1994). *The Fifth Discipline Fieldbook: Strategies and tools for building a learning organization*. London: Nicholas Brealey Publishing.

Serafeimidis, V., & Smithson, S. (2003). Information systems evaluation as an organizational institution - experience from a case study. *Information Systems Journal*, *13*, 251–274. doi:10.1046/j.1365-2575.2003.00142.x

Sharp, H., Finkelstein, A., & Galal, G. (1999). Stakeholder Identification in the Requirements Engineering Process. In *Proceedings of the 10th Int'l Workshop Database and Expert Systems Applications* (pp. 387–391).

Sherer, W., & Thrasher, S. (2005). *Contrasting CMMI and the PMBOK.* CMMI Technology Conference & User Group, November.

Sherwani, J., Ali, N., Mirza, S., Fatma, A., Memon, Y., Karim, M., et al. (2007). HealthLine: Speech-based Access to Health Information by Low-literate Users. In *Proceedings of IEEE/ACM International Conference on Information and Communication Technologies and Development*, Bangalore, India.

Siau, K., & Long, Y. (2005). Synthesizing e-government stage models - a metasynthesis based on metaethnography approach. *Industrial Management & Data Systems, 105*(4), 443–458. doi:10.1108/02635570510592352

Siavellis, P. M. (1999). Continuidad y Transformación del Sistema de Partidos en una 'Transición Modelo'. In P. Drake & I. Jäksic (Eds.), *El Modelo chileno. Democracia y desarrollo en los noventa* (pp. 223-259). Santiago, Chile: LOM.

Silva, P. (2004). Doing Politics in a Depoliticised Society: Social Change and Political Deactivation in Chile. *Bulletin of Latin American Research, 23*(1), 63–78. doi:10.1111/j.1470-9856.2004.00096.x

Simon, C., & Olbrich, S. (2005). The influence of legal constraints on business process modelling In *Proceedings of The eGovernment Workshop '05 (eGOV05)*, September 13 2005, Brunel University, UK.

Sin, H. (2007). *Personal Interview, Gobierno en Línea.* Bogotá, Colombia.

Skelcher, C. (1992), *Managing for Service Quality*, Longman, Harlow.

Smeltzer, L. R., Fann, G. L., & Nikolaïssen, V. N. (1988). Environmental scanning practices in small business. *Journal of Small Business Management, 226*(3), 52–62.

Smithson, S., & Tsiavos, P. (2004). Re-constructing information systems evaluation. In C. Avgerou, C. Ciborra, & F. Land (Eds.), *The Social Study of Information and Communication Technology: Innovation, Actors and Contexts* (pp. 207-230). Oxford, UK: Oxford University Press.

Snellen, I. Th. M., & van de Donk, W. (Eds.). (1998). *Public Administration in an Information Age. A Handbook.* Amsterdam: IOS Press.

Sotiropoulos, D. A. (2004). Southern European Public Bureaucracies in Comparative Perspective. *West European Politics, 27*(3), 405–422. doi:10.1080/0140238042000228077

Sparrow, J. (2005). Classification of different knowledge management development approaches of SMEs. *Knowledge Management Research and Practice, 3*, 136–145. doi:10.1057/palgrave.kmrp.8500053

Spender, J. C. (1998). The dynamics of individual and organisational knowledge. In C. Eden & J. C. Spender (Eds.), *Managerial and Organizational Cognition: Theory, Methods and Research* (pp. 11-39). London: Sage.

Stacey, R. D. (2000). *Complexity and management: Fad or Radical Challenge to Systems Thinking?* London: Routledge.

Stacey, R. D. (2003). *Strategic Management and organisational Dynamics (4th ed.).* Harlow, UK: Prentice-Hall.

Stamoulis, D., Gouscos, D., Georgiadis, P., & Martakos, D. (2001). Revisiting public information management for effective e-government services. *Journal of Information Management & Computer Security, 9*(4), 146–153. doi:10.1108/09685220110400327

Stats, I. W. (n.d.). *Internet Statistics on Albania.* Retrieved March 25, 2009, from http://www.internetworldstats.com/euro/al.htm

Stemberger, M. I., & Jaklic, J. (2007). Towards E-government by business process change—A methodology for public sector. *International Journal of Information Management, 27*, 221–232. doi:10.1016/j.ijinfomgt.2007.02.006

Stokes, P. (2008). *The Viability of Societies: Governance and Complexity Today.* Saarbruken, Germany: VDM Verlag.

Storey, D. (2004). Exploring the link, among small firms, between management training and firm performance: A comparison between the UK and other OECD countries. *International Journal of Human Resource Management*, *15*(1), 112–130. doi:10.1080/0958519032000157375

Stump, E. (1989). *Dialectic and its place in the development of medieval logic*. Ithaca, NY: Cornell University Press.

Subsecretaría de Transportes (SUBTEL). (2009). *Estadísticas e Indicadores por Tipo de Servicio. Internet, Series Conexiones Internet*. Retrieved March 26, 2009, from http://www.subtel.cl/prontus_subtel/site/artic/20070212/pags/20070212182348.html#T1

Sudhir, V., Srinivasan, G., & Muraleedharan, V. R. (1997). Planning for sustainable solid waste management in urban India. *System Dynamics Review*, *13*, 223–246. doi:10.1002/(SICI)1099-1727(199723)13:3<223::AID-SDR127>3.0.CO;2-Q

Sundbo, J. (2001). *The strategic Management of Innovation*. Cheltenham, UK: Edward Elgar.

Swedberg, R., & Granovetter, M. (2001). Introduction. In M. Granovetter & R. Swedberg (Eds.), *The Sociology of Economic Life (2nd Ed.)*. Oxford, UK: Westview Press.

Sydänmaanlakka, P. (2002). *An Intelligent Organization*. Oxford, UK: Capstone.

Tan, C., & Pan, S. (2003). Managing e-transformation in the public sector: An e-government study of the Inland Revenue Authority of Singapore (IRAS). *European Journal of Information Systems*, *12*, 269–281. doi:10.1057/palgrave.ejis.3000479

Tapscott, D., & Williams, A. (2007). *Wikinomics: How Mass Collaboration Changes Everything*. London: Atlantic.

The Federal Enterprise Architecture (FEA). (n.d.). *Ontology*. Retrieved from http://web-services.gov/fea-rmo.html van Dooren, W. (2006). *Performance Measurement in the Flemish Public Sector: A Supply and Demand Approach*. PhD dissertation, Faculteit Sociale Wetenschappen - Onderzoekseenheid: Instituut voor de Overheid [IO], K.U.Leuven

Thompson, D., Rust, R., & Rhoda, J. (2005). The business value of e-government for small firms. *International Journal of Service Industry Management*, *16*(4), 385–407. doi:10.1108/09564230510614022

Thomsen, S. R. (1996). @ work in cyberspace: Exploring practitioner use of the PRForum. *Public Relations Review*, *22*(2), 115–131.

Toffler, A. (1992). *Power Shift: Knowledge, Wealth and Violence at the Edge of the 21St Century*. London: Bantam Books.

Tomaney, J., & Ward, N. (2001). Locating the region. An introduction. In J. Tomaney & N. Ward (Eds.), *A Region in Transition. North East England at the Millennium* (pp. 1-22). Aldershot, UK: Ashgate.

Torres, L., Pina, V., & Acerete, B. (2006). E-Governance Developments in European Union Cities: Reshaping Government's Relationship with Citizens. *Governance: An International Journal of Policy, Administration, and Institutions*, *19*(2), 227–302.

Turke, R. E. (2008). *Governance: Systemic Foundation and Framework*. Berne, Switzerland: Physica-Verlag.

Turner, F. (2006). *From Counterculture to Cyberculture: Stewart Brand, the Whole Earth Network, and the Rise of Digital Utopianism*. Chicago: University of Chicago Press.

UK Department of Health. (2002). *Getting the best from Best Value: experience of applying Best Value in social care*. London: Department of Health (Social Services Inspectorate).

Ulrich, W. (1983). *Critical Heuristics of Social Planning: A New Approach to Practical Philosophy*. Berne, Switzerland: Haupt.

UNDP. (2006). *Desarrollo Humano en Chile. Las Nuevas Tecnologías: ¿Un Salto al Futuro?* Santiago, Chile: UNDP.

UNESCO. (2005). *E-Governance capacity building*. Retrieved December 2008, from http://portal.unesco.

org/ci/en/ev.php-URL_ID=4404&URL_DO=DO_TOPIC&URL_SECTION=201.html

Unicode Consortium. (n.d.). Retrieved from http://www.unicode.org/

Unified Medical Language System. (n.d.). Retrieved from http://www.nlm.nih.gov/research/umls/

United Nations E-government survey. (2008). *From E-government to Connected Governance* (Doc. Nr. ST/ESA/PAD/SER.E/112,). New York: Department of Economic and Social Affairs Division for Public Administration and Development Management.

United Nations. (2005). *Global E-Government Readiness Report: From E-Government to E-Inclusion*. New York: United Nations.

United Nations. (2008). *Public Administration and e. government network*. UNPAN Reports.

Universidad Católica (UC) y Cámara Chilena de la Construcción. (CCS). (2006). *Monitoreando el futuro digital: resultados encuesta WIP-Chile, 2006*. Retrieved March 16, 2007, from http://www.wipchile.cl/estudios/WIP_Chile_2006___informe.pdf

USAID Press Release Tirana. (n.d.). *A Modern Home for E-Governance: New IT Agency will host e-government services*. Retrieved May 19, 2008, from http://albania.usaid.gov/shfaqart/255/92/Press_Release.htm

Valenti, P., Anta, R., & Bendersky, M. (2003). *Manual. Gob. Estrategias de Gobierno Electrónico en los países de la Región 1: La definición de un modelo de análisis y estudio de casos*. Inter-American Development Bank, Washington.

Valenzuela, A., & Valenzuela, J. S. (1983). Los Orígenes de la Democracia. Reflexiones Teóricas sobre el Caso de Chile. *Estudios Públicos*, *12*, 7–39.

Valenzuela, J. S. (1995). Orígenes y Transformaciones del Sistema de Partidos en Chile. *Estudios Públicos*, *58*, 5–80.

Van Audenhove, L., Burgelman, J., Nulens, G., & Cammaert, B. (1999). Information Society in the developing world: A critical assessment. *Third World Quarterly*, *20*(2), 387–404. doi:10.1080/01436599913811

Vanderstraeten, R. (2005). System and environment: Notes on the autopoiesis of modern society. *Systems Research and Behavioral Science*, *22*(6). doi:10.1002/sres.662

Vial, A. (2005). Nosotros los ciudadanos a la intemperie. In *Proceedings of the Nueva Sociedad N 195*, Caracas, Venezuela (pp. 42-53).

Vision-2019 (2005). *2019 Visión Colombia II Centenario (Colombia s vision 2019 executive summary report) - Propuesta para Discusión - Resumen Ejecutivo*. Retrieved October 2007, 2007, from http://www.dnp.gov.co/paginas_detalle.aspx?idp=366

Volberda, H. W., & Lewin, A. Y. (2003). Co-evolutionary dynamics within and between firms: from evolution to co-evolution. *Journal of Management Studies*, *40*(8), 2105–2130. doi:10.1046/j.1467-6486.2003.00414.x

Von Foerster, H. (1981). Notes on an epistemology for living things. In H. Von Foerster (Eds.), *Observing Systems* (pp. 257-271). Salinas, CA: Intersystems Publications.

Wainwright, H., et al. (2007). *Networked politics. Rethinking political organisation in an age of movements and networks* (Work in progress). Amsterdam: The Transnational Institute. Retrieved from http://www.tni.org/docs/200701261425586570.pdf

Walsh, J. P., & Ungson, G. R. (1991). Organizational memory. *Academy of Management Review*, *16*, 57–91. doi:10.2307/258607

Ward, J., & Griffiths, P. (2002). *Strategic Planning for Information Systems*. Chichester, UK: John Wiley and Sons.

Ward, S., & Gibson, R. (2003). On-line and on Message? Candidate Websites in the 2001 General Election. *British Journal of Politics and International Relations*, *5*(2), 188–205. doi:10.1111/1467-856X.00103

Ward, S., Lusoli, W., & Gibson, R. (2002). Virtually Participating: A Survey of Online Party Members. *Information Polity*, *7*, 199–215.

Weber, M. (1920). *The Theory of Social and Economic Organization*. New York: The Free Press

Weber, M. (1946). Bureaucracy. In H. Gerth & C. W. Mills (Eds.), *Max Weber: Essays in Sociology* (pp. 196-244). New York: Oxford University Press.

Webster, F. (2002). *Theories of the Information Society* (2nd ed.). London: Routledge.

Wenger, E. (1999). *Communities of Practice: Learning, Meaning, and Identity*. Cambridge, UK: Cambridge University Press.

West, D. (2004). E-government and the transformation of service delivery and attitudes. *Public Administration Review, 64*(1), 15–27. doi:10.1111/j.1540-6210.2004.00343.x

West, D. M. (2003). *State and Federal e-Government in the United States, 2003*. Providence, RI: Center for Public Policy, Brown University.

Wickham, J. (1997). Where is Ireland in the global information society? *The Economic and Social Review, 28*(3), 277–294.

Wiener, N. (1965). *Cybernetics or Control and Communication in the Animal and the Machine (2nd ed.)*. Boston, MA: MIT Press.

Wikimedia Foundation. (2009). *Planned Economy. Wikipedia*. Retrieved July 20, 2009, from http://en.wikipedia.org/wiki/Planned_economy

Wilson, J. Q. (1993). *Bureaucracy*. New York: Free Press.

Wong, K. Y., & Aspinwall, E. (2005). An empirical study of the important factors for knowledge-management adoption in the SME sector. *Journal of Knowledge Management, 9*(3), 64–82. doi:10.1108/13673270510602773

World Bank. (2003). *Doing Business in 2004: Understanding Regulation*. A World Bank Publication.

World Bank. (2009). *Doing Business 2009 Report: Albania*. A World Bank Publication.

Wresch, W. (1996). *Disconnected: Have and Have-nots in the Information Age*. New Brunswick, NJ: Rutgers University Press.

Ya, A. N., & Bretschneider, S. (2007). The Decision to Contract Out: A Study of Contracting for E-Government Services in State Governments. *Public Administration Review, 67*(3), 531–544. doi:10.1111/j.1540-6210.2007.00735.x

Yildiz, M. (2007). E-government Research. Reviewing the Literature, Limitations, and Ways Forward. *Government Information Quarterly, 24*(3), 646–665. doi:10.1016/j.giq.2007.01.002

Yin, R. K. (1994), *Case Study Research, Design and Methods (2nd ed.)*, Sage Publications, Beverly Hills, CA.

Yolles, M. (2006). Organizations as Complex Systems: An Introduction to Knowledge Cybernetics. Yolles, M. (Ed). *ISCE Book Series: Managing the Complex Series.* Editor(s): M. Lissack, K. Richardson. ISCE Research.

Yu, E. (1995). *Modelling strategic relationships for process reengineering*. Ph.D. Thesis, University of Toronto, Department of Computer Science.

Zahariadis, N. (1995). *Markets, States, and Public Policy. Privatization in Britain and France*. Ann Arbor, MI: University of Michigan Press.

Zahariadis, N. (1999). Ambiguity, time and multiple streams. In P. A. Sabatier (Ed.), *Theories of the policy process* (pp. 73-93). Boulder, CO: Westview Press.

Zarza, M. (2009). *Las redes sociales siguieron la amenaza de bomba y la explosión al minuto*. Retrieved February 2009, from http://www.abc.es/20090209/nacional-terrorismo/redes-sociales-siguieron-amenaza-200902090959.html

Zed Digital. (2008). *El fenómeno de las redes sociales. Percepción, usos y publicidad. Zed digital*. Retrieved January 2009, from http://www.zeddigital.es/Estudio de redes sociales_20_11_2008.pdf

About the Contributors

José-Rodrigo Córdoba-Pachón (MA, PhD) is a senior lecturer and director of the Master of Science in Business Information Systems at the school of Management, Royal Holloway, University of London in the UK. His research interests are in the improvement of technology use people; the use of systemic thinking, and electronic government. Prior to his master and PhD experiences in the University of Hull in the UK, he worked as a project analyst and project manager in Colombia, his native country. He has published work in journals including the *European Journal of Information Systems (EJIS),* the *Journal of the Operational Research Society*, the *Information Systems Journal* , *Systems Research and Behavioural Science,* and *Organization*. Together with Teresa Torres-Coronas from Universitat Rovira I Virgili, he set up *ELIS* research group on electronic government and sustainability in 2004. He continues contributing to this group and is also an active member of the Operational Research Society and the British Computer Society in the UK. Currently he lectures in topics of information systems and complex project management. His first single authored book titled *Systems Practice in the Information Society* is to be released in Autumn 2009.

Alejandro Elias Ochoa-Arias is professor at the University of Los Andes, Venezuela and senior researcher on Interpretive Systemology and director of the Centro de Investigaciones en Sistemología Interpretativa at the University of Los Andes. Current research interests includes civil society, development, political philosophy and organizational studies, free knowledge and free technology as well as the use and implications of technology in running public affairs. Alejandro has been visiting researcher at the University of Hull (UK) and the University of Cuyo (Mendoza, Argentina). He also co-edited a book titled *Community Operational Research: OR and Systems Thinking for Community Development* with Gerald Midgley (Springer, 2004). He has several publications on refereed journals and has been invited speaker at many institutions in Venezuela and elsewhere. Alejandro was a member of the Directive Board of a Regional Development Office at Venezuela and member of the Directive Board of the National Council for Free Technology in Venezuela. He got his master's degree in information management at Lancaster University (UK) and his PhD at the University of Hull (UK). Currently, he is developing a study on the holistic design of the city at Mérida, Venezuela.

***

Ahmed Al-Maimani, is currently reading for a PhD in management at the University of Hull in the UK. He holds an MBA from the same university obtained in 2000. His dissertation topic was the impacts of information systems on socio-economic decision making and he has pursued this now by looking at the organisation of e-government projects and infrastructures in Oman. Ahmed has hold several distinguished positions in the Omani government offices including the Ministry of Commerce and Industry, the ministry of Finance, the office of the deputy primer minister for financial and economic affairs and advisory roles to his majesty the Sultan. He has also being member of executive government committee boards and has been commended for his services on several occasions.

Eduardo Araya Moreno is assistant professor at the Public Affairs Institute of the University of Chile. He is currently writing a PhD dissertation on ICT using by political parties in Chile, Costa Rica and Uruguay. He has published papers on state features in the informational society, features of personnel training programs in topics related to e-government, and e-government in Chile. He acts as editor of *Estado, Gobierno, Gestión Pública. Revista Chilena de Administración Pública.*

Alexandre Barbosa has 30 years of experience in the IT and telecom industries in Brazil. As a specialist in IT strategies he has worked for the largest Brazilian telecom operators and he has been working in large-scale projects of electronic government in the Brazilian governmental agencies. He holds a PhD degree in Business Administration (FGV-EAESP), a master's degree in computer science (UFMG) and in business administration (Bradford University), and a Bachelor of Science Degree in electrical engineering (PUC-MG). He also holds postgraduate diplomas in corporate finance and marketing (FGV). He lectures at FGV (on postgraduate courses in the areas of business policy, marketing, applied information and communication technologies).

Diego Barría is a PhD researcher at the Department of Latin American Studies of Leiden University in the Netherlands, where he is writing a dissertation on public sector change in Chile, between 1880 and 1931. He previously has published on topics such as e-government in Chile, Union participation in policy-making processes in Latin America, and administrative reforms in historical perspective.

Gustavo Campos is a MPA student at University of Chile. He holds a BA in Public Administration (University of Chile). He has been teaching-assistant at the School of Government and Public Management of the University of Chile, in topics such as Political Theory and Public Management. He currently works at the National Treasury of Chile.

Juan-Gabriel Cegarra (PhD) is associate professor of Business Management in the Polytechnic University of Cartagena (Spain). His research interests focus on eGovernment, Knowledge Management and Marketing Orientation. He has published several articles in internationally recognised journals including *Small Business Journal, International Journal of Information* Management and the *Journal of the Operational Research Society.* He has been awarded a visiting fellowship by the British Academy of Management to conduct research on knowledge management and succession in UK small and medium enterprises (SMEs). He is a member of the ELIS group (eGovernment for local integration with sustainability).

Rajneesh Chowdhury is senior consultant with Pricewaterhouse Coopers in New Delhi, India. Rajneesh has completed a BA in sociology, MA in sociology and MSc in management systems, and is a qualified Member of the British Chartered Management Institute. His areas of interest include critical systems thinking, multimethodology and change management. He has been involved in various projects in the fields of organization design, large scale change programs, organization structuring and information systems. Rajneesh has published and presented widely in the field of systems thinking and its application in management consulting. He started his career in healthcare consulting in the UK and currently works in a range of areas including manufacturing, services, pharmaceuticals and healthcare. Rajneesh was also nominated to the British Ministry of Trade & Industry in 2005 for the prestigious accolade of Business Leader of Tomorrow for making outstanding contribution to the field of healthcare informatics.

J. Ignacio Criado (PhD) is a Lecturer in the Department of Political Science and International Relations, Universidad Autónoma de Madrid. From 2001, he has published several chapters and peer-reviewed articles on e-government issues, specially about the local and regional tiers in comparative perspective. His books *Building the Local e-Administration* (published in Spanish, *Construyendo la e-Administración Local*. Madrid, EuroGestión Pública, 2004) and *Between Utopian Dreams and Pessimistic Visions* (published in Spanish, *Entre Sueños Utópicos y Visiones Pesimistas*. Madrid, Instituto Nacional de Administración Pública, 2009) both received research awards by the Spanish Ministry for Public Administrations. His research deals with different topics on ICTs and Public Administration, public sector reform policies, quality management, and leadership in the cyberspace. He is editorial board member for *International Journal of Public Sector Management*. He may be contacted at www.jicriado.com.

Jacinto Dávila. has a PhD in logic and artificial intelligence (1997) and an MSc on information technology from Imperial College in the UK. He graduated in 1991 as a systems engineer from Universidad de los Andes in Merida, Venezuela (ULA). Since November 1997 he is a lecturer and researcher at the Centre for Modelling and Simulation (CESIMO) at ULA. Jacinto has been co-ordinator of graduate studies in Computer Science (1998-2000), co-ordinator of graduate studies in modelling and simulation (1998-2005), and executive director of the Foundation for Science and Technology Development in Mérida, Venezuela (2008-2009). He is also co-developer of the multi-agent simulation software *Galatea*,(http://galatea.sourceforge.net/ , accessed in July 2009) and leader of the *ULAnux/ULAnix* project (http://ulanix.wiki.sourceforge.net/, accessed in July 2009).

Eduardo-Henrique Diniz is electronic engineer, has a MSc and a PhD in business administration, was visiting scholar at University of California, Berkeley from 1996 to 1998, and at HEC Montreal in 2007. Professor at FGV-EAESP, Sao Paulo, Brasil since 1999, is member of the GVcemf, Center for Studies in Microfinance at FGV, Eduardo has researched on banking technology and electronic government, having published several papers and articles on both subjects. Eduardo Diniz is also chief-editor of Revista de Administração de Empresas (RAE), the main academic journal in business in Brazil.

Angela Espinosa (BSc computer and systems engineer Los Andes, Colombia, 1981, PhD organisational cybernetics, Aston, 1995). Angela was born in Bogotá, Colombia and graduated as a computer and systems engineer in 1981. She worked originally as an information systems manager, in private and public enterprises, and then as the director of the Secretariat of Information and Systems of the

Colombian President's Office (1990 – 1992). From 1993 to 2002, she taught systems and cybernetics in Los Andes University and provided consultancy for both the private and public sectors. Since 2002 she has been researching and teaching within the Scarborough Management Centre of the Hull Business School and acting as its research co-ordinator. In 2007 she received two EPSRC grants to do research on Emergence and Complexity, in particular in topics of self-organisation in sustainability and regeneration programs.

Marcelo Fernandez holds a Bachelor of Mechanical Engineering and a Bachelor of Laws, both at University of São Paulo. He currently works at the Secretariat of Finance of State of São Paulo as Electronic Invoice Project leader, being responsible for a $15 MM budget and the coordination of 20 collaborators.

Amanda Gregory (PhD) is head of Management Systems in the Business School, University of Hull, UK. Prior to joining the Business School, Amanda was the research worker on a national project funded by the Leverhulme Trust to design evaluation procedures for Councils for Voluntary Service (CVS). This work provided the basis for her doctorate, awarded in 1995, entitled 'Organisational Evaluation: A Complementarist Approach'. This led to Amanda holding several prominent academic positions including being the deputy editor of the journal *Systems Research and Behavioral Science*. Amanda's current consultancy and research interests relate to participatory forms of evaluation, the use of systems approaches to support management decision making, action learning, and lean systems thinking. As regards project work, she has focused on evaluation in the not-for-profit sector and the facilitation of community initiatives.

Alvaro Junqueira holds a Master Degree in business administration at Fundação Getulio Vargas of São Paulo - FGV/SP and a Bachelor of Science Degree in informatics at Federal University of Rio de Janeiro. He also studied one semester at the University of Calgary, Canada, as an exchange student and researcher. Alvaro has particular interest in the process of implementation of electronic government projects and programs, especially when multiple agencies are involved. He has written several articles for Public Administration journals and Congresses, such as RAP - Revista da Administração Pública, BALAS and ENANPAD. Actually he works at the Secretariat of Finance of State of São Paulo as its Information Tecnology Director being responsible for a $50 MM annual budget and the coordination of 300 collaborators.

Endrit Kromidha graduated with First Class Honours, both in international economics at the Budapest Business School – Faculty of International Management and Business in Hungary and in European business administration at the University of Lincoln in UK. He obtained a Master degree in European economic studies from the University of Tirana – Faculty of Economics in Albania, in a joint program with Otto Friedrich (Bamberg) University in Germany. Endrit Kromidha has a special interest on e-government in developing countries, which is also the topic of his PhD at Royal Holloway University of London. He has also been professionally involved in banking, leasing, NGOs and think-tanks for research, marketing, business administration and lecturing.

Cecilia Loureiro-Koechlin (MBA, PhD) is a researcher at the Building the Research Information Infrastructure (BRII) project of the Oxford University Library Services in the UK. Cecilia did her PhD in the field of Information Systems looking at the social aspects of software development. Through an online ethnographic approach she analysed software developers views on their own work. She is interested in the social components of information systems and software such as groupware, social software, virtual learning environments (VLEs)and virtual research environments (VREs), as well as in the use of structuration theory as a framework to investigate software development and use. She has taken part in several projects related to the use of information technology in education and has published several journal articles and reports. Additional interests are online communities of practice, online pragmatics, online ethnography and content analysis. She writes a blog in her free time (http://mutant-clk.blogspot.com/, accessed in July 2009).

Deepankar Medhi is a professor in the Department of Computer Science and Electrical Engineering at the University of Missouri-Kansas City, USA. He received a BSc (Hons) in mathematics from Cotton College, Gauhati University, India, MSc in mathematics from the University of Delhi, India, and his PhD in computer sciences from the University of Wisconsin-Madison, USA. Prior to joining UMKC in 1989, he was with the AT&T Bell Laboratories. He was a visiting professor at the Technical University of Denmark and a visiting research fellow at Lund Institute of Technology, Sweden. He served as a Fulbright senior specialist. His research interests are resilient network design, network routing and architecture, and e-system for public health. He has published over eighty papers, and co-authored the books, *Routing, Flow, and Capacity Design in Communication and Computer Networks* (2004) and *Network Routing: Algorithms, Protocols, and Architectures* (2007), both published by Morgan Kaufmann Publishers (an imprint of Elsevier Science).

Federico Monteverde is systems analyst graduated from the Engineering School of the Universidad de la Repúbica Oriental del Uruguay. He obtained post-graduate degrees in social networks analysis and public policy analysis from the Universidad Bolivariana de Chile. He is currently a Master's candidate in electronic government at the Universidad Tecnológica Metropolitana del Estado de Chile. Monteverde is electronic government researcher and practitioner. He is a frequent lecturer in electronic government and public administration conferences and published several articles and papers in well-known national and international reviews. He is consultant for the government of Uruguay at the national electronic government agency, where he is involved in the whole process of electronic government policies. Since April, 2009, he is director of the Data Protection Authority of Uruguay.

Kevin Orr is the director of the Centre for Management and Organisational Learning, at Hull University Business School, UK. His research interests are in the areas of public sector leadership and learning. His approach to research draws upon elements of political sociology, interpretivism, organisation studies, critical management studies, and discourse analysis. He has published work on political leadership and governance, local government traditions and local government crisis and change. He has developed extensive networks of public sector practitioners and has experience of applied research for organisations which include local councils, the Scottish Parliament, the British Council, and the Society of Local Authority Chief Executives (UK). He is a member of the International Editorial Advisory Board of Management Learning and the Executive Board of Organizational Learning, Knowledge and Capabilities (www.olkc.net).

Francisco-Javier Palm-Rojas is a systems engineer with a master's degree in mathematics applied to engineering. He teaches in the Statistics and Computing Institute of Universidad de Los Andes, Mérida, Venezuela. Also, he is a long time free software developer and activist, and an active researcher in modelling and simulation, free/libre open source geomatics, and Python Collaborative Software Development. His areas of interest are open knowledge, collaborative learning, geospatial analysis and simulation. He was member of SOLVE (Free Software Venezuelan Association http://solve.net.ve) Board of Directors. Also, he is the president of SPIRALIA: Networking Learning Solutions (http://spiralia.net.ve) and active member of ENDOLOGICA: Open Informatics and Professional Services Cooperative for Endogenous Development. More information about him and his work can be found at http://fpalm.net.ve.

María-Angela Petrizzo-Páez is a politician with a master's degree in management sciences from the Universidad de Los Andes (Mérida, Venezuela). She is a long time free software and free knowledge advanced user and activist, and an active member of the Researchers Promotion Program (PPI in Spanish) by the Venezuelan Ministerio Para del Poder Popular para la Ciencia, Tecnología, Industrias Intermedias y Telecomunicaciones, since 2008. Her areas of interests are citizenship, social and political networks, endogenous development, organizational learning, e-government, e-democracy and information and communication technologies for the citizenship learning and development She is the SOLVE (Free Software Venezuelan Association http://solve.net.ve) general coordinator. Also, she is the general manager of SPIRALIA: Networking Learning Solutions (http://spiralia.net.ve) and active member of ENDOLOGICA, Open Informatics and Professional Services Cooperative for Endogenous Development. She lectured at Universidad de Los Andes in Mérida, Venezuela, Misión Sucre and Universidad Tecnológica Metropolitana from Santiago, Chile in areas like e-government, new public management, citizenship participation, conflicts resolution among others. More information about her and her work can be found at http://petrizzo.org.ve.

Elias Pimenidis is a senior lecturer at the University of East London in the UK. The core of his research work focuses on e-business and e-government development projects. His interest in security issues is on online transactions and e-services and in particular, on the effect, these could have on e-government implementations. Other research interests include the evaluation of web services, knowledge management systems and the use of computer games for educational purposes. He is also a visiting lecturer at the Informatics laboratory of the Agricultural University of Athens, Greece.

Otávio Prado is graduated in geology at Instituto de Geociências - USP, graduate in history at Faculdade de Filosofia Letras e Ciências Humanas - USP, master's and PHD in public administration at Fundação Getúlio Vargas (FGV/EAESP). Has experience in administration, focusing on Public administration, acting on the following subjects: electronic government, urban planning, environmental geology, environmental management and geotechnical cartography. Researcher of the Centro de Estudos em Administração Pública e Governo (CEAPG) of FGV/EAESP.

Ana-Magaly Reyes-Hill has an MBA from Universidad de los Andes in Merida, Venezuela (ULA, 2001) and previous to that she graduated as systems engineer from the same university. She has been general manager from HidroAndes community service, IT chief manager at the Health Ministry in Caracas, Venezuela, IT Consultant at PAHO (Panamerican Health Organization) in Caracas, IT Manager and Planning Director at the Free Trade Zone of Mérida (Zona Libre Cultural, Científica y

Tecnológica de Mérida). She has also been lecturer at Hotel/College in Mérida and Universidad de las Fuerzas Armadas (UNEFA). Since 2008 she is a freelance consultant for planning and organizational modelling and development.

Ioannis Savvas is a Phd candidate in the field of e-government at the Informatics Laboratory of the Agricultural University of Athens. He is an executive in the Region of Central Macedonia one of the 13 regional government authorities in Greece. He is involved in projects exploring the potential of e-government implementations in improving the Greek public administration's efficiency. His interests include requirements engineering, knowledge management, and intelligent system applications in PA. He holds an MSc in "informatics and administration" from AUTh and he is graduate of the Greek National School of Public Administration. His research is supported and funded by a scholarship from the Greek State Scholarship Foundation.

Alexander B. Sideridis is professor in computer science and head of the Informatics Laboratory of the Agricultural University of Athens. He earned his first degree at the University of Athens and his MSc and PhD from Brunel University. He has been the project leader of more than 30 successful national and international projects and has published more than 180 scientific articles in management and decision support systems, computer networking related to local and central administration activities, advanced computational numerical modeling, informatics & impact of computers in society and agricultural informatics

Index

E

N

O

P

R